Polymer-Based Drug Delivery:

Bridging Science and Therapeutics

Md. Abul Kalam Azad
University College MAIWP International, Malaysia

Mireia Mallandrich Miret
University of Barcelona, Spain

Sukalyan Kumar Kundu
Jahangirnagar University, Bangladesh

IGI Global
Scientific Publishing
Publishing Tomorrow's Research Today

Vice President of Editorial	Melissa Wagner
Director of Acquisitions	Mikaela Felty
Director of Book Development	Jocelynn Hessler
Production Manager	Mike Brehm
Cover Design	Jose Rosado

Published in the United States of America by
IGI Global Scientific Publishing
701 East Chocolate Avenue
Hershey, PA, 17033, USA
Tel: 717-533-8845
Fax: 717-533-7115
Website: https://www.igi-global.com E-mail: cust@igi-global.com

Library of Congress Cataloging-in-Publication Data

Names: Azad, Md. Abul Kalam, 1990- editor | Mallandrich Miret, Mireia, 1977- editor | Kundu, Sukalyan Kumar, 1975- editor
Title: Polymer-based drug delivery : bridging science and therapeutics / edited by Md. Abul Kalam Azad, Mireia Mallandrich Miret, Sukalyan Kumar Kundu.
Description: Hershey, PA : IGI Global Scientific Publishing, [2026] | Includes bibliographical references and index. | Summary: "This book intends to bridge the gap between basic scientific principles and their clinical applications, addressing both the theoretical and practical aspects of polymer-based drug delivery systems"-- Provided by publisher.

Identifiers: LCCN 2025025992 (print) | LCCN 2025025993 (ebook) | ISBN 9798337310473 hardcover | ISBN 9798337310480 paperback | ISBN 9798337310497 ebook
Subjects: LCSH: Polymeric drug delivery systems--Case studies
Classification: LCC RS201.P65 P635 2025 (print) | LCC RS201.P65 (ebook)
LC record available at https://lccn.loc.gov/2025025992
LC ebook record available at https://lccn.loc.gov/2025025993

British Cataloguing in Publication Data
A Cataloguing in Publication record for this book is available from the British Library.

Table of Contents

Detailed Table of Contents

Polymer-mediated drug delivery has completely transformed the field of modern therapeutics, providing controlled drug release, improved bioavailability, and targeted drug delivery. This chapter provides an overview of polymer-drug interaction fundamentals, and it updates the field with recent developments in biodegradable, stimuli-responsive, and nanostructured materials in the field of polymer-drug interactions. Polymer-based approaches can be used for precise and application-specific designs to prevent side effects by bridging the gap between the material science research and clinical application. Major advances in polymer engineering, formulation strategies, and drug release rates that contribute to the enhancement of therapeutic outcomes are highlighted. Final perspectives, such as smart polymers or personalized drug delivery strategies, are discussed to heighten the awareness of the transformational capacity of polymer-based systems in pharmaceutical sciences.

Biodegradable polymers play a pivotal role in modern drug delivery by enabling sustained and controlled release, reducing systemic toxicity, and improving patient adherence. These materials degrade into non-toxic byproducts, eliminating the need for surgical removal and supporting various formulations like nanoparticles, hydrogels, and implants. This review explores natural and synthetic polymers, their degradation mechanisms, and innovations such as stimuli-responsive systems and targeted delivery. Applications span oncology, infectious diseases, cardiovascular therapy, diabetes, and tissue engineering. Key challenges include consistent drug loading, initial burst control, and manufacturing scalability. The paper also highlights safety, regulatory considerations, and future directions involving smart materials and biosensor integration. Biodegradable polymers continue to shape the future of personalized and regenerative medicine.

For the administration of therapeutic agents there is need of delivery agents which will be nontoxic and amended with the desired physicochemical properties. Recently, a considerable amount of work has been reported on the use of natural polymers as drug delivery agent. In this book chapter, an overview of the natural polysaccharides (chitosan, hyaluronic acid, cellulose and bacterial exopolysaccharide) is presented. Although bacterial exopolysaccharides showed features as drug carrier, still there use as drug delivery agent is limited due to certain factors like composition, yield, cost of purification etc. Researchers are trying various formulations of this polysaccharides by altering the chemical groups, size and shape modifications, reducing the toxicity and increasing the controlled drug release propery . Herein, this book chapter aims to provide the current trends and perspectives in the utilisation of the polysaccharides originated from natural sources.

The drug delivery properties of polymer-based hydrogels make them attractive because they combine biocompatibility, biodegradability, and high-water retention capabilities. These delivery systems provide controlled drug release mechanisms that enable precise targeted delivery and reduce the side effects. Drug delivery systems include oral administration, injectable delivery, implantable systems, and wound healing applications. The development of polymer-based hydrogels faces ongoing challenges because of drug loading restrictions and difficulties with release precision, stability, and scalability. The combination of smart hydrogels with bioinspired designs and 3D printing fabrication methods provides potential solutions. This review examines hydrogel varieties and drug delivery methods, while exploring their clinical applications and future directions for developing personalized and efficient therapeutic systems.

This chapter offers a detailed overview of hydrogel-based drug delivery systems, emphasizing their potential as precision-engineered platforms for controlled therapeutic release. It examines two main drug incorporation strategies—physical entrapment and chemical conjugation—highlighting their mechanistic and kinetic differences and suitability in various physiological contexts. A key focus is the relationship between drug diffusion kinetics and the mechanical stability of hydrogel matrices. The discussion underscores how modifying hydrogel composition and structure can balance robustness with predictable release profiles. Additionally, the chapter addresses essential material properties such as biocompatibility and biodegradability, which are vital for safe and effective in vivo use. By integrating recent advancements, it provides insights into the rational design of next-generation hydrogels that enable targeted, sustained, and stimuli-responsive drug delivery, aligning with the goals of precision medicine.

Fatema Tuz Zohera, International Islamic University Malaysia, Malaysia
Abul Kalam Azad, University MAIWP International, Malaysia

Alginate-based microbeads are promising carriers for targeted drug delivery due to their biocompatibility, biodegradability, and gel-forming capacity. Formed via ionotropic gelation with divalent cations, these microbeads protect labile drugs and enable sustained, controlled, and site-specific release. Tuning parameters like the M/G ratio and crosslinking density allows precise control of drug release kinetics and mechanical properties. Recent advances—including co-polymers, nanoparticles, and stimuli-responsive systems—have improved targeting, drug retention, and stability. Applications span gastrointestinal, colonic, and tumor-targeted therapies. With low toxicity and regulatory acceptability, alginate microbeads are suitable for oral, injectable, and localized administration. This review outlines recent developments, challenges, and future directions, highlighting alginate microbeads as a versatile platform in precision drug delivery.

Nasir khan, Gomal University, DI Khan, Pakistan
Sajid Raza, Ibadat International University Islamabad, Pakistan
Muhammad Saad, Ibadat International University Islamabad, Pakistan
Muhammad Noman, Ibadat International University Islamabad, Pakistan
Muhammad Adnan, My University Islamabada, Pakistan
Abul Kalam Azad, University MAIWP International, Malaysia

This chapter contains stimuli responsive polymers (SRPs) or smart materials, change their physical and chemical properties in response to external stimuli such as pH, temperature, or light, enables controlled, targeted drug delivery, improving therapeutic efficacy and decreasing side effects. Ocular, nasal, pulmonary, transdermal, and oral drug delivery systems utilizing SRPs are tailored to specific therapeutic needs. There are challenges such as biocompatibility, stability, and industrial scale production. The integration of nanotechnology with SRPs improves their performance by allowing for more precise control of drug release. Despite challenges, SRPs are promising solutions for treating complex diseases and are expected to revolutionize drug delivery systems by making them safer, more efficient, and patient specific.

Smart polymers represent a transformative approach in CNS drug delivery, offering controlled, targeted, and responsive release of therapeutic agents. This chapter explores the design and application of stimuli-responsive and biodegradable polymers tailored to overcome the physiological barriers of the CNS, particularly the BBB. Stimuli-responsive polymers, engineered to respond to internal (pH, redox potential, enzymes) or external (temperature, magnetic field, light) triggers, enable on-demand drug release at the desired site. Concurrently, biodegradable polymers reduce long-term toxicity and eliminate the need for surgical removal, making them ideal for chronic CNS therapies. Finally, the chapter addresses recent advances, current limitations, including stability, scalability, and regulatory challenges, and outlines future directions for research. Integrating material science with neuropharmacology, smart polymers offer a promising platform to revolutionize CNS therapeutics through precision drug delivery.

Polymeric microneedles (MNs) constitute an advanced technology in transdermal drug delivery that overcomes the limitations of traditional patches bypassing the stratum corneum to facilitate precise and minimum penetrative administration. By employing biodegradable polymers, such as PLGA, PCL, and hyaluronic acid, MNs offer enhanced biocompatibility, sustainability, and controlled drug release, addressing various medical applications, including vaccines, cosmeceuticals, and gene therapy. Significant advancements include biosensor integration for real-time monitoring, nanotechnology for improved drug solubility and stability, and hydrogel systems for sustained drug release. Despite these advantages, challenges persist in addressing skin irritation, brittleness, and immunogenicity while ensuring product consistency through 3D printing. Advancements in regulatory compliance, machine

Preface

The field of drug delivery continues to evolve at an unprecedented pace, with polymer-based systems emerging as one of the most promising and versatile platforms for therapeutic innovation. *Polymer-Based Drug Delivery: Bridging Science and Therapeutics* has been conceived as a comprehensive and insightful resource that reflects the latest advances and future directions in this rapidly developing domain.

This volume brings together contributions from leading researchers and practitioners who are pioneering novel polymer systems designed to enhance the targeted, sustained, and responsive delivery of therapeutic agents. From the fundamentals of polymer design and synthesis to the application of advanced drug delivery platforms—such as hydrogels, microneedles, and biodegradable carriers—this book aims to capture the breadth and depth of polymer-based strategies currently shaping the landscape of modern therapeutics.

Our central goal in assembling this reference work is to bridge the critical gap between scientific theory and clinical application. By combining discussions on molecular properties, formulation methodologies, and real-world implementation across various delivery routes, this book strives to foster a holistic understanding that supports innovation across academia, research, and industry. Readers will encounter not only the current state of the art but also the technical challenges and knowledge gaps that remain to be addressed. In doing so, this work aims to inspire future research trajectories that prioritize patient safety, drug efficacy, and translational success.

This book is especially intended for researchers, graduate students, postdoctoral fellows, and professionals working in pharmaceutical sciences, biomedical engineering, polymer chemistry, and nanomedicine. Whether engaged in drug design, formulation science, or therapeutic development, readers will find valuable insights that can inform their research, refine their approaches, and catalyze breakthroughs in targeted drug delivery.

As editors, we are proud to present this collection of scholarship that not only celebrates the progress made in polymer-based drug delivery, but also serves as a catalyst for the next generation of innovations. We extend our gratitude to the contributors for their invaluable work, and we look forward to seeing how this knowledge will be applied to improve health outcomes around the globe.

ORGANIZATION OF THE BOOK

In Chapter 1, "Polymer-Based Drug Delivery Bridging Science and Therapeutics," the authors introduce the foundational principles of polymer-mediated drug delivery and the key advancements that have shaped this field. The chapter offers a comprehensive view of how polymers, especially biodegradable and stimuli-responsive types, have transformed drug formulation and therapeutic precision. By bridging material science and clinical application, this overview underscores the role of smart and personalized delivery systems in optimizing therapeutic outcomes and minimizing adverse effects.

Chapter 2, "Biodegradable Polymers in Modern Therapeutics: A Revolution in Sustained Drug Release," explores the versatile utility of biodegradable polymers in modern medicine. This chapter presents an in-depth review of both natural and synthetic polymers, shedding light on their degradation mechanisms and therapeutic applications ranging from oncology to cardiovascular care. The discussion extends to regulatory and manufacturing challenges, while highlighting the future of smart materials and biosensor-enhanced systems in advancing sustained and patient-centered treatments.

In Chapter 3, "Natural Polysaccharides as Drug Delivery Agent: Current Trends and Perspectives," the spotlight turns to naturally derived polymers, particularly chitosan, hyaluronic acid, cellulose, and bacterial exopolysaccharides. This chapter discusses the promise and limitations of using natural polysaccharides as drug carriers, emphasizing current innovations in modifying their structures to enhance drug release, reduce toxicity, and improve therapeutic efficacy. It offers a valuable perspective on how nature-inspired solutions are contributing to safer and more sustainable drug delivery approaches.

Chapter 4, "Polymers in Hydrogels Fabrication and Their Versatile Application in Drug Delivery," delves into the unique properties of polymer-based hydrogels, highlighting their biocompatibility, biodegradability, and moisture-retention characteristics. The authors examine diverse administration routes and the engineering strategies used to overcome formulation challenges such as drug loading and release control. This chapter also points to future advancements involving smart hydrogels, bioinspired designs, and 3D printing, reinforcing the role of hydrogels in personalized and efficient therapeutic systems.

In Chapter 5, "Integrating Physical Entrapment, Chemical Conjugation, Biocompatibility, and Mechanical-Diffusive Balance in Advanced Hydrogel Systems for Controlled Drug Delivery," the authors explore the nuanced design of advanced hydrogels for precision medicine. The chapter compares physical and chemical drug incorporation techniques while examining how these strategies impact mechanical stability and diffusion kinetics. Key considerations such as biocompatibility, degradation, and material composition are critically analyzed, setting the stage for next-generation hydrogel platforms tailored for predictable, responsive, and sustained drug delivery.

Chapter 6, "Alginate-Based Microbeads: A Promising Platform for Targeted Drug Delivery Systems," presents alginate microbeads as a robust and versatile vehicle for site-specific drug delivery. With their ability to encapsulate labile compounds and facilitate controlled release, alginate-based systems offer significant promise across gastrointestinal, colonic, and oncological therapies. The authors highlight innovations in crosslinking, co-polymers, and stimuli-responsiveness while addressing key challenges such as scalability and formulation precision, positioning alginate microbeads as a key material in precision therapeutics.

In Chapter 7, "Stimuli-Responsive Polymers in Drug Delivery," the authors examine smart polymer systems that adapt to environmental stimuli such as pH, temperature, or light to regulate drug release. The chapter reviews the use of these materials in various delivery routes—including ocular, nasal, and transdermal—while discussing the technological and manufacturing hurdles that must be overcome. By integrating nanotechnology, these systems are moving toward higher precision, lower toxicity, and broader application potential in the treatment of complex diseases.

Chapter 8, "Designing Smart Polymers for CNS Drug Delivery Stimuli: Responsive and Biodegradable Systems," focuses on the critical need for advanced drug delivery methods targeting the central nervous system (CNS). The chapter explores how smart, biodegradable polymers are engineered to overcome the blood-brain barrier through stimuli-triggered mechanisms, thereby offering precise, on-demand therapeutic delivery. The authors address challenges of stability, regulatory considerations, and clinical translation, while emphasizing the synergy between materials science and neuropharmacology in shaping the future of CNS therapies.

Finally, Chapter 9, "Polymer-Based Microneedles in Transdermal Drug Delivery Systems," presents microneedle technology as a transformative advancement in transdermal drug delivery. Utilizing biocompatible and biodegradable polymers, microneedles enable pain-free, controlled delivery while bypassing skin barriers. This chapter explores the integration of biosensors, nanotechnology, and hydrogels for enhanced performance, and discusses the regulatory and manufacturing complexities of bringing microneedle systems into clinical practice. It underscores the potential of this technology in personalized medicine and next-generation therapeutic platforms.

CONCLUSION

As we conclude this volume, *Polymer-Based Drug Delivery: Bridging Science and Therapeutics*, we are reminded of the extraordinary progress that has been made at the intersection of polymer science and pharmaceutical innovation. The chapters presented in this book reflect not only the current state of knowledge but also the boundless potential that polymer-based systems hold in transforming how therapeutic agents are designed, delivered, and experienced by patients.

From biodegradable matrices and stimuli-responsive platforms to advanced hydrogels and microneedle technologies, the contributions herein demonstrate how materials once limited to theoretical exploration have evolved into clinically significant solutions. These advancements promise improved bioavailability, targeted delivery, and minimized side effects—key pillars in the advancement of personalized medicine.

Yet, with progress comes responsibility. The future of drug delivery must remain mindful of regulatory challenges, patient safety, scalability, and environmental impact. We hope this collection will serve as both a scholarly reference and a springboard for further research, encouraging the scientific community to continue pushing boundaries while keeping therapeutic efficacy and patient-centered design at the core.

We extend our deepest appreciation to the authors and researchers whose work has shaped this book. Their contributions reflect the collaborative spirit and interdisciplinary rigor necessary for breakthroughs in this field. It is our hope that readers—from students to seasoned professionals—find inspiration and insight within these pages, and that the innovations explored here ultimately lead to better treatments and improved quality of life for patients worldwide.

CONCLUSION

Chapter 1
Polymer-Based Drug Delivery

Bably Khatun
https://orcid.org/0000-0001-7064-298X
Institute of Advanced Study in Science and Technology, India

Shagufta Rizwana
Royal Global University, India

Nur Jalal Mondal
https://orcid.org/0009-0005-1481-3322
Material Nanochemistry Laboratory, Physical Sciences Division, Institute of Advanced Study in Science and Technology, Guwahati, India & Academy of Scientific and Innovative Research, Ghaziabad, India

Devasish Chowdhury
Material Nanochemistry Laboratory, Physical Sciences Division, Institute of Advanced Study in Science and Technology, Guwahati, India & Academy of Scientific and Innovative Research, Ghaziabad, India

ABSTRACT

Polymer-mediated drug delivery has completely transformed the field of modern therapeutics, providing controlled drug release, improved bioavailability, and targeted drug delivery. This chapter provides an overview of polymer-drug interaction fundamentals, and it updates the field with recent developments in biodegradable, stimuli-responsive, and nanostructured materials in the field of polymer-drug interactions. Polymer-based approaches can be used for precise and application-specific designs to prevent side effects by bridging the gap between the material science research and clinical application. Major advances in polymer engineering, formulation strategies, and drug release rates that contribute to the enhancement of

DOI: 10.4018/979-8-3373-1047-3.ch001

therapeutic outcomes are highlighted. Final perspectives, such as smart polymers or personalized drug delivery strategies, are discussed to heighten the awareness of the transformational capacity of polymer-based systems in pharmaceutical sciences.

INTRODUCTION

Drug delivery systems (DDS) are now at the forefront of biomedical research due to the pursuit of effective therapeutic delivery. By maximizing efficacy and reducing systemic toxicity, these systems seek to maximize the delivery of therapeutic agents to particular bodily locations. Due to their exceptional adaptability and tunability, polymers have become essential components in the development of DDS. Their dominance in this field is highlighted by their ability to be engineered for controlled release, protection of bioactive molecules, and improvement of patient compliance through decreased dosing frequency. In his 1998 review, Robert Langer emphasized how polymers enable controlled release, which transforms the way that chronic illnesses like cancer and diabetes are treated. His research established the foundation for contemporary DDS, which emphasizes materials that are compatible with human physiology.

Key Features of Polymeric DDS

Biocompatibility and Biodegradability

The effectiveness of polymer-based drug delivery systems (DDS) is determined by their intrinsic qualities of biocompatibility and biodegradability. A biodegradable polymer breaks down gradually into non-toxic breakdown products that the body can safely metabolize or eliminate, while a biocompatible polymer does not interact with the host system in a way that causes inflammation or detrimental immune responses. One of the most commonly employed examples is poly(lactic-co-glycolic acid) (PLGA), which hydrolytically degrades to lactic acid and glycolic acid—both of which are normal metabolites that feed into the Krebs cycle (Danhier et al., 2012). PLGA degradation rate can be controlled by varying the proportion of lactic to glycolic acid and is thus amenable to a range of controlled-release uses, for example, in Lupron Depot® for hormone treatment. Equivalently, polylactic acid (PLA), which is biodegradable and derived from renewable sources and breaks down into lactic acid, is applied in implants as well as injectable fillers (Garlotta, 2001). Another aliphatic polyester, polycaprolactone (PCL), biodegrades at a slower rate and is ideally suited for long-term implantable systems (Dash & Konkimalla, 2012).

- Natural polymers with high enzymatic biodegradability and biocompatibility include silk fibroin, gelatin, chitosan, and alginate. A chitin derivative, chitosan is biodegradable via lysozyme and mucoadhesive, making it appropriate for use in nasal and oral drug delivery systems (Agnihotri et al., 2004). Alginate, a brown seaweed derivative, is used to make hydrogels for drug and cell encapsulation that break down in mild physiological conditions (Lee & Mooney, 2012). Tissue engineering scaffolds frequently use gelatin, a collagen derivative that breaks down with proteases (Van den Bulcke et al., 2000). For ocular and implantable DDS, silk fibroin, which is taken from silkworm cocoons, is appealing due to its mechanical durability and gradual enzymatic degradation (Kundu et al., 2013).

Mechanical Strength and functionalizability

Because of their high mechanical strength, reversibility and chemical functionalizability, synthetic polymers such as polyethylene glycol (PEG), poly(N-isopropylacrylamide) (PNIPAM), and poly(ethylene oxide)-based block copolymers are ideal for the development of complex drug delivery systems. For instance, PEG is a biocompatible, water-soluble polymer that is frequently used in the PEGylation process to extend the duration of therapeutic molecules' circulation. This allows for surface modification with targeting ligands, antibodies, or peptides to enable site-specific delivery in addition to improving solubility and reducing immunogenicity (Knop et al., 2010). Furthermore, different branching and molecular weights can be incorporated into synthetic polymers to alter mechanical characteristics like tensile strength and elasticity, which are crucial for hydrogels and implantable devices. For example, PEG hydrogels have been extensively studied in the delivery of ocular and subcutaneous drugs because they exhibit soft-tissue mimicking mechanical strengths and surface-modifiable characteristics (Tibbitt & Anseth, 2009).

Stimuli-Responsive Delivery

Smart polymers have emerged as a revolutionary class of materials that can undergo physical or chemical changes in response to specific environmental triggers such as pH, temperature, enzymes, or redox conditions. This allows for the controlled and localized release of drugs, especially in pathological microenvironments like tumors or inflamed tissues. A foundational study by Langer (1998) highlighted the use of such polymers for site-specific delivery, catalyzing a paradigm shift in drug delivery research. Thermo-responsive polymers like PNIPAM exhibit a sharp phase transition at body temperature, allowing for drug encapsulation at room temperature and release upon reaching physiological temperatures (Gil & Hudson, 2004).

pH-sensitive polymers such as poly (methacrylic acid) or poly (β-amino esters) are engineered to swell or degrade in acidic tumor microenvironments or endosomes, releasing their payload selectively at the disease site (Binauld & Stenzel, 2012). These polymers enable high therapeutic efficacy with minimal systemic toxicity and are being actively explored for applications ranging from cancer therapy to insulin delivery.

TYPES OF POLYMERS

Polymers play a pivotal role in drug delivery systems, offering tailored solutions based on their origin and properties. These materials are broadly categorized into *natural* and *synthetic* polymers, each with distinct advantages.

Advances in polymer chemistry have revolutionized drug delivery, enabling precision, safety, and patient compliance. These materials continue to shape modern medicine, from natural polymers mimicking biological systems to synthetic variants offering engineering flexibility.

Polymeric Drug Delivery Systems can be classified into Natural Polymers and Synthetic Polymers (Table 1)

Table 1. Classification of polymeric drug delivery systems

Type	Examples	Advantages	Limitations / Challenges
Natural Polymers	Chitosan, Alginate, Gelatin (Fang et al., 2022)	Inherent bioactivity (e.g., antimicrobial chitosan) - Low immunogenicity (e.g., alginate gels well with calcium ions)	Batch-to-batch variability - Thermal/mechanical instability (e.g., gelatin sensitivity)
Synthetic Polymers	PLGA, PEG, PCL, PMMA (Fang et al., 2022)	Tunable properties (e.g., PLGA degradation by copolymer ratio) - FDA-approved for multiple applications	Possible cytotoxicity or inflammatory responses (e.g., PMMA if impurities are present)

Natural Polymers

1. **Chitosan**: Derived from chitin, chitosan is celebrated for its biodegradability, mucoadhesive properties (ability to adhere to mucous membranes), and capacity to enhance drug permeability across biological barriers. For instance, it has been used in nasal and oral drug delivery to improve bioavailability (Agnihotri, Mallikarjuna, & Aminabhavi, 2004).

2. **Alginate**: This ion-sensitive polymer forms gels in the presence of calcium ions, making it ideal for wound healing and cell encapsulation. Its gentle gelation process protects sensitive therapeutics, such as proteins, during delivery (Lee & Mooney, 2012).
3. **Gelatin**: A collagen-derived, thermo-responsive polymer, gelatin transitions between sol and gel states with temperature changes, enabling its use in injectable scaffolds for tissue engineering (Lien, Ko, & Huang, 2009).

Synthetic Polymers

1. **PLA/PLGA**: Poly(lactic acid) and poly(lactic-co-glycolic acid) are FDA-approved polymers that degrade into biocompatible byproducts (lactic and glycolic acids). PLGA's tunable degradation rate makes it a cornerstone in controlled-release formulations, such as long-acting injectables (Makadia & Siegel, 2011).
2. **PEG**: Polyethylene glycol's "stealth" effect reduces immune recognition, prolonging circulation time. PEGylation (attachment of PEG chains) is widely used to enhance the stability of nanomedicines, such as mRNA vaccines (Knop et al., 2010).
3. **PCL**: Polycaprolactone's slow degradation (months to years) suits it for implants requiring sustained release, such as contraceptive devices (Woodruff & Hutmacher, 2010).

Key Properties

1. **Biodegradability**: Polymers must degrade *in vivo* to avoid surgical removal. For example, PLGA's hydrolysis-driven breakdown ensures safe elimination (Anderson & Shive, 2012).
2. **Biocompatibility**: Non-toxic polymers prevent adverse immune reactions. Chitosan's natural origin minimizes inflammatory responses, as shown in preclinical studies (Kean & Thanou, 2010).
3. **Molecular Weight (MW)**: Higher MW polymers degrade slower but may compromise drug loading efficiency. A balance is critical, as demonstrated in PLGA-based nanoparticles (Fredenberg et al., 2011).
4. **Hydrophilic/Hydrophobic Balance**: This balance governs interactions with biological fluids and drug encapsulation. For instance, hydrophobic polyesters like PLA encapsulate lipophilic drugs, while hydrophilic PEG enhances water solubility (Gref et al., 1994).

Functionalization and Conjugation

1. **Targeting Ligands**: Attaching ligands like folic acid enables active tumor targeting, as cancer cells overexpress folate receptors (Sudimack & Lee, 2000). Similarly, antibody-conjugated polymers (e.g., trastuzumab-PLGA) improve specificity for HER2-positive breast cancer (Bobo et al., 2016).
2. **Stimuli-Responsive Moieties**: pH-sensitive linkers (e.g., hydrazone bonds) release drugs in acidic tumor microenvironments, while redox-sensitive systems (e.g., disulfide bonds) respond to intracellular glutathione levels (Mura, Nicolas, & Couvreur, 2013).
3. **PEGylation**: Beyond stealth effects, PEGylation reduces renal clearance and prevents protein adsorption. For example, PEGylated liposomal doxorubicin (Doxil®) exhibits prolonged circulation compared to free drug (Harris & Chess, 2003).

MECHANISMS OF DRUG RELEASE IN POLYMER-BASED DELIVERY SYSTEMS

Diffusion-controlled systems are among the most widely used strategies in polymer-based drug delivery. These systems facilitate drug release through the passive movement of drug molecules across a polymer barrier. In matrix systems, the drug is uniformly distributed within a non-degradable polymer matrix, and release occurs as the drug diffuses through the polymer to reach the external environment. This mechanism offers prolonged, steady-state release and is commonly utilized in subcutaneous implants and ophthalmic inserts, such as those made from silicone. One of the earliest applications was in hormone regulation using silicone matrices (Folkman & Long, 1964). In contrast, reservoir systems encapsulate a drug core within a rate-limiting polymer membrane. These systems ensure a more constant, zero-order drug release due to the maintained concentration gradient across the membrane. Transdermal patches (e.g., nicotine or fentanyl patches) are classic examples, offering non-invasive, long-term delivery with reduced dosing frequency (Prausnitz & Langer, 2008).

Erosion-controlled systems rely on the degradation of the polymer matrix to facilitate drug release and can be classified into surface erosion and bulk erosion mechanisms. In surface-eroding polymers like polyanhydrides, hydrolysis occurs exclusively at the polymer's outer layer, resulting in a predictable, linear erosion profile. This enables precise control over drug release rates, making them suitable for implantable devices used in chemotherapy. Tamada and Langer (1993) demonstrated the utility of such systems in localized cancer treatments where temporal control is essential. On the other hand, bulk-eroding systems, such as those using poly(lactic-co-glycolic acid) (PLGA), degrade throughout the entire polymer matrix. Water infiltrates the structure, leading to swelling and eventual polymer breakdown (Figure 1). While this allows encapsulation of both hydrophilic and hydrophobic drugs, it also poses a risk of burst release. Nevertheless, PLGA remains extensively used in injectable microspheres and depot formulations, especially for vaccines, due to its biodegradable and biocompatible nature (Makadia & Siegel, 2011).

Figure 1. Mechanism of release of the drug

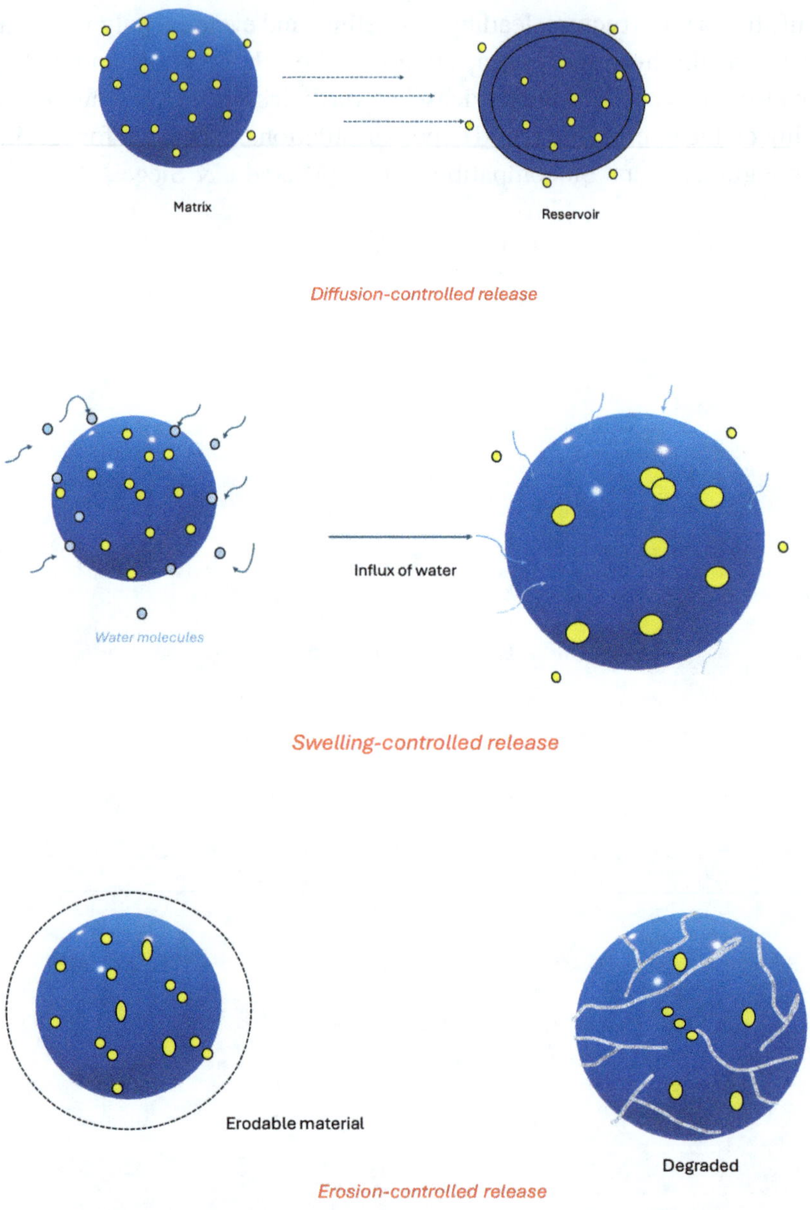

Stimuli-responsive systems represent a more advanced class of drug delivery mechanisms, offering site-specific and on-demand drug release. These smart polymers respond to changes in the biological environment, such as pH, temperature, or enzymatic activity, enabling precise targeting of pathological tissues. For example, pH-sensitive systems use polymers like chitosan that swell or dissolve in acidic environments, such as the tumor microenvironment or stomach, thus selectively releasing their payload where needed (Agnihotri et al., 2004). Similarly, temperature-responsive polymers like poly(N-isopropylacrylamide) (PNIPAM) exhibit a lower critical solution temperature (LCST) around 32°C. Above this temperature, PNIPAM undergoes a coil-to-globule transition, resulting in the expulsion of encapsulated drugs. This property is harnessed in combination with hyperthermia therapies, where local heating at tumor sites induces drug release (Qiu & Park, 2001). These innovative mechanisms mark a significant step toward personalized and responsive medicine (Figure 2).

Figure 2. Process factors of release

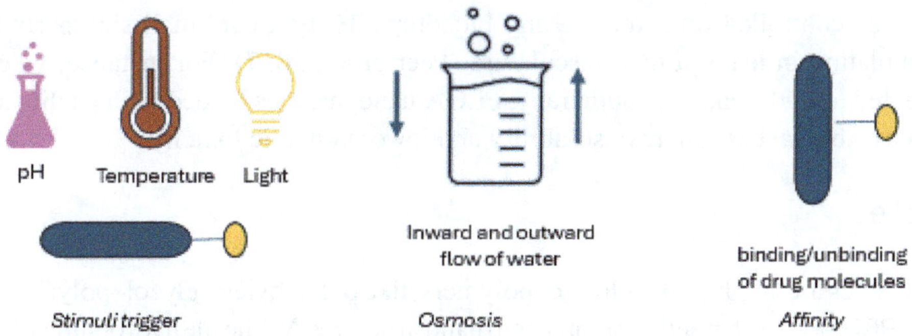

Understanding these mechanisms allows scientists to design drug delivery systems tailored to clinical needs. Diffusion and erosion control provide sustained release, while stimuli-responsive materials enable precision targeting. Together, these strategies enhance therapeutic efficacy and patient compliance.

POLYMER ARCHITECTURES AND DESIGN STRATEGIES FOR ADVANCED DRUG DELIVERY SYSTEMS

Polymer-based drug delivery systems (DDS) have revolutionized modern therapeutics by enabling precise control over drug release, improving bioavailability, and minimizing off-target effects. Their architectures and morphologies are tailored to meet specific clinical needs, ranging from passive tumor targeting to localized drug administration. Below, we elaborate on key polymer designs and surface modification strategies critical to their success.

Forms and Morphologies

Nanoparticles (10–200 nm)

Submicron carriers of polymeric nanoparticles are frequently employed for passive tumor targeting through the Enhanced Permeability and Retention (EPR) effect. While biodegradable polymers like poly(lactic-co-glycolic acid) (PLGA) guarantee controlled drug release and biocompatibility, their small size permits accumulation in leaky tumor vasculature (Peer et al., 2017). For instance, when compared to traditional chemotherapy, PLGA nanoparticles loaded with paclitaxel (such as Abraxane®) improve solubility and lower systemic toxicity.

Micelles

Micelles are amphiphilic block copolymers, like polyethylene glycol–poly(lactic acid) (PEG–PLA), that self-assemble into nanostructures. While the hydrophilic PEG shell extends the duration of circulation, the hydrophobic core encapsulates lipophilic medications. According to Kataoka et al. (2012), this design works especially well for administering medications like doxorubicin that are not very soluble in water.

Dendrimers

Hyperbranched polymers with functional surface groups and a high degree of symmetry are called dendrimers. High drug loading capacity and easy conjugation with targeting ligands are made possible by their multivalent surfaces. For example, poly(amidoamine) (PAMAM) dendrimers have been used to minimize immunogenicity while delivering anticancer medications such as methotrexate (Tomalia, 2005).

Hydrogels

For prolonged drug release in topical or implanted applications, hydrogels: three-dimensional networks that swell in aqueous environments—are perfect. Localized delivery of proteins or vaccines is made possible by temperature-sensitive hydrogels, such as poly(N-isopropylacrylamide) (PNIPAAm), which go through sol-gel transitions at body temperature (Hoare & Kohane, 2014).

Nanocapsules

By encasing medications in a polymeric membrane, these core-shell structures enable controlled release kinetics. Lipid-polymer hybrid nanocapsules, for instance, have been developed to shield siRNA from deterioration while promoting cellular uptake (Alvarez-Lorenzo & Concheiro, 2013).

Nanofibers

For transdermal patches or wound dressings, the electrospun nanofibers' high surface area-to-volume ratio makes them perfect. Polycaprolactone (PCL) nanofibers can be blended with medications, such as antibiotics, to provide sustained release at infection sites (Sill & von Recum, 2008).

Surface Modifications for Targeting

Passive Targeting

Passive targeting uses the inherent characteristics of tumors and their surroundings to deliver therapeutic agents exactly where they are required. The EPR effect passively accumulates nanoparticles by taking advantage of tumours' poor lymphatic drainage and leaky vasculature. Numerous FDA-approved nanomedicines, such as Doxil® (liposomal doxorubicin), are based on this approach (Peer et al., 2017).

Active Targeting

Active drug targeting uses ligands to attach to particular receptors or antigens on target cells in order to deliver a medication to a particular site in the body, such as a tumour or diseased tissue.

Antibody–Drug Conjugates (ADCs)

A monoclonal antibody (mAb) and a cytotoxic drug (payload) are combined via a linker in antibody-drug conjugates (ADCs), a form of targeted cancer treatment. ADCs like trastuzumab emtansine (Kadcyla®) combine monoclonal antibodies (e.g., anti-HER2) with cytotoxic drugs to selectively kill cancer cells overexpressing specific receptors (Beck et al., 2017).

Folate-Targeting

Cancer cells often overexpress folate receptors; conjugating nanoparticles with folic acid enhances uptake in ovarian and breast cancers (Low & Kularatne, 2009).

Transferrin-Conjugates

Transferrin receptors are abundant on blood-brain barrier (BBB) endothelial cells and cancer cells. Transferrin-coated nanoparticles improve brain delivery of drugs like doxorubicin (Ulbrich et al., 2011).

The versatility of polymer architectures, from dendrimers to hydrogels, enables tailored drug delivery solutions for diverse medical challenges. Coupling these designs with surface modifications like antibody conjugation or folate targeting further enhances specificity and therapeutic outcomes. Continued innovation in polymer science promises to address unmet clinical needs, including multidrug-resistant cancers and neurological disorders.

APPLICATIONS

Cancer Therapy

Polymer-based DDSs have made considerable progress in cancer therapy by improving treatment efficiency and reducing adverse side effects. One significant advancement is the usage of polymer drug conjugates, where the anticancer agent is chemically linked to polymers like polyglutamic acid, poly (N-(2-hydroxypropyl) meth acrylamide (HPMA) and polyethylene glycol (PEG). PEGylation, which involves attaching PEG molecules to a drug or nanoparticles, enhances water solubility, prolongs their presence in the bloodstream, and reduces the immune response (Parashar et al., 2024). These systems also assistance from the improved permeability and retention (EPR) effect, a passive targeting method which exploits the leaky vasculature in solid tumors, allowing these huge drug carriers to accumulate more in tumor tissues

(Jahan et al., 2025). In clinical uses, PEGylated liposomal doxorubicin (Doxil®) has been accepted for situations like multiple myeloma, ovarian cancer, and Kaposi's sarcoma, offering reduced heart toxicity and superior pharmacokinetics compared to encapsulated doxorubicin (Gabizon et al., 2025). Correspondingly, Abraxane®, which consists of paclitaxel bound to albumin nanoparticles, increases drug delivery to tumors without depending on dangerous solvents, giving rise to safer and more efficient treatment of breast and lung cancer (Barinholtz, 2012).

Antimicrobial and Antiviral Delivery

The rising frequency of viral outbreaks and antimicrobial resistance hinders the development of revolutionary delivery methods. Polymers like chitosan, a naturally occurring polysaccharide, have biocompatibility and intrinsic antimicrobial properties, making them exceptional possibilities for the delivery of antiviral and antibiotic medicines. The positive charge of chitosan increases its ability to bind with negatively charged bacterial membranes and mucosal surfaces, thus enhancing retention time and drug absorption at the sites of infection. Advanced smart polymers that are capable of stimuli-responsive drug release demonstrate a pioneering approach, where alterations in temperature, enzymatic or pH conditions activate the localized release of antimicrobials specifically at the site of infections. For example, pH-responsive polymers specifically discharge drugs in acidic environments associated with infections, hence decreasing exposure and potential side effects. These targeted delivery systems improve treatment efficacy and mitigate the risk of resistance by confirming therapeutic drug levels are sustained where they are best required (Žigrayová et al., 2023).

Gene and RNA Delivery

Delivering genetic materials like DNA, siRNA, and mRNA poses discrete challenges, including cellular uptake, stability, and intercellular transport. Polymeric transporters produce polyplexes through electrostatic interactions with nucleic acids, safeguard them from cellular uptake and enzymatic degradation. Lipid-polymer hybrid nanoparticles combine the biocompatibility of lipids with the robustness of polymers, permitting the effective delivery of gene editing tools and mRNA. The efficiency of mRNA vaccines against COVID-19, which use lipid nanoparticles (LNP) carriers to encapsulate and safeguard mRNA, has represented a noteworthy progress in nucleic acid therapeutics (Niculescu et al., 2021). These platforms warrant effective delivery to target cells, triggering immune response or gene expression with remarkable specificity and regulated release of payloads (Li et al.,

2024). Polymer-based gene delivery continues to grow with applications in genetic disorders, cancer immunotherapy, and regenerative medicine (Hou et al., 2021).

Ocular, Oral, and Transdermal Delivery

Administering medicines through oral, ocular, and transdermal routes encounters problems such as swift clearance, restricted permeability, and enzymatic breakdown. Mucoadhesive polymers like carbomers and chitosan increase drug retention on mucosal surfaces such as those found in the oral cavity or eyes by making adhesive connections that facilitate prolonged bioavailability and better drug absorption. Increasingly, thin polymer patches and films are used for controlled drug release, increasing patient compliance by weakening the frequency of dosages (Lammerding et al., 2024). The transdermal route meaningfully profits from polymeric microneedle arrays, which easily penetrate the stratum corneum barrier and deliver medicines straight into the dermis. These microneedles are prepared from biodegradable polymers that dissolve, delivering vaccines, analgesics marking a major shift towards less aggressive and more patient-friendly drug administration (Tsung et al., 2023).

CURRENT CHALLENGES AND LIMITATIONS

Scale Up and Industrial Translation

Polymer-based DDSs have displayed major potential in attaining controlled release, augmenting therapeutic effectiveness and targeted delivery. However, the transition from lab-scale synthesis to large-scale industrial production poses a considerable hurdle to commercialization. Methods such as microfluidics or nanoprecipitation used for creating hydrogels and nanoparticles often encounter scalability disputes owing to variations in process dynamics, mixing, and thermal gradients. Uniformity is negotiated by variability between batches in stability, polymer characteristics and drug loading, which necessitates severe quality control and sophisticated PAT/QbD frameworks (Sarma et al., 2025). The cost associated with costly reagents, GMP compliance, low synthesize and the need for cold chain storage makes industrialization financially challenging. Current infrastructure often lacks the specialized equipment necessary for nanoparticle production and real-time analytical monitoring. Documentation of stimuli-responsive or dynamic systems, as well as rigorous validation following process changes, are regulatory requirements. Inadequate process documentation and regulatory awakened further restrict knowledge transfer from academia to industry. Furthermore, despite promising preclinical results, intellectual property restrictions and licensing delays hamper commercialization. To realize polymer-based DDS's clinical and commercial promise, process automation, modular manufacturing platforms, regulatory-aligned R&D, and active collaboration across sectors are required (Ghavimi et al., 2021).

Stability and Storage Issues

Polymer-based DDS face significant stability challenges during storage, which can impact regulatory compliance, therapeutic efficacy, and safety. Functional groups intended for targeting or stimuli-responsiveness may decrease over time, disturbing performance. Another concern is the drug leakage during storage, particularly in nano formulations with large surface-to-volume ratios, which simplify diffusion. Lyophilization is frequently employed to extend shelf life, but it requires careful optimization of cryo/lyoprotectants to preserve particle integrity. Reconstitution must constantly replicate the novel formulation, which can be thought-provoking. Light, oxygen, and temperature are environmental factors that often require costly packaging solutions, storage and specialized conditions. Tackling these difficulties is important for maintaining DDS shelf life, quality, enhancing, and ensuring economic and clinical viability (Kumar et al., 2025).

Regulatory Hurdles

Polymer-based DDS present significant regulatory challenges due to their structural complexity and lack of standardized evaluation procedures. Unlike traditional pharmaceuticals, polymeric DDS ranging from nanoparticles to hydrogels can fall under confusing regulatory categories, sometimes being classified as combination goods that necessitate multi-agency control. This results in variable review timelines and unclear approval standards. Inadequate analytical standards exist for measuring crucial quality parameters such as drug loading, particle size, and degradation. Toxicology investigations must consider both the active medication and the polymer carrier, but validated models for predicting long-term safety and immunogenicity are lacking. Stability standards specified by ICH Q1A (R2) are mostly hard to achieve due to colloidal, novel polymer or chemical instability. The design of clinical trials must account for carrier pharmacokinetics, novel hazards and informed consent for emerging materials. Regulatory differences across regions (e.g., EMA, PMDA, Health Canada) hamper global commercialization. Besides, post-market changes in materials or processes may need additional validation (Baryakova et al., 2023). To address these tasks, robust characterization tools, early regulatory interaction, and harmonized international frameworks are needed.

Toxicity and Immunogenicity Concerns

Polymer-based DDS face considerable biocompatibility issues, including immunogenicity and toxicity, that can delay clinical translation. Surface characteristics, such as hydrophobicity and ligands, influence immune recognition and clearance. Modified surfaces can trigger opsonization or anti-drug antibody (ADA) responses, diminishing therapeutic efficiency with subsequent treatments. Certain degradation products from polymers may be toxic or reactive, leading to oxidative stress and cellular damage. The "PEG dilemma" illustrates how PEGylation extends circulation time, but it can also provoke anti-PEG immunological responses and hypersensitivity. Monomers, residual solvents, and catalysts left from manufacturing also contribute to toxicity, requiring strict cleaning protocols. Moreover, animal models may not precisely predict complicating risk assessment and human responses. To tackle these trials, thoughtful polymer design, immunological profiling and comprehensive material screening that highlight inert, degradable or immune evasive properties are vital for the effective and safe development of DDS (Arabiyat, 2024).

FUTURE TRENDS AND EMERGING TECHNOLOGIES

Stimuli-Responsive and Smart Polymers

Stimulus-responsive and smart polymers are a significant improvement in DDS, allowing for targeted, accurate and controlled drug release. These devices react to specific physiological or environmental stimuli such as temperature, enzymes, pH, redox potential or light, enabling spatiotemporal drug delivery in complicated biological contexts. Multi- stimuli responsive polymers are especially intriguing because they provide synergistic responsiveness, such as redox and pH sensitive systems that are stable in the circulation but break down quickly in the reducing and acidic tumor microenvironment. This improves site-specific delivery while minimizing off target effects. Bioinspired smart polymers replicate natural biological systems, which improve biocompatibility and efficiency (Pradhan et al., 2025). Zwitterionic polymers, for example mimic cell membrane neutrality to minimize immune clearance and extend circulation. Enzyme-responsive conjugates can selectively release medicines in the presence of disease-associated enzymes such as matrix metalloproteinase (MMPs) or elastase. Furthermore, logic-gated or signal responsive polymers are being created for precision medicine, with medicine released only in reaction to complicated molecular inputs. These new materials have the potential to change treatments for cancer, inflammation and autoimmune illnesses by delivering intelligent, adaptive delivery platforms that are adapted to the dynamic physiological environment (Liechty et al., 2010).

3D Printed Drug Delivery Systems

3D printing is renovating drug distribution by patient specific system to be built layer by layer with pharmaceutical-grade polymers. This technique has significant advantages over traditional approaches, such as personalized drug release patterns, customizable doses and the possibility of on demand manufacture in hospitals or pharmacies. It promotes personalized therapy by incorporating unique patient demands such as genetics and age. Fused deposition modeling (FDM) is commonly utilized for multi drug poly pills. Stereolithography (SLA) for high-resolution structures, such as inkjet printing and microneedles for heat sensitive medicines (Majrashi et al., 2024). Each approach requires polymers that strike a compromise between biocompatibility, printability, and drug kinetics. Recent breakthroughs have integrated stimuli-responsive, smart polymers into 3D printed electronics. For example, hydrogels containing thermo-responsive PNIAAm release medicines in reaction to fever and multi-material printing enables hybrid devices with embedded sensors or responsive behavior (Nizam et al., 2024). However, regulatory problems continue. Current frameworks are not sufficiently optimized for decentralized or patient-specific manufacturing. Ensuring quality, polymer drug compatibility, and consistency is also a challenge. Despite this, the FDA's clearance of spritam® demonstrates clinical feasibility. Looking ahead, AI-driven design, tissue-specific implants, and portable 3D printers have the potential to improve personalized medicine by providing decentralized, scalable solutions for global health care (Aftab et al., 2025).

Artificial Intelligence in DDS Design

Artificial intelligence (AI) is transforming the design of polymer-based DDS by allowing for personalized, data-driven, and efficient formulation strategies. AI simplifies polymer selection, models complex release kinetics, and predicts drug polymer compatibility, considerably decreasing the need for trial-and-error experiments. Random forest, deep learning models, and neural networks are popular methods for predicting drug release profiles, encapsulation efficiency and degradation rate. AI also allows for personalized DDS by combining patient data (e.g., genetic profiles, electronic health records) to modify drug dose and release. Closed-loop systems powered by AI can adjust drug delivery in real time in response to physiological indications (Kapoor et al., 2024). In smart polymer creation, AI-powered generative models search enormous chemical spaces for stimuli-responsive polymers with specific behaviors such as temperature or pH-triggered release. Furthermore,

AI works with 3D printing and robotic platforms to optimize print geometries and automate formulation processes using digital twins and predictive control (Gholap et al., 2024). Despite issues such as data quality, regulatory ambiguity, and interpretability, AI is reshaping the DDS market. As explainable models and standardized data sets emerge, AI will play an increasingly important role in the development of patient-specific, intelligent, and adaptable medicine delivery solutions.

Personalized and Precision Medicine

Personalized and precision medicine transforms DDS by adapting therapies to individual patient features such as genetics, biomarkers, and disease patterns. Polymer-based DDSs have distinct advantages in this paradigm because of their adjustable characteristics, biocompatibility, and adaptability. These systems can be designed with controlled release, targeted delivery, and patient-specific degradation rates. Polymeric DDs can be tailored to specific subpopulations by the integration of genomes and biomarkers data, such as nanoparticles targeting HER2-positive breast cancer or folate receptor-overexpressing tumours. Smart polymers improve precision by allowing for site-specific, stimulus-responsive medicine release in response to pH, temperature, or enzyme activity. Personalized implants and 3D printed medicine devices provide customized geometries and multi-drug delivery for complex treatment regimens, particularly in elderly or chronically unwell patients. However, implementation problems include high production costs, regulatory complications, and data integration issues. Future improvements will use AI for predictive modelling, wearable sensors for real-time monitoring, and bio printing to incorporate DDS in personalized scaffolds. As medicine progresses towards predictions and preventions, polymeric DDS will be critical in providing individualized medicines with increased efficacy and safety. These versatilities make them valuable tools in the future generation of patient-centered healthcare solutions (Alzoubi et al., 2023).

CONCLUSION

Polymer-based DDSs have changed the landscape of treatments by allowing for regulated, prolonged, and targeted drug delivery release. Their adaptable chemical and physical features allow the precise design of carriers capable of overcoming biological barriers, enhancing drug solubility, and protecting labile medicines from lowering systemic toxicity and degradation. These systems are particularly effective in treating complicated diseases, including infectious, cancer, and autoimmune diseases, where traditional medicines often failed due to side effects and lack of specificity. The capability to control site-specific delivery and drug release kinetics using nanoparticles, micelles, polymer matrices, or hydrogels signifies a watershed moment in pharmaceutical science. Additionally, the improvement of stimuli-responsive and smart polymers has shaped novel prospects for real-time and adaptive drug administration, ushering in a future of more patient-centered and effective treatment methods. Polymer-based DDS symbolizes the intersection of numerous scientific disciplines, including biomedical engineering, polymer chemistry, nanotechnology, material science, clinical medicine, and molecular biology. This multidisciplinary basis not only quickens invention but also increases polymeric systems' adaptability to a wide range of clinical needs. Several FDA-approved formulations, including Doxil® (liposomal doxorubicin) and Abraxane® (albumin-bound paclitaxel), have displayed improved therapeutic indices and patient outcomes, representing the translational potential of these systems. Moreover, improvements in artificial intelligence, additive manufacturing, and precision medicine are pushing the progress of next-generation polymeric DDS that are more personalized, smarter, and capable of cooperative individual diversity in treatment response and disease progression. These integrative methods reveal polymer-based DDS's readiness to transfer from bench to bedside in the era of precise and personalized treatments. Despite preclinical results and scientific advances, many hurdles inhibit polymer-based DDS from being broadly adopted in clinical practice. These include long-term safety, regulatory approval, scalability, and reproducibility. As these systems get more complex, with personalized designs, stimulus-responsive features or AI-guided formulations, well-known regulatory frameworks may struggle to measure their performance using normal standards. To accelerate the translation process, industry stakeholders, academic researchers, regulatory agencies, and clinicians must work together. Establishing standardized methods for efficiency, safety, and quality control, as well as open communication lines between regulators and developers, will be essential. In addition, international consortia and public-private partnerships can help with adaptive regulatory approaches, pooled resources, and real-world data integration. These collaborative works will not only quicken the clinical translation of polymer DDS but also guarantee its safe and effective application in a range of healthcare settings.

REFERENCES

Aftab, M., Ikram, S., Ullah, M., Khan, N., Naeem, M., Khan, M. A., O'g'li, R. B. B., Qizi, K. S. S., Ugli, O. O. E., Abdurasulovna, B. M., & Qizi, O. K. A. (2025). Recent trends and future directions in 3D printing of biocompatible polymers. *Journal of Manufacturing and Materials Processing*, *9*(4), 129. DOI: 10.3390/jmmp9040129

Agnihotri, S. A., Mallikarjuna, N. N., & Aminabhavi, T. M. (2004). Recent advances on chitosan-based micro- and nanoparticles in drug delivery. *Journal of Controlled Release*, *100*(1), 5–28. DOI: 10.1016/j.jconrel.2004.08.010 PMID: 15491807

Alvarez-Lorenzo, C., & Concheiro, A. (2013). *Handbook of biodegradable polymeric materials and their applications*. American Scientific Publishers.

Alzoubi, L., Aljabali, A. A., & Tambuwala, M. M. (2023). Empowering Precision Medicine: The impact of 3D printing on personalized Therapeutic. *AAPS PharmSciTech*, *24*(8), 228. Advance online publication. DOI: 10.1208/s12249-023-02682-w PMID: 37964180

Anderson, J. M., & Shive, M. S. (2012). Biodegradation and biocompatibility of PLA and PLGA microspheres. *Advanced Drug Delivery Reviews*, *64*, 72–82. DOI: 10.1016/j.addr.2012.09.004

Arabiyat, S. A. (2024). Prospectus and Concerns of Immunomodulatory Nanotechnologies and Nanoparticles Biocompatibility and Toxicity. In *Nanotechnology Based Microbicides and Immune Stimulators* (pp. 165–189). https://doi.org/DOI: 10.1007/978-981-96-0397-8_10

Baryakova, T. H., Pogostin, B. H., Langer, R., & McHugh, K. J. (2023). Overcoming barriers to patient adherence: The case for developing innovative drug delivery systems. *Nature Reviews. Drug Discovery*, *22*(5), 387–409. DOI: 10.1038/s41573-023-00670-0 PMID: 36973491

Beck, A., Goetsch, L., Dumontet, C., & Corvaïa, N. (2017). Strategies and challenges for the next generation of antibody–drug conjugates. *Nature Reviews. Drug Discovery*, *16*(5), 315–337. DOI: 10.1038/nrd.2016.268 PMID: 28303026

Binauld, S., & Stenzel, M. H. (2012). Acid-degradable polymers for drug delivery: A decade of innovation. *Chemical Communications*, *49*(21), 2082–2102. DOI: 10.1039/c2cc36589h PMID: 23320254

Bobo, D., Robinson, K. J., Islam, J., Thurecht, K. J., & Corrie, S. R. (2016). Nanoparticle-based medicines: A review of FDA-approved materials and clinical trials to date. *Pharmaceutical Research*, *33*(10), 2373–2387. DOI: 10.1007/s11095-016-1958-5 PMID: 27299311

Fang, Z., Hu, J., Yuan, Q., Zhou, W., & Fan, L. (2022). Biological responses to PMMA-based bone cements and potential improvements. *Materials (Basel)*, *15*(1), 145. DOI: 10.3390/ma15010145

Folkman, J., & Long, D. M. (1964). The use of silicone rubber as a carrier for prolonged drug therapy. *The Journal of Surgical Research*, *4*(3), 139–142. DOI: 10.1016/S0022-4804(64)80040-8 PMID: 14130164

Fredenberg, S., Wahlgren, M., Reslow, M., & Axelsson, A. (2011). The mechanisms of drug release in poly(lactic-co-glycolic acid)-based drug delivery systems—A review. *International Journal of Pharmaceutics*, *415*(1–2), 34–52. DOI: 10.1016/j.ijpharm.2011.05.049 PMID: 21640806

Gabizon, A. A., Gabizon-Peretz, S., Modaresahmadi, S., & La-Beck, N. M. (2025). Thirty years from FDA approval of pegylated liposomal doxorubicin (Doxil/Caelyx): An updated analysis and future perspective. *BMJ Oncology*, *4*(1), e000573. DOI: 10.1136/bmjonc-2024-000573 PMID: 39885941

Ghavimi, S. A., Gehret, P. M., Aronson, M. R., Schipani, R., Smith, K. W., Borek, R. C., Germiller, J. A., Jacobs, I. N., Zur, K. B., & Gottardi, R. (2021). Drug delivery to the pediatric upper airway. *Advanced Drug Delivery Reviews*, *174*, 168–189. DOI: 10.1016/j.addr.2021.04.004 PMID: 33845038

Gholap, A. D., Uddin, M. J., Faiyazuddin, M., Omri, A., Gowri, S., & Khalid, M. (2024). Advances in artificial intelligence for drug delivery and development: A comprehensive review. *Computers in Biology and Medicine*, *178*, 108702. DOI: 10.1016/j.compbiomed.2024.108702 PMID: 38878397

Gil, E. S., & Hudson, S. M. (2004). Stimuli-responsive polymers and their bioconjugates. *Progress in Polymer Science*, *29*(12), 1173–1222. DOI: 10.1016/j.progpolymsci.2004.08.003

Gref, R., Minamitake, Y., Peracchia, M. T., Trubetskoy, V., Torchilin, V., & Langer, R. (1994). Biodegradable long-circulating polymeric nanospheres. *Science*, *263*(5153), 1600–1603. DOI: 10.1126/science.8128245 PMID: 8128245

Harris, J. M., & Chess, R. B. (2003). Effect of PEGylation on pharmaceuticals. *Nature Reviews. Drug Discovery*, *2*(3), 214–221. DOI: 10.1038/nrd1033 PMID: 12612647

Hoare, T. R., & Kohane, D. S. (2014). Hydrogels in drug delivery: Progress and challenges. *Polymer*, *49*(8), 1993–2007. DOI: 10.1016/j.polymer.2008.01.027

Hou, X., Zaks, T., Langer, R., & Dong, Y. (2021). Lipid nanoparticles for mRNA delivery. *Nature Reviews. Materials*, *6*(12), 1078–1094. DOI: 10.1038/s41578-021-00358-0 PMID: 34394960

Jahan, A., Lonare, M. K., Khatoon, S., & Devi, K. K. (2025). Polymer-Based Drug Delivery Systems: Design and Characterization. In *Methods in pharmacology and toxicology* (pp. 111–131). https://doi.org/DOI: 10.1007/978-1-0716-4554-3_6

Kapoor, D. U., Sharma, J. B., Gandhi, S. M., Prajapati, B. G., Thanawuth, K., Limmatvapirat, S., & Sriamornsak, P. (2024). AI-driven design and optimization of nanoparticle-based drug delivery systems. *Science. Engineering and Health Studies*, *24010003*, 24010003. Advance online publication. DOI: 10.69598/sehs.18.24010003

Kataoka, K., Harada, A., & Nagasaki, Y. (2012). Block copolymer micelles for drug delivery: Design, characterization, and biological significance. *Advanced Drug Delivery Reviews*, *64*, 37–48. DOI: 10.1016/j.addr.2012.09.013 PMID: 11251249

Kean, T., & Thanou, M. (2010). Biodegradation, biodistribution and toxicity of chitosan. *Advanced Drug Delivery Reviews*, *62*(1), 3–11. DOI: 10.1016/j.addr.2009.09.004 PMID: 19800377

Knop, K., Hoogenboom, R., Fischer, D., & Schubert, U. S. (2010). Poly(ethylene glycol) in drug delivery: Pros and cons as well as potential alternatives. *Angewandte Chemie International Edition*, *49*(36), 6288–6308. DOI: 10.1002/anie.200902672 PMID: 20648499

Kumar, P., Goel, A., Dash, B., Rishishwar, P. S. J., Suryawanshi, M. V., Rishishwar, S., & Asokan, N. (2025). Innovations in drug delivery systems for Biologics: Enhancing stability and targeted delivery for Next-Generation therapeutics. *Chinese. Journal of Applied Physiology (Bethesda, Md.)*, *20250001*, e20250001. Advance online publication. DOI: 10.62958/j.cjap.2025.001 PMID: 39748757

Lammerding, L. C., Arora, A., Braun, S., & Breitkreutz, J. (2024). Spatial separation of different drug substances in one microneedle array patch by combining inkjet printing and micromolding technology. *International Journal of Pharmaceutics*, *125102*. Advance online publication. DOI: 10.1016/j.ijpharm.2024.125102 PMID: 39708846

Langer, R. (1998). Drug delivery and targeting. *Nature*, *392*(6679, Suppl), 5–10. DOI: 10.1038/32380 PMID: 9579855

Lee, K. Y., & Mooney, D. J. (2012). Alginate: Properties and biomedical applications. *Progress in Polymer Science, 37*(1), 106–126. DOI: 10.1016/j.progpolymsci.2011.06.003 PMID: 22125349

Li, J., Zhang, Y., Yang, Y., & Sun, T. (2024). Advancing mRNA therapeutics: The role and future of nanoparticle delivery systems. *Molecular Pharmaceutics, 21*(8), 3743–3763. DOI: 10.1021/acs.molpharmaceut.4c00276 PMID: 38953708

Liechty, W. B., Kryscio, D. R., Slaughter, B. V., & Peppas, N. A. (2010). Polymers for drug delivery systems. *Annual Review of Chemical and Biomolecular Engineering, 1*(1), 149–173. DOI: 10.1146/annurev-chembioeng-073009-100847 PMID: 22432577

Lien, S.-M., Ko, L.-Y., & Huang, T.-J. (2009). Effect of pore size on ECM secretion and cell growth in gelatin scaffold for articular cartilage tissue engineering. *Acta Biomaterialia, 5*(2), 670–679. DOI: 10.1016/j.actbio.2008.09.020 PMID: 18951858

Low, P. S., & Kularatne, S. A. (2009). Folate-targeted therapeutic and imaging agents for cancer. *Current Opinion in Chemical Biology, 13*(3), 256–262. DOI: 10.1016/j.cbpa.2009.03.022 PMID: 19419901

Majrashi, M. a. A., Yahya, E. B., & Mushtaq, R. Y., HPS, A. K., Rizg, W. Y., Alissa, M., Alkharobi, H., Badr, M. Y., & Hosny, K. M. (. (2024). Revolutionizing drug delivery: Exploring the impact of advanced 3D printing technologies on polymer-based systems. *Journal of Drug Delivery Science and Technology, 98*, 105839. DOI: 10.1016/j.jddst.2024.105839

Makadia, H. K., & Siegel, S. J. (2011). Poly(lactic-co-glycolic acid) (PLGA) as biodegradable controlled drug delivery carrier. *Polymers, 3*(3), 1377–1397. DOI: 10.3390/polym3031377 PMID: 22577513

Mura, S., Nicolas, J., & Couvreur, P. (2013). Stimuli-responsive nanocarriers for drug delivery. *Nature Materials, 12*(11), 991–1003. DOI: 10.1038/nmat3776 PMID: 24150417

Niculescu, A., Bîrcă, A. C., & Grumezescu, A. M. (2021). New applications of lipid and Polymer-Based nanoparticles for nucleic acids delivery. *Pharmaceutics, 13*(12), 2053. DOI: 10.3390/pharmaceutics13122053 PMID: 34959335

Nizam, M., Purohit, R., & Taufik, M. (2024). 3D printing in Healthcare: A review on drug printing, challenges and future perspectives. *Materials Today. Communications, 40*, 110199. DOI: 10.1016/j.mtcomm.2024.110199

Parashar, A. K., Saraogi, G. K., Jain, P. K., Kurmi, B., Shrivastava, V., & Arora, V. (2024). Polymer-drug conjugates: Revolutionizing nanotheranostic agents for diagnosis and therapy. *Discover. Oncology*, *15*(1), 641. Advance online publication. DOI: 10.1007/s12672-024-01509-9 PMID: 39527173

Peer, D., Karp, J. M., Hong, S., Farokhzad, O. C., Margalit, R., & Langer, R. (2017). Nanocarriers as an emerging platform for cancer therapy. *Nature Nanotechnology*, *2*(12), 751–760. DOI: 10.1038/nnano.2007.387 PMID: 18654426

Pradhan, T., Chelike, D. K., Roy, D., Pramanik, T., & Dolui, S. (2025). Stimuli-Responsive Multiacceptor Conjugated Polymers: Recent trend and future direction. *ACS Polymers Au*, *5*(2), 62–79. Advance online publication. DOI: 10.1021/acspolymersau.4c00082 PMID: 40226348

Prausnitz, M. R., & Langer, R. (2008). Transdermal drug delivery. *Nature Biotechnology*, *26*(11), 1261–1268. DOI: 10.1038/nbt.1504 PMID: 18997767

Qiu, Y., & Park, K. (2001). Environment-sensitive hydrogels for drug delivery. *Advanced Drug Delivery Reviews*, *53*(3), 321–339. DOI: 10.1016/S0169-409X(01)00203-4 PMID: 11744175

Sarma, L., Singh, S., Anjali, N., & Dutta, D. (2025). Challenges and future directions for Next-Generation Drug Delivery Systems. In *Methods in pharmacology and toxicology* (pp. 571–596). https://doi.org/DOI: 10.1007/978-1-0716-4554-3_25

Sill, T. J., & von Recum, H. A. (2008). Electrospinning: Applications in drug delivery and tissue engineering. *Biomaterials*, *29*(13), 1989–2006. DOI: 10.1016/j.biomaterials.2008.01.011 PMID: 18281090

Sudimack, J., & Lee, R. J. (2000). Targeted drug delivery via the folate receptor. *Advanced Drug Delivery Reviews*, *41*(2), 147–162. DOI: 10.1016/S0169-409X(99)00062-9 PMID: 10699311

Tamada, J. A., & Langer, R. (1993). Erosion kinetics of hydrolytically degradable polymers. *Proceedings of the National Academy of Sciences of the United States of America*, *90*(3), 552–556. DOI: 10.1073/pnas.90.2.552 PMID: 8421690

Tibbitt, M. W., & Anseth, K. S. (2009). Hydrogels as extracellular matrix mimics for 3D cell culture. *Biotechnology and Bioengineering*, *103*(4), 655–663. DOI: 10.1002/bit.22361 PMID: 19472329

Tomalia, D. A. (2005). Birth of a new macromolecular architecture: Dendrimers as quantized building blocks for nanoscale synthetic polymer chemistry. *Progress in Polymer Science*, *30*(3–4), 294–324. DOI: 10.1016/j.progpolymsci.2005.01.007

Tsung, T., Tsai, Y., Lee, H., Chen, Y., & Lu, D. (2023). Biodegradable Polymer-Based Drug-Delivery systems for ocular diseases. *International Journal of Molecular Sciences*, *24*(16), 12976. DOI: 10.3390/ijms241612976 PMID: 37629157

Ulbrich, K., Hekmatara, T., Herbert, E., & Kreuter, J. (2011). Transferrin- and transferrin-receptor-antibody-modified nanoparticles enable drug delivery across the blood-brain barrier (BBB). *European Journal of Pharmaceutics and Biopharmaceutics*, *71*(2), 251–256. DOI: 10.1016/j.ejpb.2008.08.021 PMID: 18805484

Woodruff, M. A., & Hutmacher, D. W. (2010). The return of a forgotten polymer—Polycaprolactone in the 21st century. *Progress in Polymer Science*, *35*(10), 1217–1256. DOI: 10.1016/j.progpolymsci.2010.04.002

Žigrayová, D., Mikušová, V., & Mikuš, P. (2023). Advances in antiviral delivery systems and Chitosan-Based polymeric and nanoparticulate antivirals and antiviral carriers. *Viruses*, *15*(3), 647. DOI: 10.3390/v15030647 PMID: 36992356

Chapter 2
Biodegradable Polymers in Modern Therapeutics: A Revolution in Sustained Drug Release

Fatema Tuz Zohera
https://orcid.org/0009-0006-4718-5390
International Islamic University Malaysia, Malaysia

Wan Mohd Azizi Wan Sulaiman
University MAIWP International, Malaysia

Abul Kalam Azad
https://orcid.org/0000-0003-4874-1169
University MAIWP International, Malaysia

ABSTRACT

Biodegradable polymers play a pivotal role in modern drug delivery by enabling sustained and controlled release, reducing systemic toxicity, and improving patient adherence. These materials degrade into non-toxic byproducts, eliminating the need for surgical removal and supporting various formulations like nanoparticles, hydrogels, and implants. This review explores natural and synthetic polymers, their degradation mechanisms, and innovations such as stimuli-responsive systems and targeted delivery. Applications span oncology, infectious diseases, cardiovascular therapy, diabetes, and tissue engineering. Key challenges include consistent drug loading, initial burst control, and manufacturing scalability. The paper also highlights safety, regulatory considerations, and future directions involving smart materials and biosensor integration. Biodegradable polymers continue to shape the future of personalized and regenerative medicine.

DOI: 10.4018/979-8-3373-1047-3.ch002

INTRODUCTION

Traditional drug delivery strategies often encounter challenges such as inconsistent bioavailability, the need for frequent administration, and undesired systemic effects. These limitations highlight the necessity for innovative delivery platforms that enable prolonged and controlled release of therapeutic compounds. In this context, biodegradable polymers have emerged as a pivotal advancement, allowing extended drug availability, improved clinical outcomes, and better adherence to treatment regimens. Because they decompose into non-toxic, metabolizable byproducts, they eliminate the need for surgical retrieval and reduce the risk of long-term side effects. This review examines the diverse applications and advancements of biodegradable polymers in therapeutics, encompassing their types, degradation mechanisms, innovations, and real-world clinical usage.

Overview of Drug Delivery Challenges

The field of drug delivery has faced significant hurdles in achieving efficient, targeted, and patient-compliant therapeutic systems. Traditional dosage forms often lack the ability to maintain therapeutic drug levels in plasma for prolonged durations. Consequently, many medications require frequent administration, leading to poor patient compliance and variable drug exposure. Additionally, issues such as rapid systemic clearance, instability of therapeutic agents, non-specific targeting, and dose-dependent toxicity further complicate effective drug delivery. Novel drug delivery platforms are thus needed to overcome these limitations, ensuring optimal therapeutic outcomes with minimal side effects (Langer & Peppas, 2003).

Need for Sustained Release Systems

Sustained or controlled drug release systems are engineered to maintain drug plasma concentrations within the optimal therapeutic window over prolonged periods, thereby enhancing treatment outcomes and patient adherence. Unlike conventional dosage forms that cause rapid fluctuations in drug levels, these systems provide a steady release rate, minimizing the risks of sub-therapeutic dosing or toxicity associated with peak-trough variations (Siepmann & Siepmann, 2012). This controlled release is especially critical for drugs with narrow therapeutic indices, where maintaining a precise concentration range is essential to avoid adverse effects or therapeutic failure. Additionally, sustained delivery improves bioavailability by facilitating gradual absorption, which can be particularly beneficial for poorly soluble or unstable drugs that degrade quickly under physiological conditions. Various technologies have been developed to realize these goals, including biodegradable implants that provide localized delivery over weeks or months, micro- and nano-carriers that enhance cellular uptake and targeted release, and injectable depots that offer minimally invasive administration with prolonged drug presence. The design of these platforms often involves tailoring polymer properties, such as degradation rate and hydrophilicity, to modulate drug release kinetics according to specific therapeutic needs. Collectively, these advanced delivery systems contribute to improved pharmacokinetics, reduced dosing frequency, enhanced patient compliance, and optimized therapeutic efficacy (Siepmann & Siepmann, 2012).

Role of Biodegradable Polymers in Modern Therapeutics

Biodegradable polymers are a class of materials capable of breaking down within the physiological environment into non-toxic, metabolizable byproducts such as water, carbon dioxide, and naturally occurring acids. This intrinsic biodegradability eliminates the necessity for surgical removal after the therapeutic function is completed, significantly improving patient compliance and safety (Makadia & Siegel, 2011). Their ability to encapsulate a wide range of therapeutic agents—ranging from small molecules to proteins and nucleic acids—and release them in a controlled, sustained, or stimuli-responsive manner makes them exceptionally suitable for modern drug delivery systems. Among synthetic biodegradable polymers, **poly(lactic-co-glycolic acid) (PLGA)**, **polylactic acid (PLA)**, and **polycaprolactone (PCL)** have garnered substantial attention due to their well-characterized degradation kinetics, mechanical properties, and established safety profiles. PLGA, in particular, is FDA-approved and extensively used in injectable microspheres and implants for controlled drug release owing to its tunable lactic to glycolic acid ratio that directly affects degradation rates and drug release profiles (Danhier et al., 2012). In addition to synthetic variants, natural biodegradable polymers such as **chitosan**, **alginate**, **gelatin**, and **collagen** offer remarkable **biocompatibility**, **low immunogenicity**, and inherent bioactivity. Chitosan, derived from chitin, exhibits mucoadhesive properties and the ability to transiently open tight junctions, enhancing paracellular drug transport across epithelial barriers (Mao et al., 2005). Alginate, obtained from brown algae, forms ionically crosslinked gels under mild conditions, making it ideal for encapsulating sensitive biologicals like proteins and cells. Together, synthetic and natural biodegradable polymers have transformed the landscape of drug delivery by facilitating the development of long-acting injectables, targeted delivery platforms, and implantable depots. Their flexibility in **physicochemical tuning, drug loading capacity**, and **release profiles** allows for the design of personalized and site-specific therapies with improved efficacy, reduced systemic toxicity, and enhanced patient adherence (Jain, 2000).

CLASSIFICATION OF BIODEGRADABLE POLYMERS

Natural Biodegradable Polymers

Natural polymers are derived from renewable biological sources. These include polysaccharides and proteins such as chitosan, gelatin, and alginate. Chitosan, derived from chitin, exhibits excellent mucoadhesive properties and promotes wound healing. Gelatin, a collagen derivative, is used in vaccine delivery and tissue scaffolding due to its biocompatibility and tunable gelation. Alginate, sourced from brown seaweed, forms hydrogels in the presence of divalent cations like calcium and is frequently used in encapsulation. While natural polymers are inherently biocompatible and non-immunogenic, they often suffer from variability in batch quality and mechanical instability, requiring modification or blending with synthetic polymers for improved performance (Ulery et al., 2011).

Synthetic Biodegradable Polymers

Synthetic polymers offer the advantage of precise control over molecular weight, composition, and degradation kinetics. They are often more consistent than natural polymers and can be engineered to meet specific delivery requirements.

Polylactic Acid (PLA)

PLA is a hydrophobic aliphatic polyester synthesized from lactic acid. It degrades slowly through hydrolysis of its ester bonds, making it suitable for long-term drug release applications such as implants. PLA is approved by the FDA for medical use and has been used in orthopedic fixation devices, sutures, and sustained drug delivery systems.

Polyglycolic Acid (PGA)

PGA is a hydrophilic, fast-degrading polyester used primarily in surgical sutures and tissue engineering. Its high crystallinity leads to rapid hydrolysis, making it suitable for short-term therapeutic interventions. However, its degradation produces glycolic acid, which may cause localized pH changes.

Poly(lactic-co-glycolic acid) (PLGA)

PLGA is a copolymer of PLA and PGA, combining the hydrophobicity of PLA and the hydrophilicity of PGA. The degradation rate of PLGA can be tailored by altering the lactic to glycolic acid ratio. PLGA microspheres and nanoparticles are widely used for controlled release of proteins, peptides, and small-molecule drugs. Its safety profile and ability to encapsulate a wide range of therapeutics make it one of the most used biodegradable polymers (Makadia & Siegel, 2011).

Polycaprolactone (PCL)

PCL is a semi-crystalline, hydrophobic polymer with very slow degradation properties. It is suitable for long-term implants and delivery systems requiring prolonged release. Due to its low melting point, it is easily processable, and its degradation products are non-toxic.

MECHANISM OF BIODEGRADATION AND DRUG RELEASE

Hydrolytic and Enzymatic Degradation

Biodegradable polymers degrade via two primary mechanisms: hydrolysis and enzymatic action. Hydrolysis involves the cleavage of ester bonds in the polymer backbone through water absorption, commonly observed in synthetic polymers like PLGA and PLA. Enzymatic degradation, more common in natural polymers, involves enzymes such as lysozymes breaking down the polymer chains. The degradation process leads to polymer erosion, either bulk or surface erosion, which directly influences the release of the encapsulated drug (Anderson & Shive, 1997).

Kinetics of Drug Release

Drug release from biodegradable polymers can be governed by diffusion, erosion, or a combination. Initially, drug molecules near the polymer surface are released by diffusion. As degradation proceeds, polymer erosion allows for the release of embedded drug molecules. Controlled release is achieved by tuning polymer properties, such as crystallinity, porosity, and hydrophilicity. Mathematical models like the Higuchi and Korsmeyer-Peppas equations are often used to study and predict release kinetics.

Factors Affecting Degradation and Release Profiles

Several factors influence the degradation rate and release profile of drugs. These include polymer molecular weight, crystallinity, copolymer ratio, environmental pH, temperature, and presence of enzymes. For instance, PLGA with a higher glycolic acid content degrades faster due to increased hydrophilicity. The size and shape of the delivery system also affect release, with smaller particles degrading faster due to increased surface area.

ADVANCEMENTS IN BIODEGRADABLE POLYMER SYSTEMS

Smart and Stimuli-Responsive Polymers

Smart polymers can alter their structure or properties in response to external stimuli such as pH, temperature, light, or enzymes. These systems allow for targeted and on-demand drug release, improving site-specific delivery and reducing systemic side effects. For example, pH-sensitive polymers release drugs in acidic tumor microenvironments or within lysosomes.

Nanoparticles and Microparticles

Biodegradable polymer-based nanoparticles and microparticles improve drug solubility, stability, and targeting. They protect sensitive drugs from enzymatic degradation and facilitate controlled release. Nanoparticles made from PLGA have shown significant promise in delivering anticancer drugs, proteins, and nucleic acids (Danhier et al., 2012).

Injectable Hydrogels and Implantable Devices

Injectable hydrogels form gels in situ upon administration, acting as drug reservoirs. These hydrogels can encapsulate small molecules, proteins, or cells for localized delivery. Implantable devices made from biodegradable polymers provide long-term delivery, releasing drugs over weeks to months (Vermonden et al., 2012).

Surface Modification and Targeted Delivery

Functionalization of polymer surfaces with targeting ligands, such as antibodies or peptides, enhances cell-specific delivery. Surface engineering can also improve mucoadhesion, cellular uptake, and circulation time, offering more precise and efficient therapeutic outcomes.

Extended Mucoadhesive Delivery via Biodegradable Polymers

An emerging strategy in sustained drug release is the use of **mucoadhesive biodegradable polymers**, which adhere to mucosal surfaces and extend the residence time of drug formulations at the site of absorption. Polymers such as **chitosan**, **carbopol**, and **alginate** have demonstrated strong mucoadhesive properties in addition to being biodegradable, making them excellent candidates for **oral, buccal, nasal, and vaginal drug delivery systems** (Prajapati et al., 2013). These systems enhance drug absorption by maintaining prolonged contact with the mucosa, allowing for **controlled and localized release** while bypassing hepatic first-pass metabolism. Moreover, the use of mucoadhesive biodegradable matrices ensures gradual degradation and drug diffusion, reducing systemic side effects and improving bioavailability for drugs with short half-lives or poor solubility. This approach offers a dual advantage—**sustained release kinetics and improved patient compliance**—making it particularly valuable for chronic therapies, hormone replacement, and mucosal vaccines.

THERAPEUTIC APPLICATIONS

Oncology

Biodegradable polymers play a pivotal role in oncology by enabling targeted delivery of chemotherapeutic agents. Nanoparticles and microparticles formulated from polymers like PLGA can deliver drugs such as paclitaxel and doxorubicin directly to tumor tissues, enhancing efficacy while minimizing systemic toxicity. These carriers exploit the enhanced permeability and retention (EPR) effect to accumulate in tumor sites and release drugs in a controlled manner, reducing side effects and improving patient outcomes.

Infectious Diseases

In the treatment of chronic infections, biodegradable sustained-release systems ensure prolonged therapeutic drug levels at the infection site. Antibiotics such as ampicillin and rifampicin have been encapsulated in polymeric microspheres to treat diseases like osteomyelitis and tuberculosis, improving compliance by minimizing the frequency of administration. These systems help overcome issues of poor bioavailability and short drug half-life often associated with conventional formulations.

Cardiovascular Disorders

Biodegradable drug-eluting stents are used to locally release anti-proliferative agents such as sirolimus to prevent restenosis after angioplasty. Unlike permanent metallic stents, these polymeric stents degrade over time, leaving behind a healed arterial lumen with no foreign material, thus reducing long-term risks such as inflammation or late thrombosis. Their temporary presence supports vascular healing and reduces the need for prolonged antiplatelet therapy.

Diabetes and Chronic Illnesses

Chronic conditions like diabetes benefit significantly from biodegradable polymer-based formulations. PLGA microspheres are used to create long-acting injectable forms of antidiabetic agents like exenatide, providing sustained drug release over weeks. This minimizes the burden of daily injections, enhances therapeutic compliance, and maintains better glycemic control, which is critical in managing disease progression and complications.

Regenerative Medicine and Tissue Engineering

Biodegradable polymers such as PLGA, PCL, and natural biopolymers like collagen and chitosan are used to fabricate scaffolds that support tissue regeneration. These scaffolds provide a temporary structure for cell attachment, migration, and proliferation while gradually degrading into biocompatible byproducts. Applications include bone regeneration, neural tissue repair, and dermal wound healing, making them essential components in regenerative therapies (Ulery et al., 2011).

TOXICITY, BIOCOMPATIBILITY, AND REGULATORY CONSIDERATIONS

Safety and Toxicity Assessment

Safety evaluation of biodegradable polymers centers on their degradation products, immunogenicity, and the effects of prolonged exposure. PLGA, for instance, breaks down into lactic and glycolic acids, which are naturally metabolized via the Krebs cycle. Preclinical studies examine both acute and chronic toxicity to confirm systemic safety and compatibility (Makadia & Siegel, 2011).

FDA and EMA Guidelines

The FDA and EMA require comprehensive data on biocompatibility, degradation kinetics, and pharmacokinetics for regulatory approval. Developers must adhere to Good Manufacturing Practices (GMP), with stringent testing protocols tailored to each product. This ensures efficacy, reproducibility, and patient safety in clinical use.

Clinical Trials and Marketed Products

Several pharmaceutical products utilizing biodegradable polymers have gained regulatory approval, underscoring their clinical viability and safety. Notably, Lupron Depot, which uses PLGA microspheres for sustained leuprolide acetate delivery, and Zoladex, a goserelin acetate implant, have demonstrated long-term efficacy in hormone-related therapies. These formulations exemplify successful translation from research to market, offering patient-friendly dosing and reduced side effects. Their approval validates the potential of polymer-based systems in chronic disease management (Suri et al., 2007).

CHALLENGES AND LIMITATIONS

Drug Loading and Encapsulation Efficiency

Encapsulating hydrophilic or high-molecular-weight biomolecules, such as proteins and peptides, in biodegradable polymers poses significant challenges due to low entrapment efficiency and potential denaturation during solvent-based processing. These molecules often exhibit rapid diffusion, leading to an undesirable initial burst release that compromises sustained delivery. Addressing this requires optimization of polymer matrices, emulsification techniques, and stabilizing excipients to enhance loading efficiency and release control.

Stability and Shelf Life

Biodegradable formulations are susceptible to hydrolytic and oxidative degradation during storage, which can alter their physicochemical properties and compromise drug release profiles. Such instability can reduce therapeutic efficacy and shelf life, especially in moisture- or temperature-sensitive environments. To maintain formulation integrity, advanced packaging solutions—such as moisture-barrier materials and inert atmosphere storage—are essential. These measures ensure product stability from manufacturing to administration.

Scale-up and Manufacturing Barriers

Scaling up biodegradable polymer-based drug delivery systems from laboratory to commercial production presents significant challenges in maintaining batch-to-batch reproducibility, ensuring sterilization without compromising polymer integrity, and controlling production costs. Achieving regulatory approval also necessitates validated, scalable manufacturing protocols that meet stringent quality and safety standards. As noted by Suri et al. (2007), these hurdles must be addressed to ensure successful clinical translation and market entry.

KEY ROLES OF BIODEGRADABLE POLYMERS IN SUSTAINED DRUG DELIVERY

Controlled and Sustained Drug Release

Biodegradable polymers facilitate controlled and sustained drug release, maintaining therapeutic drug levels within the desired plasma concentration range over extended periods. This steady release minimizes the peaks and troughs commonly observed with conventional formulations, thereby reducing side effects and enhancing efficacy. By avoiding rapid drug clearance and frequent dosing, these systems improve patient adherence to treatment regimens. Such polymers are especially beneficial for drugs with narrow therapeutic indices, where consistent plasma levels are critical. As highlighted by Makadia and Siegel (2011), these systems offer a reliable approach to optimizing pharmacokinetic profiles and therapeutic outcomes.

Elimination of Surgical Removal

Biodegradable polymers such as PLGA and PLA degrade through hydrolysis of their ester linkages, producing non-toxic byproducts like lactic acid and glycolic acid, which are naturally metabolized via the Krebs cycle. This intrinsic biodegradability eliminates the need for surgical retrieval of the delivery system after drug release, enhancing patient compliance and reducing medical costs. The degradation rate can be tailored by adjusting the polymer's molecular weight and monomer composition. This feature makes them ideal for applications requiring temporary implants or long-acting injectable depots. As noted by Danhier et al. (2012), such polymers ensure safe and efficient drug delivery with minimal residual impact on the body.

Customizable Degradation and Drug Release Kinetics

By adjusting the composition and molecular weight of biodegradable polymers, researchers can finely tune the degradation rate and drug diffusion properties to achieve desired release kinetics. This customization allows for precise control over how long and how much drug is released, matching the pharmacological needs of various therapies. Such design flexibility is crucial for developing long-acting formulations that maintain efficacy while minimizing dosing frequency (Park et al., 2013).

Enhanced Bioavailability of Poorly Soluble Drugs

Polymer encapsulation significantly enhances the solubility and systemic availability of hydrophobic or labile drug molecules by protecting them from degradation and promoting controlled release. Biodegradable polymers form micro- or nano-scale carriers that increase drug dispersion in biological fluids. This approach improves absorption and bioavailability, particularly for poorly water-soluble compounds. Moreover, encapsulation reduces premature clearance, allowing more drug to reach the target site. As a result, therapeutic efficacy is improved with reduced dosing frequency.

Targeted and Site-Specific Drug Delivery

Functionalized biodegradable polymers, modified with targeting ligands such as peptides, antibodies, or aptamers, enable selective delivery to specific tissues or cellular receptors. This targeted approach enhances drug accumulation at the disease site while reducing off-target distribution, thereby improving therapeutic efficacy. By exploiting receptor-mediated endocytosis, these polymers facilitate intracellular drug delivery, crucial for treating conditions like cancer or inflammatory diseases.

Functionalization also allows real-time imaging and diagnostics when combined with contrast agents or markers. Overall, this strategy minimizes systemic toxicity and maximizes pharmacological outcomes.

Versatile Routes of Administration

Biodegradable polymer-based sustained release systems are versatile platforms that support various administration routes, enhancing therapeutic flexibility. These include injectable depots for long-term parenteral delivery, implantable devices for site-specific action, oral formulations for systemic effects, and transdermal patches for controlled absorption through the skin. Such adaptability broadens their clinical utility across diverse treatment settings (Park et al., 2013).

Applications in Chronic and Complex Disease Management

Long-acting drug formulations are essential for managing chronic diseases such as cancer, cardiovascular disorders, diabetes, and neurodegenerative conditions by maintaining therapeutic drug levels over extended periods and improving patient compliance (Patel et al., 2014). In cancer treatment, biodegradable implants provide sustained release of chemotherapeutics, reducing toxicity and enhancing targeted delivery (Feng et al., 2017). For cardiovascular diseases, sustained-release antihypertensive drugs ensure steady plasma concentrations, improving blood pressure control and reducing adverse events. Long-acting formulations in diabetes enable less frequent dosing of insulin and glucose-lowering agents, enhancing glycemic control and minimizing hypoglycemia risk (Bailey & Day, 2013). In neurodegenerative disorders, polymeric systems facilitate continuous delivery of neuroprotective agents across the blood-brain barrier, improving therapeutic outcomes (Zhao et al., 2021). Overall, these formulations provide stable pharmacokinetics, reduced dosing frequency, and better efficacy, making them vital in chronic disease management.

Commonly Used Biodegradable Polymers

Biodegradable polymers commonly employed in drug delivery include synthetic types such as poly(lactic-co-glycolic acid) (PLGA), polylactic acid (PLA), and polycaprolactone (PCL), alongside natural polymers like chitosan, alginate, and gelatin. These materials offer diverse degradation rates and biocompatibility profiles, enabling tailored therapeutic applications.

Clinically Approved Formulations and Products

Several biodegradable polymer-based formulations have received FDA approval, exemplified by Lupron Depot® (leuprolide acetate in PLGA microspheres) and Zoladex® (goserelin acetate implant). These products highlight the clinical viability and sustained-release advantages of polymer-based drug delivery systems.

Reduction of Dose-Dependent Toxicity and Side Effects

Sustained release drug delivery systems maintain steady plasma drug concentrations, thereby preventing the high peak levels associated with conventional dosing that often lead to toxicity. By minimizing these plasma level spikes, such systems reduce adverse effects and improve the overall safety profile of therapeutic agents. This controlled release also enhances patient tolerability and compliance, making treatment more effective over time (Makadia & Siegel, 2011).

FUTURE PERSPECTIVES

Emerging Polymer Technologies

Emerging polymer technologies are expected to transform the future landscape of drug delivery. The development of advanced biodegradable polymers focuses on improving specificity, controlled degradation, and multifunctionality. Novel materials such as dendritic polymers, block copolymers, and hyperbranched polymers offer unique architectures that facilitate higher drug loading, improved bioavailability, and programmable release profiles (Sosnik & Augustine, 2016; Li et al., 2021). These innovations are engineered to respond to biological cues, such as enzymes or reactive oxygen species, enabling site-specific and condition-specific drug release. In addition, advances in polymer chemistry have led to the creation of stimuli-responsive polymers that can alter their physical or chemical properties in response to temperature, pH, light, or magnetic fields (Zhang et al., 2020). This adaptability allows for intelligent delivery systems that can be fine-tuned according to disease pathology. For instance, thermoresponsive polymers can release their payload upon reaching a specific temperature, which is particularly useful in treating localized infections or tumors. Another major frontier is the incorporation of biodegradable polymers into nanotechnology-based systems, such as polymeric micelles, dendrimers, and nanogels, which provide enhanced stability, circulation time, and cellular uptake. These nanostructures offer the potential for combination therapy by delivering multiple drugs or therapeutic agents simultaneously, thereby increasing treatment efficacy while minimizing toxicity. Furthermore, 3D printing and biofabrication techniques are being explored for designing customized, patient-specific implants and drug-eluting devices using biodegradable polymers. These personalized solutions are tailored to match the physiological and pathological conditions of individual patients, aligning with the principles of precision medicine. Together, these emerging technologies are paving the way for next-generation drug delivery systems that are smarter, more efficient, and more patient-centric.

Personalized and Precision Medicine Integration

The integration of biodegradable polymers with personalized and precision medicine represents a significant evolution in therapeutic strategies. Personalized medicine involves tailoring treatment based on an individual's genetic profile, lifestyle, and specific disease characteristics, while precision medicine applies this concept to develop highly targeted therapies (Collins & Varmus, 2015). Biodegradable polymers can be engineered to encapsulate drugs at patient-specific doses and release rates, accommodating interpatient variability in metabolism and drug response. This is particularly advantageous for drugs with narrow therapeutic indices or those that require site-specific delivery. Smart biodegradable systems equipped with diagnostic sensors or biosensors can monitor physiological conditions and modulate drug release accordingly, a feature especially relevant in chronic diseases like diabetes and cancer (Makadia & Siegel, 2011; Ma et al., 2020). Furthermore, advances in genomics and proteomics have facilitated the identification of biomarkers that can guide the selection of polymeric carriers and targeting ligands to optimize therapy at the molecular level. Combining polymer science with data-driven medicine allows for more effective treatment regimens, reducing adverse effects and improving patient outcomes (Ginsburg & Phillips, 2018). The synergy between biodegradable polymers and personalized medicine is also being harnessed in vaccine development and immunotherapy, where polymer-based delivery vehicles can present antigens or adjuvants in a patient-specific manner to elicit tailored immune responses. The future of drug delivery lies in this convergence, enabling treatments that are not only biocompatible and effective but also individualized and adaptive.

Interdisciplinary Approaches for Next-Gen Delivery

Next-generation drug delivery systems are evolving through a fusion of disciplines including materials science, nanotechnology, and biomedical engineering. Biodegradable polymers benefit from engineering innovations such as microfluidics, enabling precise nanoparticle fabrication with tunable features (Karnik et al., 2008). Systems biology and computational modeling support the prediction of degradation kinetics and drug release profiles, improving design accuracy (Langer & Peppas, 2003). Bioinformatics enables integration of patient-specific data, fostering more tailored therapeutic strategies. The collaboration between clinicians and scientists has led to hybrid systems like theranostics, which combine diagnostics with therapy in one polymeric platform. These advances exemplify how cross-disciplinary efforts enhance drug targeting, efficacy, and safety. Continued convergence of these fields will drive innovations in responsive, patient-centered delivery platforms (Tibbitt et al., 2016).

CONCLUSION

Biodegradable polymers have transformed modern therapeutics by enabling safer, more effective, and patient-friendly drug delivery systems. These materials degrade into non-toxic byproducts after fulfilling their therapeutic role, thus eliminating the need for surgical removal and reducing long-term risks. Their intrinsic adaptability allows for the engineering of diverse delivery platforms, ranging from nanoparticles and hydrogels to implants and injectable depots, which can be tailored to meet specific clinical needs. One of the most significant contributions of biodegradable polymers is their capacity to support sustained and targeted drug release. By maintaining drug concentrations within the therapeutic window for prolonged periods, these systems improve pharmacokinetics, enhance bioavailability, and minimize dosing frequency. This, in turn, leads to better patient adherence and therapeutic outcomes, particularly in the management of chronic illnesses, cancers, and localized infections. Furthermore, their compatibility with smart technologies, such as stimuli-responsive release and ligand-mediated targeting, has broadened their application in precision and regenerative medicine. Polymers like PLGA, PLA, PCL, chitosan, and alginate have become the foundation for creating customized delivery systems aligned with an individual's genetic and pathological profile. With the ongoing advancements in material science, nanotechnology, and personalized healthcare, biodegradable polymers are expected to play a pivotal role in the next generation of drug delivery solutions. Their versatility, biocompatibility, and design flexibility position them as indispensable tools in shaping the future of safe, effective, and patient-centered therapeutics.

REFERENCES

Anderson, J. M., & Shive, M. S. (1997). Biodegradation and biocompatibility of PLA and PLGA microspheres. *Advanced Drug Delivery Reviews*, *28*(1), 5–24. DOI: 10.1016/S0169-409X(97)00048-3 PMID: 10837562

Anderson, J. M., & Shive, M. S. (1997). Biodegradation and biocompatibility of PLA and PLGA microspheres. *Advanced Drug Delivery Reviews*, *28*(1), 5–24. DOI: 10.1016/S0169-409X(97)00048-3 PMID: 10837562

Bailey, C. J., & Day, C. (2013). Traditional plant medicines as treatments for diabetes. *Diabetes Care*, *26*(8), 250–252. DOI: 10.2337/diacare.26.8.250 PMID: 2673695

Collins, F. S., & Varmus, H. (2015). A new initiative on precision medicine. *The New England Journal of Medicine*, *372*(9), 793–795. DOI: 10.1056/NEJMp1500523 PMID: 25635347

Danhier, F., Ansorena, E., Silva, J. M., Coco, R., Le Breton, A., & Préat, V. (2012). PLGA-based nanoparticles: An overview of biomedical applications. *Journal of Controlled Release*, *161*(2), 505–522. DOI: 10.1016/j.jconrel.2012.01.043 PMID: 22353619

Feng, L., Mumper, R. J., & Ruan, G. (2017). Biodegradable polymeric nanocarriers for cancer drug delivery. *Current Pharmaceutical Design*, *23*(8), 1173–1185. DOI: 10.2174/1381612822666161227144912

Ginsburg, G. S., & Phillips, K. A. (2018). Precision medicine: From science to value. *Health Affairs*, *37*(5), 694–701. DOI: 10.1377/hlthaff.2017.1624 PMID: 29733705

Jain, R. A. (2000). The manufacturing techniques of various drug-loaded biodegradable poly(lactide-co-glycolide) (PLGA) devices. *Biomaterials*, *21*(23), 2475–2490. DOI: 10.1016/S0142-9612(00)00115-0 PMID: 11055295

Karnik, R., Gu, F., Basto, P., Cannizzaro, C., Dean, L., Kyei-Manu, W., Langer, R., & Farokhzad, O. C. (2008). Microfluidic platform for controlled synthesis of polymeric nanoparticles. *Nano Letters*, *8*(9), 2906–2912. DOI: 10.1021/nl801736q PMID: 18656990

Langer, R., & Peppas, N. A. (2003). Advances in biomaterials, drug delivery, and bionanotechnology. *AIChE Journal. American Institute of Chemical Engineers*, *49*(12), 2990–3006. DOI: 10.1002/aic.690491202

Li, M., Rouaud, O., & Poncelet, D. (2021). Microencapsulation by solvent evaporation: State of the art for process engineering approaches. *International Journal of Pharmaceutics, 561*, 43–56. PMID: 18706988

Ma, P., Luo, Y., & Sun, Y. (2020). Emerging polymeric nanocarriers for precision drug delivery. *Acta Pharmaceutica Sinica. B, 10*(2), 209–220.

Makadia, H. K., & Siegel, S. J. (2011). Poly lactic-co-glycolic acid (PLGA) as biodegradable controlled drug delivery carrier. *Polymers, 3*(3), 1377–1397. DOI: 10.3390/polym3031377 PMID: 22577513

Mao, S., Sun, W., & Kissel, T. (2005). Chitosan-based formulations for delivery of DNA and siRNA. *Advanced Drug Delivery Reviews, 57*(9), 1303–1322. DOI: 10.1016/j.addr.2005.04.004 PMID: 19796660

Park, K., Skidmore, S., Hadar, J., Garner, J., Park, H., Otte, A., & Jiang, X. (2013). Injectable, long-acting PLGA formulations: Analyzing PLGA and understanding microparticle formation. *Journal of Controlled Release, 172*(2), 465–472. DOI: 10.1016/j.jconrel.2013.08.012 PMID: 31071374

Patel, R. B., Patel, M. R., Bhatt, K. K., & Patel, B. G. (2014). Extended-release matrix technology for oral delivery. *Journal of Controlled Release, 192*, 122–137. DOI: 10.1016/j.jconrel.2014.07.033

Prajapati, V. D., Jani, G. K., Moradiya, N. G., & Randeria, N. P. (2013). Pharmaceutical applications of various natural gums, mucilages and their modified forms. *Carbohydrate Polymers, 92*(2), 1685–1699. DOI: 10.1016/j.carbpol.2012.11.021 PMID: 23399207

Siepmann, J., & Siepmann, F. (2012). Modeling of drug release from delivery systems based on hydroxypropyl methylcellulose (HPMC). *Advanced Drug Delivery Reviews, 64*(2), 163–174. DOI: 10.1016/j.addr.2012.09.028 PMID: 11369079

Sosnik, A., & Augustine, R. (2016). Challenges and opportunities in the design of oral drug delivery systems for pediatric patients. *Advanced Drug Delivery Reviews, 73*, 140–155. DOI: 10.1016/j.addr.2014.05.004 PMID: 24819219

Suri, S. S., Fenniri, H., & Singh, B. (2007). Nanotechnology-based drug delivery systems. *Journal of Occupational Medicine and Toxicology (London, England), 2*(1), 16. DOI: 10.1186/1745-6673-2-16 PMID: 18053152

Suri, S. S., Fenniri, H., & Singh, B. (2007). Nanotechnology-based drug delivery systems. *Journal of Occupational Medicine and Toxicology (London, England), 2*(1), 16. DOI: 10.1186/1745-6673-2-16 PMID: 18053152

Tibbitt, M. W., Dahlman, J. E., & Langer, R. (2016). Emerging frontiers in drug delivery. *Journal of the American Chemical Society, 138*(3), 704–717. DOI: 10.1021/jacs.5b09974 PMID: 26741786

Ulery, B. D., Nair, L. S., & Laurencin, C. T. (2011). Biomedical applications of biodegradable polymers. *Journal of Polymer Science. Part B, Polymer Physics, 49*(12), 832–864. DOI: 10.1002/polb.22259 PMID: 21769165

Vermonden, T., Censi, R., & Hennink, W. E. (2012). Hydrogels for protein delivery. *Chemical Reviews, 112*(5), 2853–2888. DOI: 10.1021/cr200157d PMID: 22360637

Zhang, Y., Chan, H. F., & Leong, K. W. (2020). Advanced materials and processing for drug delivery: The past and the future. *Advanced Drug Delivery Reviews, 65*(1), 104–120. DOI: 10.1016/j.addr.2012.10.003 PMID: 23088863

Zhao, X., & Li, X. (2021). Biodegradable polymers in clinical practice: A review. *International Journal of Pharmaceutics, 610*, 121241.

Zhao, Z., Gao, P., You, Y., Wang, H., & Han, X. (2021). Polymeric nanoparticles for brain-targeted drug delivery: Current status and future perspectives. *Acta Pharmaceutica Sinica. B, 11*(8), 2289–2309. DOI: 10.1016/j.apsb.2021.03.02 PMID: 34522580

Chapter 3
Natural Polysaccharides as Drug Delivery Agent: Current Trends and Perspectives

Meghmala S Waghmode
https://orcid.org/0000-0002-5635-134X
PDEA's Annasaheb Magar Mahavidyalaya, Pune, India

ABSTRACT

For the administration of therapeutic agents there is need of delivery agents which will be nontoxic and amended with the desired physicochemical properties. Recently, a considerable amount of work has been reported on the use of natural polymers as drug delivery agent. In this book chapter, an overview of the natural polysaccharides (chitosan, hyaluronic acid, cellulose and bacterial exopolysaccharide) is presented. Although bacterial exopolysaccharides showed features as drug carrier, still there use as drug delivery agent is limited due to certain factors like composition, yield, cost of purification etc. Researchers are trying various formulations of this polysaccharides by altering the chemical groups, size and shape modifications, reducing the toxicity and increasing the controlled drug release propery . Herein, this book chapter aims to provide the current trends and perspectives in the utilisation of the polysaccharides originated from natural sources.

INTRODUCTION

A natural polysaccharide-based drug delivery system utilizes naturally occurring polysaccharides to deliver drugs to specific sites in the body in a controlled and targeted manner (Thirumalaisamy et al.,2021). These systems are designed to improve therapeutic efficacy, reduce side effects, and enhance patient compliance

DOI: 10.4018/979-8-3373-1047-3.ch003

by controlling the release rate, improving drug stability, and enabling site-specific delivery (Kumar et al., 2004). Natural polysaccharides are macromolecules derived from biological sources such as plants, animals, or microorganisms (Kumar et al., 2004). Polysaccharides are biodegradable, biocompatible, and often non-toxic, making them ideal for biomedical applications. Common examples of natural polymers for drug delivery includes polysaccharides, proteins and glycosaminoglycans with forms as well as advantages is illustrated in Figure 1.

Figure 1. Examples of natural polysaccharides with advantages and formulations for the drug delivery

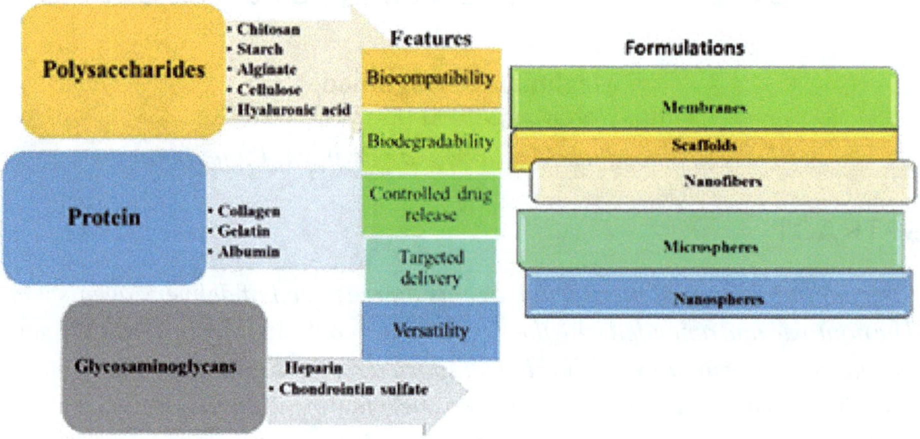

Biocompatibility, biodegradability, controlled release, targeted delivery, and adaptability are important characteristics of drug delivery systems based on natural polymers (Doppalapudi et al.,2015; Srivastava et al.,2015; Atanase, 2021).). Polymers can be made into films, hydrogels, micelles, scaffolds, nanoparticles, and more. Diffusion-controlled release, degradation-controlled release, stimuli-responsive release, and targeted delivery are important drug delivery techniques (Holowka et al.,2014).

Properties of natural polysaccharides to serve as drug delivery agent, is given in Table 1.

Table 1. Key features of chitosan, hyaluronic acid, and cellulose as drug carrier

Property	Chitosan	Hyaluronic Acid (HA)	Cellulose
Biocompatibility	Safe for use in humans and animals	Excellent natural component	Good; inert and non-toxic
Biodegradability	Degraded by lysozymes into harmless products	Enzymatically degraded by hyaluronidases	Slowly biodegradable
Hydrophilicity	Derivatization is needed.	Highly hydrophilic	Hydrophilic
Targeting capability	Good	Binds CD44 receptor; tumor targeting	General inert matrix; can be functionalized for targeting
Mechanical strength	Good	Moderate; depends on crosslinking	Good; forms flexible films and scaffolds
Modifiability	Chemically modifiable (e.g., crosslinking)	Chemically modifiable (e.g., crosslinking)	Easily chemically modified (e.g., derivatives)

Due to their non-invasive nature, wound dressings are the most often used method of promoting wound healing in diabetic patients. Alginates, foams, hydrogels, hydrocolloids, hydroactives, films, and hydrofibers are examples of wound dressings. Because they are flexible, easy to apply, permit some moisture evaporation, act as a barrier against external contamination, and provide wound bed inspection without dressing removal, polymer-based wound dressing films are beneficial (Savencu et al.,2021). For the effective wound dressing preparation selection of proper polymer and active pharmaceutical ingredients is needed.

Scaffolds created via tissue engineering were typically utilised as models for the body's damaged tissues or organs to regenerate. Tissue engineering scaffolds must be clinically operable, mechanically stable, biocompatible, and biodegradable to satisfy the body's requirements (Yang et al.,2019). Additionally, the scaffolds ought to help with nutrition delivery, migration, proliferation, and cell adhesion. Numerous methods, including electrospinning, phase separation, rapid prototyping, and self-assembly, have been developed thus far to generate tissue engineering scaffolds (Yang et al.,2019).

Over the past 20 years, there has been a significant increase in the number of novel delivery platforms for the treatment of cancer that are based on natural and synthetic polymers with improved pharmacokinetic and therapeutic potential (Ding et al.,2024). Through a variety of processes, including as physical adsorption, chemical conjugation, and/or internal loading, polymers can increase delivery, promote and extend circulation, enable selective targeting, and offer the regulated release of cargos. A promising future for efficient medication delivery with the ability to go past the challenges faced is provided by biomimetic and bio-inspired systems based on natural polymers (Ding et al.,2024). Natural polymers, including chitosan, albumin, silk fibroin, cellulose, hyaluronic acid, and collagen, have been employed for many years to enhance drug delivery efficacy (Fazal et al.,2023).

Research into a more effective remedial treatment is still difficult because many therapeutic methods are not resolutive hence researchers are trying to modify natural polymers and its combination with in use drugs. Reaching therapeutic concentrations at the exact biological location is the ultimate objective of polymer-based drug delivery systems. In this book chapter, chitosan, hyaluronic acid, cellulose, and bacterial exopolysaccharides sources, properties, and its role in drug delivery and limitations are discussed.

CHITOSAN

Chitosan is a natural polymer obtained by alkaline deacetylation of chitin. It is nontoxic, biocompatible, and biodegradable which makes it a good candidate for conventional and novel drug delivery systems. Chitosan, a cationic polysaccharide derivative of chitin has been employed in different routes of administration including nasal, ocular, intravenous, oral, mucosal, etc. (Fazal et al.,2023). Chitosan forms colloidal particles and entraps bioactive molecules through several mechanisms, including chemical crosslinking, ionic crosslinking, and ionic complexation. A possible alternative of chitosan by the chemical modification also has been useful for the association of bioactive molecules to polymer and controlling the drug release profile. Because of the high affinity of chitosan for cell membranes, it has been used as a coating agent for liposome formulations.

Chitosan is the only polycation in nature and its charge density depends on the degree of acetylation and pH of the media (Aranaz et al.,2021). The solubility of the polymer depends on the acetylation degree and molecular weight. Chitosan oligomers are soluble over a wide pH range, from acidic to basic ones (Domard, 1987).On the contrary, chitosan samples with higher molecular weight are only soluble in acidic aqueous media even at high deacetylation degrees. This lack of solubility at neutral and basic pH has hindered the use of chitosan in some applications under neutral physiological conditions (i.e., pH 7.4). This is the reason behind the preparation of chitosan derivatives with enhanced solubility (Aranaz et al.,2021). The field of chitosan chemistry involves the preparation of phosphorylated chitosan, biodegradable chitosan, trimethylated chitosan, N-succinylated chitosan, thiolatedchitosan, azidatedchitosans, sugar-modified chitosan, chitosan–dendrimer hybrid, cyclodextrin-linked chitosan crown-ether-bound chitosan and enzymatically modified chitosan (Kumar et al.,2004; Aranazet al.,2021).Phosphorylated chitosan has wide application in the tissue engineering due to its transfection rate (Zhu et al.,2010; Aranaz et al.,2021). Chitosan has a cationic character because of its primary amino groups. These primary amino groups are responsible for properties such as controlled drug release, mucoadhesion, in situ gellation, transfection, permeation enhancement, and efflux pump inhibitory properties (Bernkop-Schnürch, and Dünnhaupt, S. 2012). Formulation of chitosan and anticancer drug doxorubicin is schematically presented in Figure 2.

Figure 2. Flow chart of chitosan-based nanopartidcle formulaton

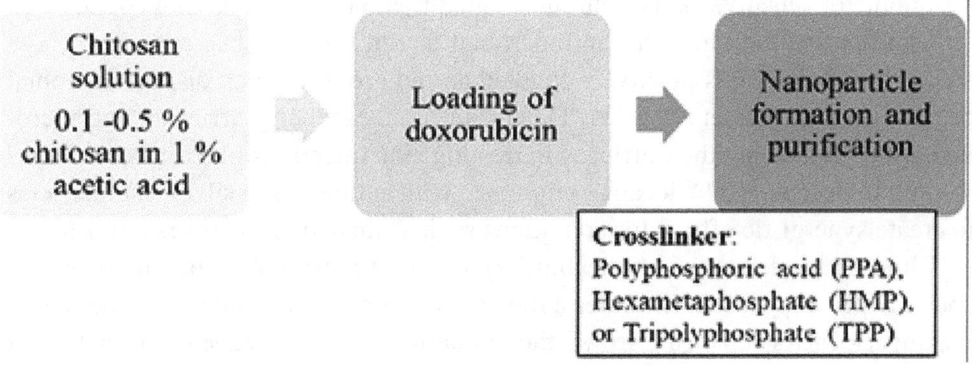

Serious drawbacks of the current treatment for diabetes mellitus, which involves subcutaneous insulin delivery and oral anti-diabetic medications, include patient noncompliance and sporadic hypoglycemia. If the oral route is offered, it would be the most practical and favoured option. Because they are hydrophilic, biocompatible, biodegradable, and nontoxic, natural polymers are particularly interesting. Chitosan (CS), one of the natural polymers utilised for oral insulin delivery, is being investigated extensively because of its advantageous biological characteristics and simplicity of chemical modification (Sonia, and Sharma, 2012). Furthermore, several benefits like safety, biodegradability, accessibility, and affordability support the ongoing development of a potential CS-based insulin delivery system. Alginate/chitosan nanoparticles were prepared by ionotropic pre-gelation of an alginate core followed by chitosan polyelectrolyte complexation, investigated to be effective for oral insulin delivery (Sarmento et al.,2007). Chitosan–6-mercaptonicotinic acid (chitosan–6-MNA) (thiolated chitosan), quaternized chitosan derivatives, chitosan cyclodextrin complexes, pegylated chitosan has been reported as a promising excipient for the oral delivery of insulin (Chaudhury, and Das, 2011; Millotti et al.,2014). An overview of recent patented methods for delivering insulin without injection highlights the importance of chitosan due to its mucoadhesive properties, which can shield insulin from enzymatic degradation, increase insulin retention time, and open the inter-epithelial tight junction to promote systemic insulin transport (Wong et al.,2009). Chitosan can be used to improve the multi-particulate matrix's and insulin's physicochemical stability. The positive charges of the dosage form required to induce mucoadhesiveness are retained and/or enhanced by the addition of chitosan coat or by co-formulating chitosan with cationic gelatin or electrolytes that offer solidified or partially crosslinked structures (Wong et al.,2009).

Chitosan is non-toxic, biocompatible, and biodegradable, minimizing the risk of adverse reactions in dermatological applications (Ali, & Ahmed, 2018). Its natural origin (from crustacean shells or fungi) ensures compatibility with skin tissue, making it suitable for sensitive skin or chronic wound treatments (Ali, & Ahmed, 2018).

Chitosan exhibits inherent antimicrobial activity against bacteria, fungi, and some viruses due to its positively charged amino groups, which disrupt microbial cell membranes (Dai et al.,2011). This makes it an excellent carrier for antimicrobial drugs, enhancing their efficacy in treating skin infections like acne or infected wounds (Dai et al.,2011). It can be combined with antibiotics or silver nanoparticles to create synergistic effects for conditions such as diabetic foot ulcers or burns.

Chitosan accelerates wound healing by promoting cell proliferation, angiogenesis, and collagen deposition. It creates a moist environment conducive to tissue repair, making it ideal for delivering growth factors, anti-inflammatory agents, or analgesics to wound sites. Chitosan-based dressings (e.g., films, hydrogels) protect wounds from external contaminants while delivering therapeutic agents (Dai et al.,2011). Chitosan's mucoadhesive properties allow it to adhere to the skin or mucosal surfaces, prolonging drug residence time and enabling sustained release. This is particularly useful for delivering drugs like corticosteroids or retinoids for psoriasis or eczema (Chuah et al.,2023). Chitosan can be formulated into nanoparticles, micelles, or liposomes to enhance the penetration of hydrophobic or hydrophilic drugs through the skin barrier (Ali, & Ahmed, 2018). Its positive charge interacts with negatively charged skin components, improving drug permeation (Bernkop-Schnürch, & Dünnhaupt,2012).

Chitosan's ability to open tight junctions in the skin enhances the delivery of large molecules, such as peptides or proteins. Chitosan's solubility and molecular weight can be modified to tailor drug delivery systems.

It can be processed into hydrogels, films, nanoparticles, or scaffolds, offering flexibility for various dermatological needs, such as anti-aging creams, wound dressings, or transdermal patches (Ali, & Ahmed, 2018).Chitosan forms protective films on the skin, maintaining hydration and acting as a barrier against environmental irritants. This property enhances the delivery of moisturizing agents or anti-aging compounds, such as hyaluronic acid or vitamin C, in cosmetic dermatology.

Chitosan's anti-inflammatory properties support its use in delivering drugs for inflammatory skin conditions like dermatitis or rosacea. (Rajitha et al.,2016). It can reduce local inflammation while carrying therapeutic agents. Its antioxidant capabilities protect encapsulated drugs from degradation, improving stability in formulations (Muthu et al.,2021). Although rare, potential allergenicity from crustacean-derived chitosan may limit its use in some patients, necessitating fungal-derived alternatives. Production costs and variability in chitosan's molecular weight or deacetylation degree can affect reproducibility in large-scale applications. Recently chitosan-based nanoparticles were utilised to improve the pharmacotherapeutic efficacy, pharmacokinetic profiles, and solubility characteristics of pharmaceuticals (Biswas et al.,2025).

HYALURONIC ACID (HA)

Two saccharide units of glucuronic acid and N-acetylglucosamine are alternatingly connected to form hyaluronic acid (Hyaluronan), a linear macromolecular mucopolysaccharide (Alaniz et al., 2016; Huang and Huang, 2018). It has been shown that hyaluronic acid and drug conjugates have the dual benefits of receptor-mediated endocytosis and tumour site aggregation (Ossipov, 2010; Huang, and Huang, 2018). Increased bioactive uptake through receptor-mediated targeting, extended bioactive retention in the synovium, decreased proinflammatory mediator expression, improved cartilage regeneration, decreased drug toxicity from sustained release, and better and more economical treatment are all reasons why HA-based bioactive delivery systems are useful in the treatment of arthritis (Walvekar et al., 2024).HA can adhere to mucosal surfaces, prolonging residence time and improving drug absorption when administered mucosally (oral, nasal, ocular)(Fu et al.,2023; Walvekar et al.,2024).

Hyaluronic acid is a naturally occurring gel-like substance that is particularly helpful in the growth and development of bone and joint cartilage (Huang et al., 2019). According to Jiang et al. (2010), HA is also crucial in lowering joint discomfort and inflammation brought on by damage or tissue deterioration. As a result, HA intra-articular injection has emerged as a successful treatment approach for osteoarthritis pain. Due to its strong affinity for binding water, HA also demonstrated remarkable gelling capabilities; as a result, it has emerged as one of the most appealing controlled drug release materials in a variety of biological applications

(Chang et al., 2021). Formulation of the hyaluronic acid for the delivery of doxorubicin is presented in **Figure 3.**

Figure 3. Flow chart of hydraluronic acid-based liposome formulation

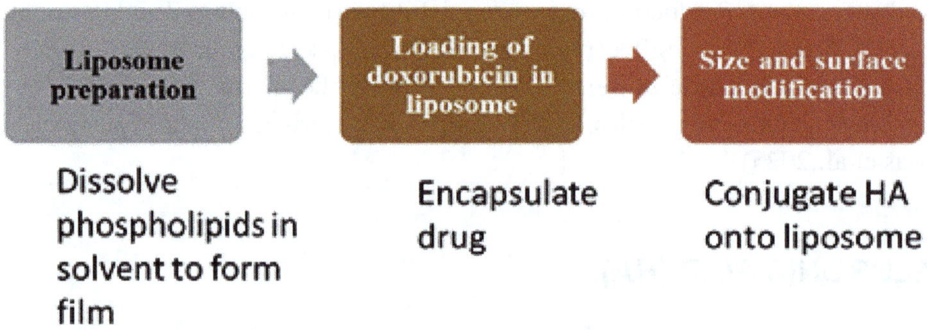

Doxorubicin (DOX) is one of the most effective chemotherapy drugs but has high toxicity to the kidneys, liver, and heart and hence researchers are trying to formulate HA based systemto serve as carrier for doxorubicin (Fu et al.,2023). Numerous hyaluronic acid (anionic polysaccharide) and its derivatives have been used in conjunction with paclitaxel to treat several cancer types, including breast, lung, and colon cancer (Fazal et al.,2023).

In-silico anti-viral activity of hydroxychloroquine (HCQ) and its hyaluronic acid-derivative (HA-HCQ) towards different SARS-CoV-2 protein molecular targets (Mpro:PDB ID 5R82 and Helicase:PDB ID 6ZSL) revealed that the HA-HCQ conjugate has multiple advantages of efficient drug delivery to its CD44 variant isoform receptors of the lower respiratory tract, highest interactive binding affinity with SARS-CoV-2 protein target (Thirumalaisamy et al.,2021). As per the study, conjugate showed strong interaction with SARS-CoV-2 Helicase with good binding energy of -23.1778 KJ/mol as compared to only hydroxychloroquine -13.6327 KJ/mol. Hence, HA-HCQ conjugate can be the promising antiviral medicine (Thirumalaisamy et al.,2021). The polymer molecule was synthesized from hyaluronic acid (HA) and hydroxychloroquine (HCQ) through disulfide bonds as redox-sensitive linkers that self-assemble into nanocapsules (HA-ss-HCQ nanocapsules) displayed active targeting and selectivity to metastatic breast cancer cells (He et al.,2022). The nanocapsules exerted significantly enhanced cellular internalization capacity, which was greatly diminished by free hyaluronic acid. Nanocapsules provides a promising strategy for active-targeting therapy in metastatic breast cancer (He et al.,2022).

Nonetheless, the human body readily breaks down hyaluronic acid (Bot et al., 2008). Consequently, a nitroxide-containing material should be added to prevent the breakdown of hyaluronic acid, or a hyaluronidase inhibitor should be introduced to stop the degradation of hyaluronic acid by blocking hyaluronidase activity (Sung et al., 2014; Huang and Huang, 2018).

HA is a natural component of skin, reducing the risk of immunogenicity or toxicity. Its biocompatibility ensures minimal adverse reactions, making it ideal for topical and transdermal drug delivery (Brown, & Jones, 2005). HA's ability to integrate with skin tissue supports its use in formulations for sensitive skin or wound healing applications. HA's high water-binding capacity (up to 1000 times its weight) helps maintain skin hydration, creating an optimal environment for drug penetration and efficacy (Brown, & Jones, 2005).This property is critical in dermatological formulations, enhancing skin barrier function and improving the delivery of active ingredients like anti-inflammatory or anti-aging compounds.

HA's viscoelastic nature allows it to form gels, creams, or films that improve the retention and controlled release of drugs on the skin surface. It enhances drug penetration by hydrating the stratum corneum, facilitating diffusion (Zhu et al.,2020).

HA can be modified (e.g., cross-linked or conjugated) to encapsulate drugs, enabling targeted delivery to deeper skin layers (Witting et al.,2015). For instance, HA-based nanoparticles can improve the delivery of hydrophobic drugs like retinoids or corticosteroids.

HA's interaction with CD44 receptors on skin cells (e.g., keratinocytes) enables receptor-mediated uptake, enhancing targeted drug delivery for conditions like psoriasis or melanoma. HA promotes wound healing by modulating inflammation, stimulating angiogenesis, and supporting cell migration and proliferation. As a drug carrier, it can deliver growth factors or antimicrobial agents directly to wound sites, improving therapeutic outcomes.HA-based dressings or hydrogels are used to treat burns, ulcers, or surgical wounds, combining drug delivery with tissue repair (Witting et al.,2015).

HA's molecular weight can be tailored (low, medium, or high) to achieve specific functions like low molecular weight HA could penetrate deeper into the skin, suitable for transdermal drug delivery, whereas high molecular weight HA could form protective films on the skin surface, ideal for topical applications (Witting et al.,2015).HA can be formulated into nanoparticles, micelles, liposomes, or hydrogels, allowing for versatile delivery systems for drugs like antibiotics, antioxidants, or chemotherapeutic agents as part of skin treatment (Brown, & Jones, 2005).

Among its many biological functions, HA plays a role in inflammation, wound healing, cell differentiation, embryological development, viscoelasticity, and more (Girish, & Kemparaju, 2007; Juncan et al.,2021). HA is widely used in anti-aging products due to its ability to improve skin elasticity and reduce wrinkles. As a carrier, it enhances the delivery of active ingredients like vitamin C, peptides, or retinoids, improving their stability and efficacy (Juncan et al.,2021). HA fillers, used in cosmetic dermatology, can incorporate drugs to reduce inflammation or infection at injection sites. HA can be conjugated with antimicrobial agents (e.g., silver nanoparticles) to treat skin infections while maintaining a moist healing environment. Its anti-inflammatory properties enhance the delivery of corticosteroids or non-steroidal anti-inflammatory drugs (NSAIDs) for conditions like eczema or dermatitis (Maniar et al.,2018).

Although HA has wide application in cosmetic industry, high-purity HA production can be expensive, impacting scalability for widespread dermatological use (Juncan et al.,2021). Researchers are exploring HA-based surface-decorated nanocarriers and hydrogels to enhance the effectiveness of rheumatoid arthritis therapy, aiming for tailored delivery, reduced side effects, and improved patient adherence (Priya et al.,2024).

CELLULOSE

Cellulose is the most abundant natural renewable polymer on earth. It is a linear polysaccharide formed by β-1-4-linked D-glucopyranose repeating units (Duan et al.,2018). Cellulose aerogels havereceived increasing research interest in biomedical fields concerning tissue engineering, controllable delivery systems, and wound dressings (Xu et al., 2018). The features of carboxymethyl cellulose (CMC), which are frequently employed in drug administration and other biomedical studies, include hydrophilicity, bio adhesiveness, pH sensitivity, non-toxicity, and the capacity to gel (Javanbakht, and Shaabani, 2019). This characteristic, which is based on its carboxylate groups, causes various situations and, as a result, makes it suitable for use in drug delivery systems (Javanbakht, and Shaabani, 2019). Three-dimensional hydrophilic CMC networks can be formed as a polymer hydrogel that is capable to absorb large amounts of biological fluids or water and release drugs at the controlled dosage. The polymer crosslinking can be achieved via polyvalent carboxylic acids such ascitric acid, by gamma irradiation, or by chloromethylene iodide (Javanbakht, and Shaabani, 2019).

Carboxymethyl cellulose showed promising characteristic as oral delivery agent. In controlled drug release systems, the CMC polymeric delivery systems could also prevent the drug from drug crystallization or protect degradation. Furthermore, the

CMC could improve drug release frequency via the rate of drug diffusion over the polymeric material, and/or via the polymeric erosion/degradation rate, or over the gels formed of the CMC by contact with water (Gupta et al.,2002). As well as, CMC could potentially to be used as responsive release agents, or as fast dissolving agents that accelerate pill dissolution. Oral formulation of the drugs with low solubility often have variable and/or poor oral bioavailability, poses great challenge to formulation scientists. CMC may be useful in amorphous solid dispersion application, which can dramatically improve solution concentration, particularly for extremely crystalline drugs, via reducing the crystallinity of drug and in the solid dispersion and after dissolution steadying the amorphous drug contrary to recrystallization (Murdande et al.,2011).

The table 2 provides examples of polysaccharides used in medication delivery, preparation, delivery systems, and applications.

Table 2. Polysaccharides (chitosan, hyaluronic acid, cellulose, and composites) used in the delivery of drugs, preparation, delivery systems and application

Polymer	Delivery vector for-	Preparation	Delivery systems	Key features	Reference
Chitosan	Antibiotic: Amoxicillin	Encapsulation in the N-2-hydroxypropyl trimethyl ammonium chloride chitosan	Nanoparticle	Sustained and slow release of drug, improve bioavailability, prolong residence time at the target site, and reduce the dosage of drug.	Zhao et al.,2022
	Anti-inflammatory drugs: Dexamethasone	Chitosan	Hydrogel	Efficient for buccal application	Pignatello et al.,2007
	Anticancer agents: Doxorubicin.	Ionic gelation	Nanoparticle	Superior cytotoxic properties against MCF-7 cells	Saeed et al.,2020
	Vaccine antigens: Hepatitis B surface antigen	Emulsion solvent evaporation technique	Microspheres	Effective oral administration of HBsAg.	Premaletha et al., 2010
	Antioxidant: Curcumin	Ionotropic gelation with tripolyphosphate or polyelectrolyte complexation with alginate	Encapsulated	Improved antioxidant property	Akolade et al.,2018
	Gene	*N*-methylene phosphonic chitosan (NMPCS)	Macromolecule	Improved luciferase expression	Zhu et al.,2010
Hyaluronic Acid	Osteoarthritis drugs: diclofenac and Dexamethasone	Hyaluronan-Loaded Liposomal Dexamethasone–Diclofenac	Nanoparticles	Efficient formulation for drug deliery	Chang et al.,2021
	Cancer therapeutics: Doxorubicin and paclitaxel	Surface modified liposome	Nanocarrier	Tumour targeting via CD44 receptor	Song et al.,2019
	Antimalarial, antiarthritisdrug:Hydroxychloroquine	HA-ss-HCQ nanocapsules	Nanocapsules	Active-targeting therapy in metastatic breast cancer	He et al., 2022

continued on following page

Table 2. Continued

Polymer	Delivery vector for-	Preparation	Delivery systems	Key features	Reference
Cellulose	Antibiotic: Amoxicillin-	Amoxicillin loaded cellulose aerogels	Aerogels	Controlled release behaviour and efficient antibacterial performance	Ye et al., 2018
	Anticancer drug: Doxorubicin	Polymeric microspheres P(MAA-co-NIPAAM-co-EGDMA) coated with modified cellulose	Microspheres	Improved water solubility	Metaxa et al.,2014
	Anticancer drug: 5-fluorouracil	Aminated-glycidylmethacrylate grafted cellulose-grafted polymethacrylic acid-succinyl cyclodextrin (Cell-g-(GMA/en)-PMA-SCD)	-	Releases drug in a controlled manner and is toxic to MCF-7 cells	Anirudhan et al.,2015
	Antiviral agent: Nevirapine	cellulose acetate butyrate nanoparticles loaded with nevirapine	Nanoparticles	Potential carrier for efficient anti-viral drug delivery.	Varshosaz et al.,2018
Chitosan and Hyaluronic acid	Anti-inflammatory agent: Dexamethasone	Dexamethasone-sodium-phosphate (DEX)-chitosan nanoparticles (CSNPs) coated with hyaluronic acid (HA)	Nanoparticle	Improved anti-inflammatory properties of drug	Alkholief et al.,2023
Bacterial Cellulose	Aspirin, and gentamicin	Polyethylenimine-Functionalized Bacterial Cellulose	Aerogels	-	Chen et al.,2018

Cellulose, a naturally abundant polysaccharide, plays a significant role in dermatological applications due to its biocompatibility, biodegradability, and versatile physicochemical properties. Its applications range from wound healing and drug delivery to tissue engineering and cosmetic formulations. Due to their distinct nanostructure, microbial cellulose membranes may find numerous other applications in regenerative medicine and wound healing, including guided tissue regeneration (GTR), periodontal therapies, and the restoration of dura mater, a membrane that envelops brain tissue (Czaja et al.,2007). There are three different types of nanocellulose occur as cellulose nanocrystals (CNC), cellulose nanofibrils (CNF) and bacterial cellulose (BC) Lin, & Dufresne, 2014).

Cellulose, particularly in the form of bacterial cellulose (BC) or nanocrystalline cellulose (NCC), is widely used in wound dressings due to its high water-holding capacity, flexibility, and ability to maintain a moist environment conducive to healing (Lin, & Dufresne, 2014)..

Bacterial cellulose (BC) mimics the extracellular matrix, promoting cell adhesion and proliferation. It is used in products like Biofill® and XCell® for treating burns, chronic wounds, and ulcers. BC's high tensile strength and barrier properties protect wounds from infections while allowing gas exchange (Czaja et al.,2007). BC provides a moist environment, absorbs exudate, and reduces pain by cooling the wound site. It also supports angiogenesis and tissue regeneration. Cellulose derivatives like hydroxypropyl methylcellulose (HPMC), carboxymethyl cellulose (CMC), and ethyl cellulose are used as carriers for controlled drug release in topical and transdermal applications (Roy et al.,2009). Cellulose derivatives act as thickeners, stabilizers, and gelling agents in creams and gels, ensuring uniform drug distribution and sustained release. For example, HPMC is used in gels for delivering anti-inflammatory drugs like diclofenac. Cellulose-based hydrogels and films provide a matrix for drugs, enabling controlled diffusion through the skin with minimum irritation (Roy et al.,2009).

Cellulose scaffolds, especially BC and cellulose nanocrystals (CNC), are used in skin tissue engineering due to their structural similarity to collagen and ability to support cell growth. BC's 3D nanofibrous structure provides a scaffold for fibroblast and keratinocyte growth, aiding in skin regeneration. It is used in artificial skin grafts and regenerative therapies (Fu et al.,2013). Cellulose can be modified with bioactive molecules (e.g., growth factors) to enhance tissue repair.

Cellulose and its derivatives are used in cosmetics for their thickening, emulsifying, and moisturizing properties. Microcrystalline cellulose (MCC) and CMC are added to creams, lotions, and masks to improve texture and stability (Klemm et al.,2011). They enhance skin hydration by forming a protective film. Cellulose beads are used as gentle exfoliants, while cellulose-based hydrogels are employed in hydrating face masks.

Cellulose can be functionalized with antimicrobial agents (e.g., silver nanoparticles, quaternary ammonium compounds) to create wound dressings or patches with infection-fighting properties (Wei et al.,2011). Functionalized cellulose releases antimicrobial agents to prevent bacterial growth, making it ideal for chronic wounds or burns (Lin, & Dufresne, 2014).

Cellulose's natural origin and biodegradability make it an environmentally friendly option for dermatological products, reducing the risk of adverse reactions compared to synthetic polymers (Lin, & Dufresne, 2014). Cellulose based hydrogels have recently been investigated as effective drug carrier formulation (Majumdar,2025).

Chitosan, hyaluronic acid, and cellulose have lots of applications in the pharmaceutical industries. Still extensive research is needed on the cost-effective formulations and finding its potency on the targeted diseases. In this book chapter covers the properties of the polysaccharides, formulations, treatments, and prospects.

BACTERIAL EXOPOLYSACCHARIDES

Bacterial exopolysaccharides (EPS) are complex sugars produced by bacteria, often used as protective layers (Saha and Datta 2022). Exopolysaccharides can be recovered from the medium culture if the organisms are grown via fermentative procedures, as they are typically released into the environment by various organisms (Tabernero, and Cardea, 2020). A variety of mechanisms, including synthase-dependent pathways, Wxz or Wzy-dependent pathways, or even the possibility that bacteria or organisms (like red seaweed) secrete an enzyme that polymerises the necessary sugar source and forms the polymer in the medium culture (extracellular synthesis with enzymes), can account for the formation of EPS (Tabernero, and Cardea, 2020). They show potential as drug carriers because they are safe for the body and can break down naturally. They can encapsulate drugs, enable controlled release, and target specific tissues, enhancing therapy effectiveness. Examples include alginate, dextran, and hyaluronic acid, each with unique applications like wound healing and cancer treatment (Saha and Datta 2022).

Different EPS have unique roles like alginate from *Azotobacter vinelandii* is used in wound healing and drug delivery for lungs and mouth (Wang et al.,2023). Dextran from *Leuconostoc mesenteroides* helps in nanoparticle drug delivery, especially for cancer treatments (Shalash et al.,2021). Hyaluronic Acid from *Streptococcus equi* is used in cancer diagnosis and eye treatments, targeting specific cells.

Bacterial EPS has the ability to form hydrogels, nanoparticles, fibers, and scaffolds makes them suitable for encapsulating and delivering therapeutic agents, enhancing drug stability, and enabling controlled release (Ahuja, et al.,2023).

Bacterial cellulose (from *Acetobacter xylinum*) has been grafted with polyethylenimine showed good drug carrier capacity for aspirin and gentamicin (Chen et al.,2018). Bacterial exopolysaccharides (EPS) are high-molecular-weight polysaccharides produced by microorganisms, particularly bacteria, as a protective layer outside their cell walls. These EPS can play a significant role in drug delivery systems due to their biocompatibility, biodegradability, and ability to encapsulate various therapeutic agents. EPS's biodegradability reduces the possibility of accumulating in the body, which makes it appropriate for medication administration (Rinaudo, M. 2006). By shielding encapsulated medications from deterioration brought on by external elements including pH, temperature, and enzyme activity, EPS can improve their stability (Saha and Datta 2022). A sustained therapeutic impact can be achieved by adjusting the structure of EPS to regulate the combined drug's release rate (Rinaudo, M. 2006). Because it forms films and is biocompatible, pullulan from *Aureobasidium pullulans* has been employed in drug delivery (Kaith et al., 2025). Gellan gum, a byproduct of *Sphingomonas elodea*, is frequently utilised in formulations that call for controlled drug release because of its gelling qualities (Abdl Aali et al., 2024).

For the efficient administration of anticancer medications, EPS can be used to coat nanoparticles (Shehata et al., 2023). Hydrogel formulations that can be utilised for localised medication administration, particularly in wound healing applications, can comprise bacterial extracellular polymeric substances (EPS) (Saha and Datta 2022). When taken orally, EPS nanosystems can improve absorption and bioavailability by shielding medications from the severe environment of the gastrointestinal tract (Sumaila et al., 2021).

Curdlan is a neutral, essentially linear, (1→3)-β-glucan which may have a few intra- or inter-chain (1→6) (McIntosh et al.,2005). Curdlan producers are *Alcaligenes* spp., *Agrobacterium* spp., *Paenibacillus* spp., *Rhizobium* spp., *Saccharomyces cerevisiae*, *Candida* spp., *Aureobasidium pullulan*, *Poria cocos,* etc. The ability of curdlan to create gels with varying properties is reflected in its name (Yotsuzuka 2001). A high-set, thermo-irreversible gel is created when curdlan aqueous solutions are heated above about 80°C and subsequently cooled, while a low-set, thermo-reversible gel is created when they are heated to 55°C.Although curdlan, a linear (1→3)-β-glucan, has potential in the pharmaceutical and medical fields, its prospective uses resulting from its nanocrystalline structure as curdlan megalosaccharides have not yet been investigated (McIntosh et al.,2005).Curdlan can combine with a variety of biologically significant bioactive compounds to generate single and triple helical structures that have antibacterial and anticancer properties (Ganie et al.,2022). Curdlan offers the opportunity to process this polysaccharide with bioactive molecules into macromolecules with specified activities because it is advantageous for enhancing solubility when compared to other polysaccharides. Curdlan conjugated with several antibacterial and anticancer drugs has been investigated as a drug delivery vehicle or as a drug-containing biomaterial. Curdlan exhibits immunoregulatory, antitumor, antiviral (against human immunodeficiency virus (HIV) and Aedes aegypti virus), prebiotic, antimalarial properties (Chaudhari et al.,2021). *Gluconacetobacter xylinus* produces curdlan pellicle, which is one of the greatest bio-based materials because of its super network structure and distinctive physiochemical characteristics and may find usage in a variety of tissue engineering and therapeutic settings (Fang et al.,2015). Research on microencapsulation and nanotechnology is required for the creation and administration of curdlan parenteral, antiviral, and anticancer treatments (Chaudhari et al.,2021). Curdlan preparations require the application of sound manufacturing practices and quality control techniques (Chaudhari et al.,2021).

Alginate is a copolymer constituted by β-d-mannuronic and α-l-guluronic acids which is obtained from the bacterium *Azotobacter vinelandii* (Tabernero and Cardea 2020).Alginate's superior gelation and biocompatibility make it may be the most widely utilised polysaccharide carrier. Alginate hydrogels have the ability to encapsulate thousands of various substances, including as proteins, growth factors, cells, etc., primarily for use in tissue engineering, wound dressing systems, and pulmonary and oral delivery (Tabernero and Cardea 2020). Their anionic nature, however, makes it difficult to produce fibre from them using electrospinning techniques. Alginate should be chemically functionalised or blended with other chemicals and polymers to increase its stiffness and mechanical qualities (Tabernero and Cardea 2020).

Dextran is an exopolysaccharide with neutral charge and has the potential to prevent blood clots. The bacterial polysaccharide dextran is made up of α-1,6-linked d-glucopyranose residues and is typically generated by enzymes from specific strains of *Streptococcus* or *Leuconostoc*.Oxidized, methacrylated photocrosslinked forms of dextran are used as drug delivery systems (Lévesque, and Shoichet, 2006; Yang et al.,2019).

Carrageenans are anionic polysaccharides with D-galactoses (β (1-4)) as main constituents. Because carrageenan polymers contain sulphate groups, they have an anionic charge (Li et al.,2014). Carrageenans come in a variety of forms, however the most commonly used in biomedical applications are the kappa, iota, and lambda types containing one, two, or three sulphate molecules (Tabernero and Cardea 2020). The characteristics of the polymer are obviously altered by the quantity of sulphate groups. Indeed, following thermal stimulation, kappa and iota carrageenans can create thermally reversible gels that release medications (Tabernero and Cardea 2020).A very adaptable polymer, carrageenan can be treated using a variety of techniques to create hydrogels (e.g., order–disorder transitions, monovalent and divalent cations, electrolyte complexes). It's crucial to note in this context that carrageenan and potassium ions combine to form a tubular structure that may be helpful in certain situations. Because of its simple hydrogel formation (also in the form of beads), it is one of the most thoroughly studied EPSs for aerogel manufacture (Tabernero and Cardea 2020).

Several bacterial EPS have been extensively studied for drug delivery, each with unique properties and applications. Ocular medication administration is a highly difficult task because of the distinct physiological and anatomical barriers. In order to decrease the frequency of instillations, scientists have been required to create new formulations that transport medications to ocular tissues at a controlled rate due to the limited ocular bioavailability (<10%) acquired from standard formulations. Natural polymers have shown promise for topical medication delivery through the constrained precorneal space and long-term release (Pahuja et al.,2012). The following table (Table 3) summarizes key EPS, their sources, properties, and examples of use:

Table 3. Examples, and formulation of the bacterial exopolysaccharides in drug delivery

EPS	Source	Applications	Example	References
Alginate	*Azotobacter vinelandii*	Wound healing	Alginate fibers with chlorhexidine gluconate (CHG) fabricated by wet blending spinning technology.	Shi et al.,2020
Dextran	*Leuconostoc mesenteroides*	colon targeting vehicles for oral delivery of drug nanoparticles.	Paclitaxel and dextran hydrogel	Yu, and Pishko, 2011
Xanthan Gum (XG)	*Xanthomonas campestris*	Hydrogel films cosmetics, drug delivery	Chitosan-XG fibers for Neomycin, Ibuprofen	Layek, 2024
Gellan Gum	*Sphingomonas paucimobilis*	Nasal drug delivery	Salbutamol sulfateand gellan gum gel and hydroxylpropyl methyl cellulose (HPMC) for nasal administration.	Salunke, and Patil, 2016
Curdlan	*Alcaligenes faecalis*	Nanofibers	Tetracycline hydrochloride in Fibers as antibacterialagent against*E. coli*	El-Naggar et al., 2016
Levan	*Halomonas smyrnensis*	microparticular system	potential carrier of antibiotic drug Vancomycin	Sezer et al.,2017
Hyaluronic Acid (HA)	*Streptococcus zooepidemicus*	Cancer treatment	Doxorubicin carrier	Rigo et al.,2022
Pullulan	*Aurebasidium pullulans*	Drug delivery	Drug carrier	Grigoras, 2019
Curdlan	*Agrobacterium sp. ATCC 31750*	Drug carrier to target *Mycobacterium smegmatis* infected macrophages	Curdlan sulphate - Chitosan nanoparticles for Rifampicin drug	Ravindran et al.,2021

The bio fabrication of naturally derived polymer scaffolds with encapsulated cells for biomedical applications (e.g., cell therapies, bone, and dental grafts, cardiovascular or musculoskeletal tissue regeneration, and wound healing) has been thoroughly studied using 3D printing, microfluidics, and electrospinning techniques (Puertas-Bartolomé et al.,2021; Pal et al.,2024).

CHALLENGES AND CURRENT PERSPECTIVES

Batch-to-batch variability, limited mechanical strength, immunogenicity, complex purification, stability issues and regulatory challenges limits the potential of polymer as drug delivery agent. Selection of the polymer depends on the aspects such inexpensive extraction procedures, environmentally friendly production processes, chemistry, surface qualities, rheology, bulk properties, high biocompatibility, and acceptable biodegradability (Al-Hussainawy, and Jasim Al-Hayder, 2024). Even with their broad range of applications, natural polymers still need a lot of work to be established on a commercial scale in innovative formulations. Polysaccharide's are widely employed in targeted and controlled-release medication delivery systems. Reports of in vivo research are uncommon, nevertheless, and the majority of studies are currently only in the in vitro experimental stage. Furthermore, there has been little progress in the creation and application of drug carriers that target chemical modification of hyaluronic acid polysaccharide and employ variety (Huang and Huang, 2018). For biopharmaceutical delivery to be continuously optimised, research on the alteration of its derivatives should be expanded. It is challenging to industrialise polysaccharides as medication carriers because of their intricate procedure (Huang and Huang, 2018).

Chitosan as a drug Carrier has challenges like poor solubility (in neutral or basic physiological conditions), mechanical weakness, variations in molecular weight and degree of deacetylation, limitedtargeting, potential immunogenicity. Current perspectives in the use of chitosan as drug carrier involves chemical modification through derivatization, nanocomposite formulations, stimuli-responsive systems, and hybrid systems.

Chemical modifications of chitosan through derivatization (e.g., quaternization, carboxymethylation) enhances solubility and stability across a broader pH range. For instance, quaternized chitosan (QCS) cryogels improve water absorption and mechanical resilience for hemostasis and wound healing (Mottaghitalab et al.,2024). Combining chitosan with nanoparticles (e.g., hyaluronic acid-coated chitosan nanoparticles) improves mechanical properties and enables targeted delivery, such as for cancer therapy via CD44 receptor targeting (Wang et al.,2017).pH-responsive chitosan/hyaluronic acid films, prepared via layer-by-layer assembly, offer controlled drug release in specific microenvironments (e.g., tumor sites) (Lu et al.,2019).Blending chitosan with other polymers like alginate or cellulose enhances mechanical strength and bioactivity, as seen in calcium ion-crosslinked chitosan-alginate hydrogels for oral site specific drug carrier (Xu et al.,2007). Chitosan's mucoadhesive properties make it promising for nasal/oral mRNA vaccine delivery, with ongoing research to optimize payload stability (Parmaksız, and Şenel, 2021).

Cellulose as drug carriers has limitations due to insolubility, limited chemical reactivity, slow degradation, high processing cost, less yield, and inefficiency in drug loading. Cellulose's high crystallinity and hydrogen bonding make it insoluble in water and most organic solvents, complicating processing into drug carriers.The hydroxyl groups on cellulose offer fewer modification sites compared to chitosan or hyaluronic acid, restricting functionalization for targeted delivery. Bacterial cellulose (BC) is highly resistant to enzymatic degradation, which can hinder its use in applications requiring biodegradability. Whilebacterial cellulose offers superior purity and biocompatibility, its large-scale production is costly due to fermentation requirements and low yields. Cellulose's dense fibrillar structure limits drug encapsulation efficiency, particularly for hydrophobic drugs. Current perspectives in the utilisation of cellulose as drug carrier involves creation of the composites, chemical derivatization, cost effective production and the nanocellulose based system.

Bacterial cellulose 3D nanofibril network can be combined with biopolymers like hyaluronic acid or chitosan to enhance drug loading and water retention, as seen in BC/HA composites for wound dressings (Li et al.,2015).Chemical derivatization (e.g., carboxymethylation) improves solubility and introduces reactive groups for drug conjugation, enabling controlled release systems.

Strategies like fructose-based fermentation and co-cultivation of bacterial cellulose and hyaluronic acid producers (e.g., *Komagataeibacter* and *Lactocaseibacillus*) reduce costs and yield biocomposite with enhanced properties (Brugnoli et al.,2025). Cellulose nanocrystals (CNCs) and nanofibers offer high surface area for drug adsorption, with applications in oral drug delivery and tissue engineering (Grishkewich et al.,2017).Bacterial cellulose's high water absorption and biocompatibility make it ideal for sustained-release matrices, such as insulin-loaded cellulosenanoparticles (Esmaeili et al.,2024).

The use of hyaluronic acid as drug carrier has challenges like weak mechanical strength, rapid degradation, expensive polymer, non-specific interactions, and processing complexity.

HA's high hydrophilicity results in weak mechanical strength, limiting its use in solid scaffolds or long-term drug release systems. Hyaluronic acid is quickly degraded by hyaluronidases in vivo, reducing the duration of drug release unless modified. Large-scale production of high-molecular-weight HA is expensive, particularly for pharmaceutical-grade applications.

HA's interaction with CD44 receptors, while useful for cancer targeting, can lead to non-specific uptake in healthy tissues expressing CD44.HA's high viscosity and polyanionic nature complicate the formation of stable nanoparticles or hydrogels without cross-linking agents.

Current perspectives in the use of hyaluronic acid as drug carrier comprises hydrogel based formulations, composites for the targeted delivery.

HA-based hydrogels, cross-linked with chitosan or synthetic polymers, improve mechanical stability and control drug release, as in HA/oxidized chitosan hydrogels for wound disinfection (Li et al.,2014).HA's affinity for CD44 receptors is leveraged for cancer therapy, with HA-coated chitosan nanoparticles delivering drugs like doxorubicin or enzyme/prodrug systems (Wang et al.,2017).Chemical modifications (e.g., thiolation, aldehyde modification) tailor HA's degradation rate and mechanical properties, supporting applications like cartilage regeneration (Mashaqbeh et al.,2024).HA combined with cationic polymers like diethylaminoethyl chitosan forms PECs for antibiotic delivery (e.g., colistin), offering high encapsulation efficiency and modified release (Dubashynskaya et al.,2021).HA reduces bacterial fouling in chitosan-based wound dressings, enhancing infection resistance without additional antimicrobial agents (Silvestro et al.,2020).

In one study, individual cells of genetically modified *Gluconacetobacter hansenii* (*G. hansenii*) demonstrated the ability to produce bacterial cellulose (BC) and hyaluronan (HA) simultaneously, potentially creating naturally composited nanofibrils. Hyaluronan synthase and UDP-glucose dehydrogenase genes, which are necessary for HA production, were introduced into the cells by manipulation. When compared to traditional in situ nanocomposites, the gel-like nanocomposite materials made by the altered *G. hansenii* shown better qualities. Using *G. hansenii* to create integrated cellulose-based nanomaterials is made easier by this genetic engineering technique.

Despite their potential, several challenges remain in the use of bacterial exopolysaccharide like variability (EPS composition), difficulty in purification, scalability, stability, and poor mechanical strength. Future research aims to address scalability, improve mechanical properties through blending, and explore novel EPS from extremophiles for enhanced stability. The integration of genetic and metabolic engineering could increase EPS production yields, making them more competitive with synthetic alternatives.

Combining biopolymers (e.g., chitosan/HA, BC/HA) leverages complementary properties, improving mechanical strength, bioactivity, and drug release control.

Advances in microbial fermentation and co-culture systems aim to reduce costs and environmental impact. Tailoring biopolymer properties (e.g., molecular weight, cross-linking) enables customized drug delivery system for specific diseases.

The novel approach in the drug carrier system is the use of bio-inspired yeast microcapsule for delivery of charged nanoparticles like quantum dots, iron oxide nanoparticle, and various fluorescent nanoparticles (Sabu et al.,2018). The long-term administration has a good safety profile compared to other delivery systems. Particles can be incorporated into yeast microcapsule by electrostatic interaction, layer-by-layer approach, and surface derivatization (Sabu et al.,2018). The primary component of yeast capsules is β-glucan, which Dectin-1 on M cells recognises (Wu et al.,2023). To encapsulate medications in yeast microcapsules, conventional mixing and spray drying techniques are employed. For the oral administration of nanotherapies to remote locations, yeast microcapsules present a viable platform (Sabu et al.,2018). The development of vaccines has benefited greatly from the use of recombinant yeast-based vaccine carriers (WYV) (Tan et al.,2022). The yeast-based drug delivery system can be administered by a variety of channels because of its distinct phagocyte-specific targeting mechanism and immunogenicity, which may be useful for the delivery of medications and vaccines (Tan et al.,2022).

CONCLUSION

Natural polymer-based drug delivery systems offer a promising platform for safe, effective, and targeted drug administration due to their biocompatibility, biodegradability, and versatility. While challenges like variability and stability need to be addressed, ongoing research and technological advancements are expanding their applications in cancer therapy, tissue engineering, and personalized medicine. By leveraging the unique properties of natural polymers, these systems are poised to play a significant role in the future of drug delivery.

Natural polysaccharides, such as chitosan, cellulose, hyaluronic acid, and bacterial exopolysaccharides, hold significant promise as drug carrier agents due to their biocompatibility, biodegradability, and tunable physicochemical properties. The addition of nanoparticles to a bioink (formulation that contain biologically active components and cells) to enhance its stiffness, shear-thinning, degradation, or stability is intriguing from the perspective of the polymer discipline. These biopolymers enable diverse applications, including targeted cancer therapy, wound healing, and vaccine delivery, by facilitating controlled drug release, mucoadhesion, and specific receptor interactions. However, challenges like poor solubility, mechanical weakness, high production costs, and regulatory hurdles limit their widespread clinical adoption. Current strategies, including chemical modifications, nanotechnology, hybrid composites, and sustainable production methods, are addressing these limitations, enhancing their functionality and cost-effectiveness. Continued research into optimizing biopolymer properties, scaling production, and validating safety through in vivo studies will be crucial for realizing their full potential in personalized and advanced drug delivery systems. The concept of the synthetic biology can be incorporated to fabricate the natural polymers to serve as drug delivery system. The interpretation of drug delivery design from an experimental to a commercial development standpoint, as well as chemical and structural stability during application and storage, are significant concerns that require more research.

ACKNOWLEDGEMENT

I acknowledge the MahaJyoti fellowship 2022.

REFERENCES

Abdl Aali, R. A. K., & Al-Sahlany, S. T. G. (2024). Gellan gum as a unique microbial polysaccharide: Its characteristics, synthesis, and current application trends. *Gels (Basel, Switzerland)*, *10*(3), 183. DOI: 10.3390/gels10030183 PMID: 38534601

Ahuja, V., Bhatt, A. K., Banu, J. R., Kumar, V., Kumar, G., Yang, Y. H., & Bhatia, S. K. (2023). Microbial exopolysaccharide composites in biomedicine and healthcare: Trends and advances. *Polymers*, *15*(7), 1801. DOI: 10.3390/polym15071801 PMID: 37050415

Akolade, J. O., Oloyede, H. O. B., Salawu, M. O., Amuzat, A. O., Ganiyu, A. I., & Onyenekwe, P. C. (2018). Influence of formulation parameters on encapsulation and release characteristics of curcumin loaded in chitosan-based drug delivery carriers. *Journal of Drug Delivery Science and Technology*, *45*, 11–19. DOI: 10.1016/j.jddst.2018.02.001

Al-Hussainawy, M. K., & Jasim Al-Hayder, L. S. (2024). Drug carriers in the delivery and release of hydroxychloroquine by biopolymer. *Chemicke Zvesti*, *78*(9), 5255–5266. DOI: 10.1007/s11696-024-03463-w

Alaniz, L., Cabrera, P. V., Blanco, G., Ernst, G., Rimoldi, G., Alvarez, E., & Hajos, S. E. (2016). Interaction of CD44 with different forms of hyaluronic acid. Its role in adhesion and migration of tumor cells. *Cell Communication & Adhesion*, *9*(3), 117–130. DOI: 10.1080/15419060214522 PMID: 12521133

Ali, A., & Ahmed, S. (2018). A review on chitosan and its nanocomposites in drug delivery. *International Journal of Biological Macromolecules*, *109*, 273–286. DOI: 10.1016/j.ijbiomac.2017.12.078 PMID: 29248555

Alkholief, M., Kalam, M. A., Raish, M., Ansari, M. A., Alsaleh, N. B., Almomen, A., Ali, R., & Alshamsan, A. (2023). Topical Sustained-Release Dexamethasone-Loaded Chitosan Nanoparticles: Assessment of Drug Delivery Efficiency in a Rabbit Model of Endotoxin-Induced Uveitis. *Pharmaceutics*, *15*(9), 2273. DOI: 10.3390/pharmaceutics15092273 PMID: 37765242

Anirudhan, T. S., Nima, J., & Divya, P. L. (2015). Synthesis, characterization and in vitro 5-fluorouracil, an anticancer drug. *Applied Surface Science*, *355*, 64–73. DOI: 10.1016/j.apsusc.2015.07.077

Aranaz, I., Alcántara, A. R., Civera, M. C., Arias, C., Elorza, B., Heras Caballero, A., & Acosta, N. (2021). Chitosan: An Overview of Its Properties and Applications. *Polymers*, *13*(19), 3256. DOI: 10.3390/polym13193256 PMID: 34641071

Atanase, L. I. (2021). Micellar drug delivery systems based on natural biopolymers. *Polymers*, *13*(3), 477. DOI: 10.3390/polym13030477 PMID: 33540922

Bernkop-Schnürch, A., & Dünnhaupt, S. (2012). Chitosan-based drug delivery systems. *European Journal of Pharmaceutics and Biopharmaceutics*, *81*(3), 463–469. DOI: 10.1016/j.ejpb.2012.04.007 PMID: 22561955

Biswas, R., Mondal, S., Ansari, M. A., Sarkar, T., Condiuc, I. P., Trifas, G., & Atanase, L. I. (2025). Chitosan and Its Derivatives as Nanocarriers for Drug Delivery. *Molecules (Basel, Switzerland)*, *30*(6), 1297. DOI: 10.3390/molecules30061297 PMID: 40142072

Bot, P. T., Hoefer, I. E., Piek, J. J., & Pasterkamp, G. (2008). Hyaluronic acid: Targeting immune modulatory components of the extracellular matrix in atherosclerosis. *Current Medicinal Chemistry*, *15*(8), 786–791. DOI: 10.2174/092986708783955554 PMID: 18393848

Brown, M. B., & Jones, S. A. (2005). Hyaluronic acid: A unique topical vehicle for the localized delivery of drugs to the skin. *Journal of the European Academy of Dermatology and Venereology*, *19*(3), 308–318. DOI: 10.1111/j.1468-3083.2004.01180.x PMID: 15857456

Brugnoli, M., Carvalho, J. P. F., Arena, M. P., Oliveira, H., Vilela, C., Freire, C. S. R., & Gullo, M. (2025). Co-cultivation of Komagataeibacter sp. and Lacticaseibacillus sp. strains to produce bacterial nanocellulose-hyaluronic acid nanocomposite membranes for skin wound healing applications. *International Journal of Biological Macromolecules*, *140208*, 140208. Advance online publication. DOI: 10.1016/j.ijbiomac.2025.140208 PMID: 39848389

Chang, M.-C., Chiang, P.-F., Kuo, Y.-J., Peng, C.-L., Chen, K.-Y., & Chiang, Y.-C. (2021). Hyaluronan-Loaded Liposomal Dexamethasone–Diclofenac Nanoparticles for Local Osteoarthritis Treatment. *International Journal of Molecular Sciences*, *22*(2), 665. DOI: 10.3390/ijms22020665 PMID: 33440880

Chaudhari, V., Buttar, H. S., Bagwe-Parab, S., Tuli, H. S., Vora, A., & Kaur, G. (2021). Therapeutic and industrial applications of curdlan with overview on its recent patents. *Frontiers in Nutrition, 8*, 646988. DOI: 10.3389/fnut.2021.646988 PMID: 34262922

Chaudhury, A., & Das, S. (2011). Recent advancement of chitosan-based nanoparticles for oral controlled delivery of insulin and other therapeutic agents. *AAPS PharmSciTech, 12*(1), 10–20. DOI: 10.1208/s12249-010-9561-2 PMID: 21153572

Chen, X., Xu, X., Li, W., Sun, B., Yan, J., Chen, C., Liu, J., Qian, J., & Sun, D. (2018). Effective drug carrier based on polyethylenimine-functionalized bacterial cellulose with controllable release properties. *ACS Applied Bio Materials, 1*(1), 42–50. DOI: 10.1021/acsabm.8b00004

Chuah, L. H., Loo, H. L., Goh, C. F., Fu, J. Y., & Ng, S. F. (2023). Chitosan-based drug delivery systems for skin atopic dermatitis: Recent advancements and patent trends. *Drug Delivery and Translational Research, 13*(5), 1436–1455. DOI: 10.1007/s13346-023-01307-w PMID: 36808298

. Czaja, W. K., Young, D. J., Kawecki, M., & Brown, R. M. (2007). The future prospects of microbial cellulose in biomedical applications. biomacromolecules, 8(1), 1-12. DOI: 10.1021/bm060620d

Dai, T., Tanaka, M., Huang, Y. Y., & Hamblin, M. R. (2011). Chitosan preparations for wounds and burns: Antimicrobial and wound-healing effects. *Expert Review of Anti-Infective Therapy, 9*(7), 857–879. DOI: 10.1586/eri.11.59 PMID: 21810057

Ding, L., Agrawal, P., Singh, S. K., Chhonker, Y. S., Sun, J., & Murry, D. J. (2024). Polymer-Based Drug Delivery Systems for Cancer Therapeutics. *Polymers, 16*(6), 843. DOI: 10.3390/polym16060843 PMID: 38543448

Domard, A. (1987). pH and cd measurements on a fully deacetylated chitosan: Application to CuII—polymer interactions. *International Journal of Biological Macromolecules, 9*(2), 98–104. DOI: 10.1016/0141-8130(87)90033-X

. Doppalapudi, S., Katiyar, S., Domb, A. J., & Khan, W. (2015). Biodegradable natural polymers. Advanced polymers in medicine, 33-66.

Duan, Y., Freyburger, A., Kunz, W., & Zollfrank, C. (2018). Cellulose and chitin composite materials from an ionic liquid and a green co-solvent. *Carbohydrate Polymers, 192*, 159–165. DOI: 10.1016/j.carbpol.2018.03.045 PMID: 29691008

Dubashynskaya, N. V., Raik, S. V., Dubrovskii, Y. A., Demyanova, E. V., Shcherbakova, E. S., Poshina, D. N., & Skorik, Y. A. (2021). Hyaluronan/diethylaminoethyl chitosan polyelectrolyte complexes as carriers for improved colistin delivery. *International Journal of Molecular Sciences*, *22*(16), 8381. DOI: 10.3390/ijms22168381 PMID: 34445088

El-Naggar, M. E., Abdelgawad, A. M., Salas, C., & Rojas, O. J. (2016). Curdlan in fibers as carriers of tetracycline hydrochloride: Controlled release and antibacterial activity. *Carbohydrate Polymers*, *154*, 194–203. DOI: 10.1016/j.carbpol.2016.08.042 PMID: 27577910

Esmaeili, J., Pirzadeh, K., Pakrooyan, M., Lukolayeh, M. E., & Kırboğa, K. K. (2024). Synthesis of cellulose nanocrystals from spinach waste for insulin delivery: Comparison to chitosan nanoparticles. *New Journal of Chemistry*, *48*(17), 7953–7963. DOI: 10.1039/D4NJ00073K

Fang, J., Kawano, S., Tajima, K., & Kondo, T. (2015). In vivo curdlan/cellulose bionanocomposite synthesis by genetically modified Gluconacetobacter xylinus. *Biomacromolecules*, *16*(10), 3154–3160. DOI: 10.1021/acs.biomac.5b01075 PMID: 26360299

Fazal, T., Murtaza, B. N., Shah, M., Iqbal, S., Rehman, M. U., Jaber, F., & Ibrahium, H. A. (2023). Recent developments in natural biopolymer based drug delivery systems. *RSC Advances*, *13*(33), 23087–23121. DOI: 10.1039/D3RA03369D PMID: 37529365

Fu, C. P., Cai, X. Y., Chen, S. L., Yu, H. W., Fang, Y., Feng, X. C., & Li, C. Y. (2023). Hyaluronic acid-based nanocarriers for anticancer drug delivery. *Polymers*, *15*(10), 2317. DOI: 10.3390/polym15102317 PMID: 37242892

Fu, L., Zhang, J., & Yang, G. (2013). Present status and applications of bacterial cellulose-based materials for skin tissue repair. *Carbohydrate Polymers*, *92*(2), 1432–1442. DOI: 10.1016/j.carbpol.2012.10.071 PMID: 23399174

Ganie, S. A., Rather, L. J., & Li, Q. (2022). Review on anti-cancer and anti-microbial applications of curdlan biomaterials. *Journal of Polymers and the Environment*, *30*(4), 1284–1299. DOI: 10.1007/s10924-021-02299-9

Girish, K. S., & Kemparaju, K. (2007). The magic glue hyaluronan and its eraser hyaluronidase: A biological overview. *Life Sciences*, *80*(21), 1921–1943. DOI: 10.1016/j.lfs.2007.02.037 PMID: 17408700

Grigoras, A. G. (2019). Drug delivery systems using pullulan, a biocompatible polysaccharide produced by fungal fermentation of starch. *Environmental Chemistry Letters, 17*(3), 1209–1223. DOI: 10.1007/s10311-019-00862-4

Grishkewich, N., Mohammed, N., Tang, J., & Tam, K. C. (2017). Recent advances in the application of cellulose nanocrystals. *Current Opinion in Colloid & Interface Science, 29*, 32–45. DOI: 10.1016/j.cocis.2017.01.005

Gupta, P., Vermani, K., & Garg, S. (2002). Hydrogels: From controlled release to pH-responsive drug delivery. *Drug Discovery Today, 7*(10), 569–579. DOI: 10.1016/S1359-6446(02)02255-9 PMID: 12047857

He, Y., Xu, Y., Huang, Y., Quang, H., Xia, X., Zhao, H., & Yu, J. (2022). Redox sensitive nano-capsules self-assembled from hyaluronic acid-hydroxychloroquine conjugates for CD44-targeted delivery of hydroxychloroquine to combat breast cancer metastasis in vitro and in vivo. *Colloids and Surfaces. B, Biointerfaces, 210*, 112249. DOI: 10.1016/j.colsurfb.2021.112249 PMID: 34864404

. Holowka, E. P., Bhatia, S. K., Holowka, E. P., and Bhatia, S. K. (2014). Controlled-release systems. *Drug delivery: materials design and clinical perspective*, 7-62.

Huang, G., & Huang, H. (2018). Application of hyaluronic acid as carriers in drug delivery. *Drug Delivery, 25*(1), 766–772. DOI: 10.1080/10717544.2018.1450910 PMID: 29536778

Huang, Y., Seitz, D., König, F., Müller, P. E., Jansson, V., & Klar, R. M. (2019). Induction of articular chondrogenesis by chitosan/hyaluronic-acid-based biomimetic matrices using human adipose-derived stem cells. *International Journal of Molecular Sciences, 20*(18), 4487. DOI: 10.3390/ijms20184487 PMID: 31514329

Javanbakht, S., & Shaabani, A. (2019). Carboxymethyl cellulose-based oral delivery systems. *International Journal of Biological Macromolecules, 133*, 21–29. DOI: 10.1016/j.ijbiomac.2019.04.079 PMID: 30986470

Jiang, D., Zou, J., Huang, L., Shi, Q., Zhu, X., Wang, G., & Yang, H. (2010). Efficacy of intra-articular injection of celecoxib in a rabbit model of osteoarthritis. *International Journal of Molecular Sciences, 11*(10), 4106–4113. DOI: 10.3390/ijms11104106 PMID: 21152323

Juncan, A. M., Moisă, D. G., Santini, A., Morgovan, C., Rus, L. L., Vonica- incu, A. L., & Loghin, F. (2021). Advantages of hyaluronic acid and its combination with other bioactive ingredients in cosmeceuticals. *Molecules (Basel, Switzerland), 26*(15), 4429. DOI: 10.3390/molecules26154429 PMID: 34361586

Kaith, A., Garg, U., Jain, N., Pandey, M., Kaul, S., Gorain, B., & Amin, M. C. I. M. (2025). Pullulan as a sustained release carrier for ocular drug delivery: A review. *International Journal of Biological Macromolecules*, *143146*, 143146. Advance online publication. DOI: 10.1016/j.ijbiomac.2025.143146 PMID: 40233906

Klemm, D., Kramer, F., Moritz, S., Lindström, T., Ankerfors, M., Gray, D., & Dorris, A. (2011). Nanocelluloses: A new family of nature-based materials. *Angewandte Chemie International Edition*, *50*(24), 5438–5466. DOI: 10.1002/anie.201001273 PMID: 21598362

Kumar, M. R., Muzzarelli, R., Muzzarelli, C., Sashiwa, H., & Domb, A. J. (2004). Chitosan chemistry and pharmaceutical perspectives. *Chemical Reviews*, *104*(12), 6017–6084. DOI: 10.1021/cr030441b PMID: 15584695

Layek, B. (2024). A Comprehensive Review of Xanthan Gum-Based Oral Drug Delivery Systems. *International Journal of Molecular Sciences*, *25*(18), 10143. DOI: 10.3390/ijms251810143 PMID: 39337626

Lévesque, S. G., & Shoichet, M. S. (2006). Synthesis of cell-adhesive dextran hydrogels and macroporous scaffolds. *Biomaterials*, *27*(30), 5277–5285. DOI: 10.1016/j.biomaterials.2006.06.004 PMID: 16793132

Li, L., Wang, N., Jin, X., Deng, R., Nie, S., Sun, L., & Gong, C. (2014). Biodegradable and injectable in situ cross-linking chitosan-hyaluronic acid based hydrogels for postoperative adhesion prevention. *Biomaterials*, *35*(12), 3903–3917. DOI: 10.1016/j.biomaterials.2014.01.050 PMID: 24507411

Li, Y., Jiang, H., Zheng, W., Gong, N., Chen, L., Jiang, X., & Yang, G. (2015). Bacterial cellulose–hyaluronan nanocomposite biomaterials as wound dressings for severe skin injury repair. *Journal of Materials Chemistry. B*, *3*(17), 3498–3507. DOI: 10.1039/C4TB01819B PMID: 32262233

Lin, N., & Dufresne, A. (2014). Nanocellulose in biomedicine: Current status and future prospect. *European Polymer Journal*, *59*, 302–325. DOI: 10.1016/j.eurpolymj.2014.07.025

Lu, B., Luo, D., Zhao, A., Wang, H., Zhao, Y., Maitz, M. F., & Huang, N. (2019). pH responsive chitosan and hyaluronic acid layer by layer film for drug delivery applications. *Progress in Organic Coatings*, *135*, 240–247. DOI: 10.1016/j.porgcoat.2019.06.012

Majumdar, S. (2025). Cellulose-based hydrogels for drug delivery applications. In *Cellulose-Based Hydrogells* (pp. 259–279). Elsevier. DOI: 10.1016/B978-0-443-22049-4.00012-6

Maniar, K. H., Jones, I. A., Gopalakrishna, R., & Vangsness, C. T.Jr. (2018). Lowering side effects of NSAID usage in osteoarthritis: Recent attempts at minimizing dosage. *Expert Opinion on Pharmacotherapy*, *19*(2), 93–102. DOI: 10.1080/14656566.2017.1414802 PMID: 29212381

Mashaqbeh, H., Al-Ghzawi, B., & BaniAmer, F. (2024). Exploring the formulation and approaches of injectable hydrogels utilizing hyaluronic acid in biomedical uses. *Advances in Pharmacological and Pharmaceutical Sciences*, *2024*(1), 3869387. DOI: 10.1155/2024/3869387 PMID: 38831895

McIntosh, M., Stone, B. A., & Stanisich, V. A. (2005). Curdlan and other bacterial (1→ 3)-β-D-glucans. *Applied Microbiology and Biotechnology*, *68*(2), 163–173. DOI: 10.1007/s00253-005-1959-5 PMID: 15818477

Metaxa, A. F., Efthimiadou, E. K., & Kordas, G. (2014). Cellulose-based drug carriers for cancer therapy: Cytotoxic evaluation in cancer and healthy cells. *Materials Letters*, *132*, 432–435. DOI: 10.1016/j.matlet.2014.06.134

Millotti, G., Laffleur, F., Perera, G., Vigl, C., Pickl, K., Sinner, F., & Bernkop-Schnürch, A. (2014). In vivo evaluation of thiolated chitosan tablets for oral insulin delivery. *Journal of Pharmaceutical Sciences*, *103*(10), 3165–3170. DOI: 10.1002/jps.24102 PMID: 25139279

Mottaghitalab, F., Yazdi, M. K., Saeb, M. R., Bączek, T., & Farokhi, M. (2024). Green and sustainable hydrogels based on quaternized chitosan to enhance wound healing. *Chemical Engineering Journal*, *152288*, 152288. Advance online publication. DOI: 10.1016/j.cej.2024.152288

Murdande, S. B., Pikal, M. J., Shanker, R. M., & Bogner, R. H. (2011). Aqueous solubility of crystalline and amorphous drugs: Challenges in measurement. *Pharmaceutical Development and Technology*, *16*(3), 187–200. DOI: 10.3109/10837451003774377 PMID: 20429826

Muthu, M., Gopal, J., Chun, S., Devadoss, A. J. P., Hasan, N., & Sivanesan, I. (2021). Crustacean waste-derived chitosan: Antioxidant properties and future perspective. *Antioxidants*, *10*(2), 228. DOI: 10.3390/antiox10020228 PMID: 33546282

Ossipov, D. A. (2010). Nanostructured hyaluronic acid-based materials for active delivery to cancer. *Expert Opinion on Drug Delivery*, *7*(6), 681–703. DOI: 10.1517/17425241003730399 PMID: 20367530

Pahuja, P., Arora, S., & Pawar, P. (2012). Ocular drug delivery system: A reference to natural polymers. *Expert Opinion on Drug Delivery*, *9*(7), 837–861. DOI: 10.1517/17425247.2012.690733 PMID: 22703523

Pal, P., Sambhakar, S., Paliwal, S., Kumar, S., & Kalsi, V. (2024). Biofabrication paradigms in corneal regeneration: Bridging bioprinting techniques, natural bioinks, and stem cell therapeutics. *Journal of Biomaterials Science. Polymer Edition*, *35*(5), 717–755. DOI: 10.1080/09205063.2024.2301817 PMID: 38214998

Paliwal, S. R., Paliwal, R., Agrawal, G. P., & Vyas, S. P. (2016). Hyaluronic acid modified pH-sensitive liposomes for targeted intracellular delivery of doxorubicin. *Journal of Liposome Research*, *26*(4), 276–287. DOI: 10.3109/08982104.2015.1117489 PMID: 26784587

. Parmaksız, S., and Şenel, S. (2021). An overview on chitosan-based adjuvant/vaccine delivery systems. *Chitosan for Biomaterials IV: Biomedical Applications*, 293-379..DOI: 10.1007/12_2021_93

Pignatello, R., Stancampiano, A. H. S., Ventura, C. A., & Puglisi, G. (2007). Dexamethasone sodium phosphate-loaded Chitosan based delivery systems for buccal application. *Journal of Drug Targeting*, *15*(9), 603–610. DOI: 10.1080/10611860701502806 PMID: 17968714

Prabaharan, M., & Mano, J. F. (2004). Chitosan-based particles as controlled drug delivery systems. *Drug Delivery*, *12*(1), 41–57. DOI: 10.1080/10717540590889781 PMID: 15801720

Premaletha, K., Licy, C. D., Jose, S., Saraladevi, A., Shirwaikar, A., & Shirwaikar, A. (2010). Formulation, characterization and optimization of Hepatitis B surface antigen (HBsAg)-loaded chitosan microspheres for oral delivery. *Pharmaceutical Development and Technology*, *17*(2), 251–258. DOI: 10.3109/10837450.2010.535824 PMID: 21108582

Priya, S., Daryani, J., Desai, V. M., & Singhvi, G. (2024). Bridging the gap in rheumatoid arthritis treatment with hyaluronic acid-based drug delivery approaches. *International Journal of Biological Macromolecules*, *271*, 132586. DOI: 10.1016/j.ijbiomac.2024.132586 PMID: 38795889

Puertas-Bartolomé, M., Mora-Boza, A., & García-Fernández, L. (2021). Emerging Biofabrication Techniques: A Review on Natural Polymers for Biomedical Applications. *Polymers*, *13*(8), 1209. DOI: 10.3390/polym13081209 PMID: 33918049

Rajitha, P., Gopinath, D., Biswas, R., Sabitha, M., & Jayakumar, R. (2016). Chitosan nanoparticles in drug therapy of infectious and inflammatory diseases. *Expert Opinion on Drug Delivery*, *13*(8), 1177–1194. DOI: 10.1080/17425247.2016.1178232 PMID: 27087148

Ravindran, R., Mitra, K., Arumugam, S. K., & Doble, M. (2021). Preparation of Curdlan sulphate-Chitosan nanoparticles as a drug carrier to target Mycobacterium smegmatis infected macrophages. *Carbohydrate Polymers*, *258*, 117686. DOI: 10.1016/j.carbpol.2021.117686 PMID: 33593559

Rigo, D., da Silva, L. M., Fischer, B., Colet, R., Dallago, R. M., & Zeni, J. (2023). Hyaluronic Acid–from Production to Application: A. *Biointerface Research in Applied Chemistry*, *13*(3), 211. DOI: 10.33263/BRIAC133.211

Rinaudo, M. (2006). Chitin and chitosan: Properties and applications. *Progress in Polymer Science*, *31*(7), 603–632. DOI: 10.1016/j.progpolymsci.2006.06.001

Roy, D., Semsarilar, M., Guthrie, J. T., & Perrier, S. (2009). Cellulose modification by polymer grafting: A review. *Chemical Society Reviews*, *38*(7), 2046–2064. DOI: 10.1039/b808639g PMID: 19551181

Sabu, C., Mufeedha, P., & Pramod, K. (2018). Yeast-inspired drug delivery: Biotechnology meets bioengineering and synthetic biology. *Expert Opinion on Drug Delivery*, *16*(1), 27–41. DOI: 10.1080/17425247.2019.1551874 PMID: 30463446

Saeed, R. M., Dmour, I., & Taha, M. O. (2020). Stable chitosan-based nanoparticles using polyphosphoric acid or hexametaphosphate for tandem ionotropic/covalent crosslinking and subsequent investigation as novel vehicles for drug delivery. *Frontiers in Bioengineering and Biotechnology*, *8*, 4. DOI: 10.3389/fbioe.2020.00004 PMID: 32039190

Saha, I., & Datta, S. (2022). Bacterial exopolysaccharides in drug delivery applications. *Journal of Drug Delivery Science and Technology*, *74*, 103557. DOI: 10.1016/j.jddst.2022.103557

Salunke, S. R., & Patil, S. B. (2016). Ion activated in situ gel of gellan gum containing salbutamol sulphate for nasal administration. *International Journal of Biological Macromolecules*, *87*, 41–47. DOI: 10.1016/j.ijbiomac.2016.02.044 PMID: 26899173

Sarmento, B., Ribeiro, A., Veiga, F., Sampaio, P., Neufeld, R., & Ferreira, D. (2007). Alginate/chitosan nanoparticles are effective for oral insulin delivery. *Pharmaceutical Research*, *24*(12), 2198–2206. DOI: 10.1007/s11095-007-9367-4 PMID: 17577641

Savencu, I., Iurian, S., Porfire, A., Bogdan, C., & Tomu ă, I. (2021). Review of advances in polymeric wound dressing films. *Reactive & Functional Polymers*, *168*, 105059. DOI: 10.1016/j.reactfunctpolym.2021.105059

Sezer, A. D., Kazak Sarılmışer, H., Rayaman, E., Çevikbaş, A., Öner, E. T., & Akbuğa, J. (2017). Development and characterization of vancomycin-loaded levan-based microparticular system for drug delivery. *Pharmaceutical Development and Technology*, *22*(5), 627–634. DOI: 10.3109/10837450.2015.1116564 PMID: 26607946

Shalash, W. K., Al-Ezzy, R. M., & Ahmaed, A. S. (2021). Antitumor activity of dextran produced from localized leuconostocmesenteroides isolates. *Biochemical and Cellular Archives*, *21*(1), 2371–2374.

Shehata, N. S., Elwakil, B. H., Elshewemi, S. S., Ghareeb, D. A., & Olama, Z. A. (2023). Selenium nanoparticles coated bacterial polysaccharide with potent antimicrobial and anti-lung cancer activities. *Scientific Reports*, *13*(1), 21871. DOI: 10.1038/s41598-023-48921-9 PMID: 38072846

Shi, S. S., Lei, S. J., & Fu, C. X. (2020). Studies of the properties of CHG-Loaded alginate fibers for medical application. *Polymer Testing*, *83*, 106141. DOI: 10.1016/j.polymertesting.2019.106141

Silvestro, I., Lopreiato, M., Scotto d'Abusco, A., Di Lisio, V., Martinelli, A., Piozzi, A., & Francolini, I. (2020). Hyaluronic acid reduces bacterial fouling and promotes fibroblasts' adhesion onto chitosan 2D-wound dressings. *International Journal of Molecular Sciences*, *21*(6), 2070. DOI: 10.3390/ijms21062070 PMID: 32197295

Song, M., Liang, Y., Li, K., Zhang, J., Zhang, N., Tian, B., & Han, J. (2019). Hyaluronic acid modified liposomes for targeted delivery of doxorubicin and paclitaxel to CD44 overexpressing tumor cells with improved dual-drugs synergistic effect. *Journal of Drug Delivery Science and Technology*, *53*, 101179. DOI: 10.1016/j.jddst.2019.101179

Sonia, T. A., & Sharma, C. P. (2012). An overview of natural polymers for oral insulin delivery. *Drug Discovery Today*, *17*(13-14), 784–792. DOI: 10.1016/j.drudis.2012.03.019 PMID: 22521664

Srivastava, A., Yadav, T., Sharma, S., Nayak, A., Kumari, A. A., & Mishra, N. (2015). Polymers in drug delivery. *Journal of Biosciences and Medicines*, *4*(1), 69–84. DOI: 10.4236/jbm.2016.41009

Sumaila, M., Marimuthu, T., Kumar, P., & Choonara, Y. E. (2021). Lipopolysaccharide nanosystems for the enhancement of oral bioavailability. *AAPS PharmSciTech*, *22*(7), 1–25. DOI: 10.1208/s12249-021-02124-5 PMID: 34595578

Sung MH, Park C, Choi JC, et al. (2014). Hyaluronidase inhibitor containing poly-gamma-glutamic acid as an effective component. US, US8916141.

Tabernero, A., & Cardea, S. (2020). Microbial Exopolysaccharides as Drug Carriers. *Polymers, 12*(9), 2142. DOI: 10.3390/polym12092142 PMID: 32961830

Takahama, R., Kato, H., Tajima, K., Tagawa, S., & Kondo, T. (2021). Biofabrication of a hyaluronan/bacterial cellulose composite nanofibril by secretion from engineered Gluconacetobacter. *Biomacromolecules, 22*(11), 4709–4719. DOI: 10.1021/acs.biomac.1c00987 PMID: 34705422

Tan, Y., Chen, L., Li, K., Lou, B., Liu, Y., & Liu, Z. (2022). Yeast as carrier for drug delivery and vaccine construction. *Journal of Controlled Release, 346*, 358–379. DOI: 10.1016/j.jconrel.2022.04.032 PMID: 35483637

Thirumalaisamy, R., Aroulmoji, V., Iqbal, M. N., Deepa, M., Sivasankar, C., Khan, R., & Selvankumar, T. (2021). Molecular insights of hyaluronic acid-hydroxychloroquine conjugate as a promising drug in targeting SARS-CoV-2 viral proteins. *Journal of Molecular Structure, 1238*, 130457. DOI: 10.1016/j.molstruc.2021.130457 PMID: 33867575

Varshosaz, J., Taymouri, S., Jafari, E., Jahanian-Najafabadi, A., & Taheri, A. (2018). Formulation and characterization of cellulose acetate butyrate nanoparticles loaded with nevirapine for HIV treatment. *Journal of Drug Delivery Science and Technology, 48*, 9–20. DOI: 10.1016/j.jddst.2018.08.020

Walvekar, P., Lulinski, P., Kumar, P., Aminabhavi, T. M., & Choonara, Y. E. (2024). A review of hyaluronic acid-based therapeutics for the treatment and management of arthritis. *International Journal of Biological Macromolecules, 130645*, 130645. Advance online publication. DOI: 10.1016/j.ijbiomac.2024.130645 PMID: 38460633

Wang, J., Liu, S., Huang, J., Ren, K., Zhu, Y., & Yang, S. (2023). Alginate: Microbial production, functionalization, and biomedical applications. *International Journal of Biological Macromolecules, 242*, 125048. DOI: 10.1016/j.ijbiomac.2023.125048 PMID: 37236570

Wang, T., Hou, J., Su, C., Zhao, L., & Shi, Y. (2017). Hyaluronic acid-coated chitosan nanoparticles induce ROS-mediated tumor cell apoptosis and enhance antitumor efficiency by targeted drug delivery via CD44. *Journal of Nanobiotechnology, 15*(1), 1–12. DOI: 10.1186/s12951-016-0245-2 PMID: 28068992

Wei, B., Yang, G., & Hong, F. (2011). Preparation and evaluation of a kind of bacterial cellulose dry films with antibacterial properties. *Carbohydrate Polymers, 84*(1), 533–538. DOI: 10.1016/j.carbpol.2010.12.017

Witting, M., Boreham, A., Brodwolf, R., Vavrova, K., Alexiev, U., Friess, W., & Hedtrich, S. (2015). Interactions of hyaluronic acid with the skin and implications for the dermal delivery of biomacromolecules. *Molecular Pharmaceutics, 12*(5), 1391–1401. DOI: 10.1021/mp500676e PMID: 25871518

Wong, T. W. (2009). Chitosan and its use in design of insulin delivery system. *Recent Patents on Drug Delivery & Formulation*, *3*(1), 8–25. DOI: 10.2174/187221109787158346 PMID: 19149726

Wu, Y., Li, P., Jiang, Z., Sun, X., He, H., Yan, P., Xu, Y., & Liu, Y. (2023). Bioinspired yeast-based β-glucan system for oral drug delivery. *Carbohydrate Polymers*, *319*, 121163. DOI: 10.1016/j.carbpol.2023.121163 PMID: 37567689

Xu, A. R., Wang, J. J., & Guo, X. (2018). Fabrication of cellulose aerogels using a green/clean procedure. *Journal of Macromolecular Science, Part B: Physics*, *57*(1), 1–7. DOI: 10.1080/00222348.2017.1396825

Xu, Y., Zhan, C., Fan, L., Wang, L., & Zheng, H. (2007). Preparation of dual crosslinked alginate–chitosan blend gel beads and in vitro controlled release in oral site-specific drug delivery system. *International Journal of Pharmaceutics*, *336*(2), 329–337. DOI: 10.1016/j.ijpharm.2006.12.019 PMID: 17223290

Yang, X., Yang, D., Zhu, X., Nie, J., & Ma, G. (2019). Electrospun and photocross-linked gelatin/dextran–maleic anhydride composite fibers for tissue engineering. *European Polymer Journal*, *113*, 142–147. DOI: 10.1016/j.eurpolymj.2019.01.059

Ye, S., He, S., Su, C., Jiang, L., Wen, Y., Zhu, Z., & Shao, W. (2018). Morphological, release and antibacterial performances of amoxicillin-loaded cellulose aerogels. *Molecules (Basel, Switzerland)*, *23*(8), 2082. DOI: 10.3390/molecules23082082 PMID: 30127283

Yotsuzuka, F. (2001). Curdlan. In Cho, S. S., & Dreher, M. L. (Eds.), *Handbook of dietary fiber* (pp. 737–757). DOI: 10.1201/9780203904220-41

Yu, X., & Pishko, M. V. (2011). Release of paclitaxel from pH sensitive and biodegradable dextran based hydrogels. *Soft Matter*, *7*(19), 8898–8904. DOI: 10.1039/c1sm05729d

Zhao, Z., Han, J., Xu, S., Jin, Z., Yin, T. H., & Zhao, K. (2022). Amoxicillin encapsulated in the N-2-hydroxypropyl trimethyl ammonium chloride chitosan and N, O-carboxymethyl chitosan nanoparticles: Preparation, characterization, and antibacterial activity. *International Journal of Biological Macromolecules*, *221*, 613–622. DOI: 10.1016/j.ijbiomac.2022.09.035 PMID: 36089095

Zhu, D., Yao, K., Bo, J., Zhang, H., Liu, L., Dong, X., & Leng, X. (2010). Hydrophilic/lipophilic N-methylene phosphonic chitosan as a promising non-viral vector for gene delivery. *Journal of Materials Science. Materials in Medicine*, *21*(1), 223–229. DOI: 10.1007/s10856-009-3849-3 PMID: 19680604

Zhu, J., Tang, X., Jia, Y., Ho, C. T., & Huang, Q. (2020). Applications and delivery mechanisms of hyaluronic acid used for topical/transdermal delivery–a review. *International Journal of Pharmaceutics*, *578*, 119127. DOI: 10.1016/j.ijpharm.2020.119127 PMID: 32036009

KEY TERMS AND DEFINITIONS

Biocompatibility: The ability of a material to survive with living tissues or creatures without causing harm is known as biocompatibility.

Biofabrication: The process of creating complex biologic products from raw materials such living cells, matrices, biomaterials, and chemicals is commonly referred to as biofabrication.

Drug delivery system: It is characterised as a formulation or a tool that makes it possible to provide a drug to the target site while enhancing its safety and effectiveness by regulating the drug's release rate, timing, and location.

Exopolysaccharide: It is high molecular weight polymer of microbial origin.

Fermentation: The chemical conversion of any organic material through microbial metabolism, which is facilitated by several enzymes, is known as fermentation.

Tissue engineering: This area of study applies concepts from biology, engineering, and medicine to develop biological replacements that can preserve, enhance, or restore tissue function.

Chapter 4
Polymers in Hydrogels Fabrication and their Versatile Application in Drug Delivery:
Hydrogels for Sustained Delivery

Sabina Nazish
https://orcid.org/0009-0008-9803-5266
Askari Institute of Health Sciences, Pakistan

Palwasha Bibi
University of Science and Technology, Pakistan

Sajid Raza
Ibadat International University, Pakistan

Rubina Anwar
Ibadat International University, Pakistan

Mohammad Adnan
Ibadat International University, Pakistan

Abul Kalam Azad
University MAIWP International, Malaysia

ABSTRACT

The drug delivery properties of polymer-based hydrogels make them attractive because they combine biocompatibility, biodegradability, and high-water retention capabilities. These delivery systems provide controlled drug release mechanisms that enable precise targeted delivery and reduce the side effects. Drug delivery systems include oral administration, injectable delivery, implantable systems, and wound healing applications. The development of polymer-based hydrogels faces ongoing challenges because of drug loading restrictions and difficulties with release

DOI: 10.4018/979-8-3373-1047-3.ch004

precision, stability, and scalability. The combination of smart hydrogels with bio-inspired designs and 3D printing fabrication methods provides potential solutions. This review examines hydrogel varieties and drug delivery methods, while exploring their clinical applications and future directions for developing personalized and efficient therapeutic systems.

1. INTRODUCTION

Soft and pliable hydrogels, as a category of smart materials, have attracted researchers' interest in biomedical applications because they possess characteristics such as high-water content, biocompatibility, tunable mechanical properties, and mimicking the natural ECM. These characteristics make hydrogels especially useful for various applications, including drug delivery, tissue engineering, and wound healing. Polymer selection is fundamental in hydrogel fabrication because the choice of polymer substantially determines the final hydrogel properties. Synthetic polymers are biocompatible, biodegradable, and show low toxicity; therefore, they are used in different biomedical applications (Huang et al., 2019).

Chitosan in Hydrogel Drug Delivery Systems

Chitosan, a natural polysaccharide obtained from chitin and widely available from the exoskeletons of crustaceans such as shrimp and crabs, is one of the most commonly used natural polymers in hydrogel drug delivery systems because of its biocompatibility, biodegradability, and multifunctional nature. Hydrogel has often been used by researchers and clinicians for the development of hydrogel drug delivery systems. The unique properties of chitosan, such as biocompatibility, biodegradability, and versatile functionality, have made it a popular choice for researchers and clinicians working in the field of hydrogel drug delivery (Duceac et al., 2019). Chitosan-based hydrogels are widely used in drug delivery systems owing to their versatility. This study specifically focused on the formation of chitosan hydrogels through interpenetrating polymer network (IPN) formation to improve the structural properties, including the swelling efficiency exhibited by hydrogels, which is essential in drug delivery applications. Moreover, the use of synthetic polymers such as poly can enhance the properties of chitosan-based hydrogels to serve different drug delivery applications.

Key Properties of Chitosan

Chitosan, a natural polysaccharide derived from crustacean shells, is widely used in drug delivery due to its biocompatibility, biodegradability, and mucoadhesive properties (Ahmed & Aljaeid, 2016). It does not trigger immune responses and degrades into safe byproducts, making it ideal for short- and long-term therapies. Its ability to adhere to mucosal tissues enhances drug residence time and efficacy. Chitosan also exhibits intrinsic antimicrobial activity, making it suitable for wound healing and infection-targeted systems. Additionally, its pH-sensitive solubility and hydrophilic nature enable the formation of hydrogels with high drug-loading capacity and controlled release.

The Role of Chitosan in Hydrogel Fabrication for Drug Delivery

Chitosan, a non-toxic, biodegradable biopolymer, is widely used in drug delivery due to its ability to form hydrogels that encapsulate various drugs, including hydrophilic drugs, peptides, and proteins. These hydrogels support sustained and controlled release through diffusion-dependent mechanisms influenced by gel structure, porosity, and swelling. Chitosan's pH sensitivity enables targeted delivery in different body regions, such as the stomach or intestines. Its antimicrobial and wound healing properties make it suitable for localized drug release at injury sites. Additionally, combining chitosan with nanoparticles enhances drug encapsulation, prolongs release duration, and improves bioavailability and site-specific delivery (Younes & Rinaudo, 2015).

ALGINATE-BASED HYDROGELS FOR CONTROLLED DRUG DELIVERY

Alginate water-soluble natural polysaccharides derived from brown seaweed have been widely accepted for drug delivery applications because of their highly biocompatible and biodegradable nature and ability to form gels in the presence of divalent ions such as calcium ions. Some of the benefits of alginate-based hydrogels for controlled and sustained drug delivery make them suitable for many pharmaceutical and biomedical applications. Alginate as a material for fabrication of drug delivery system has been widely accepted due to its properties such as ability to form hydrogels under mild conditions.Alginate hydrogels have the capability to encapsulate and stabilize drugs similarly: the release profile of the drugs in alginate hydrogels is sustained, overcoming shorter burst release problem that was associated with most drug delivery systems (Zou et al., 2020

Key Properties of Alginate: Implications for Medical and Pharmaceutical Applications

Alginate, a natural, biodegradable polysaccharide, is highly valued in medical and pharmaceutical applications due to its biocompatibility and non-toxic degradation products. It forms hydrogels through ionic crosslinking with divalent ions like calcium, creating networks suitable for drug delivery without toxic chemical agents. Alginate's hydrophilicity enables prolonged drug release, while its pH sensitivity supports targeted delivery to areas like the stomach or intestines. Additionally, the mechanical strength and viscosity of alginate gels can be adjusted by varying alginate types and crosslinker concentrations, enhancing their versatility across therapeutic applications (Huang, 2021).

Role of Alginate in Hydrogel Drug Delivery Systems

Currently, alginate is a highly suitable biopolymer for hydrogel-based drug delivery due to its ability to encapsulate a variety of therapeutic agents, including proteins, peptides, and small molecules. When cross-linked with calcium ions, alginate forms dense, porous hydrogels that enable high drug capture efficiency and sustained, controlled release. Its pH-responsive swelling and degradation allow targeted delivery to specific sites like the stomach or intestines (Thu et al., 2012). Additionally, alginate can be injected in liquid form and gelled in situ by calcium ions, forming hydrogels directly at the target site without surgery, making it ideal for minimally invasive therapeutic applications.

GELATIN IN HYDROGEL FABRICATION FOR DRUG DELIVERY

The natural polymer gelatin derived from collagen presents promising opportunities for drug delivery applications using hydrogel fabrication methods. The special features of gelatin for safe use in body tissues and its ability to break down naturally and change into gels on temperature change make it a good choice for drug delivery solutions. Hydrogels form a flexible drug delivery system owing to their water-absorbing structure composed of polymer and protein networks. Gelatin-based hydrogels possess two key features that benefit drug delivery: protection of drugs within their structure and controlled release mechanisms for timed drug delivery. Gelatin forms stronger hybrid hydrogels when combined with alginate owing to its ability to bind positively and negatively charged polymers. By adding crosslinking agents, we improved the gelatin-based hydrogel properties to make them more effective for drug delivery (Mathews, 2017).

Key Properties of Gelatin

Natural polymer gelatin acts as a strong biomaterial that offers many useful properties in the medical field. First, gelatin works well with human tissues and simplifies into simple natural materials that can naturally break down easily. According to Bupphathong et al. (2022), doctors can safely use gelatin materials within the human body during prolonged treatment periods. Gelatin exhibits a special reversible temperature-sensitive behavior. At temperatures near 30–35°C, gelatin changes its form from a flowing liquid to a bounce-like gel and turns back into a liquid upon heating above 30–35°C. This special feature enables gelatin to serve as a perfect material for injectable hydrogels that solidify upon injection. The ability of gelatin to bind moisture allows the development of hydrogels that can trap large quantities of water and drug molecules. The capacity of gelatin to attract water enables effective drug transfer and extended release from drug delivery systems. Gelatin has biological effects because collagen splits into bioactive peptides during manufacturing. Biological molecules in gelatin promote cell bonding and multiplication; therefore, the material helps medical treatments heal the body tissue. Gelatin shows a minimal immune response when used as a delivery material in extended drug delivery systems (Nordin et al., 2023).

The Role of Gelatin in Hydrogel Drug Delivery Systems

Scientists have used gelatin hydrogel platforms for versatile drug delivery because they work well across different treatment needs. This material process allows these hydrogels to store and send various drugs, including hydrophilic medicines and proteins. Gelatin-based hydrogels are versatile platforms for drug delivery, capable of encapsulating a wide range of therapeutics, including hydrophilic drugs and proteins. Their structure allows controlled, sustained release through mechanisms like swelling and biodegradation. Injectable formulations take advantage of gelatin's thermosensitivity, enabling minimally invasive, site-specific drug delivery. In wound healing, these hydrogels conform to injury sites and release healing agents effectively. Additionally, by modifying polymer compositions and integrating stimuli-responsive elements, gelatin hydrogels can achieve precise, targeted delivery, particularly useful in cancer and infection treatments. Overall, gelatin hydrogels offer multifunctional advantages across various therapeutic applications (Foox & Zilberman, 2015).

HYALURONIC ACID IN HYDROGEL DRUG DELIVERY SYSTEMS

Hyaluronic Acid exists naturally in connective tissues, skin, eyes, and synovial fluids, which drives its extensive use as a vital ingredient in hydrogel drug delivery systems. HA-based hydrogels possess natural biofriendly properties that allow them to break down safely, while serving as ideal platforms for medical treatment. This linear sugar polymer contains repeated d-glucuronic acid and N-acetyl-d-glucosamine units that participate naturally in tissue repair and cell signaling operations. HA functions as an important biological element by providing order to connective tissue networks and maintaining normal tissue conditions, while regulating moisture balance and facilitating molecule transport within tissue spaces while decreasing friction (Dicker et al., 2013)

Key Properties of Hyaluronic Acid

Medical and pharmaceutical scientists closely examine HA because this natural polysaccharide offers unique physical and biological advantages to research. Hyaluronic acid possesses two fundamental attributes: it is safe for biological tissues and is easily decomposable. Our body accepts hyaluronic acid as safe and harmless, yet it maintains its performance without causing immune system problems. Due to hyaluronidase enzyme processing, this substance is naturally converted into the safe metabolite's glucuronic acid and N-acetylglucosamine. Medical and pharmaceutical scientists look closely at hyaluronic acid because this natural polysaccharide offers unique physical and biological advantages to research. Hyaluronic acid possesses two fundamental attributes through being both safe for biological tissues and easily decomposable. Our body accepts hyaluronic acid as safe and harmless yet it maintains its performance without causing immune system problems (D'Amora et al., 2019). When the body processes hyaluronic acid with the enzyme hyaluronidase, it converts it into safe substances, including glucuronic acid and N-acetylglucosamine. The pH response of hyaluronic acid serves as the basis for the development of drug delivery methods. The response of hyaluronic acid to pH levels creates opportunities to deliver drugs in targeted pH environments in the stomach and intestine.

The Role of Hyaluronic Acid in Hydrogel Drug Delivery

Hyaluronic acid performs well for drug delivery because its special physical traits allow products to interact with multiple cell receptors. According to Jacob et al., HA-based hydrogels serve as an adaptable medium for enclosing multiple types of therapeutic drugs (Jacob et al., 2021). Controlling drug release from HA-based hydrogels depends on various factors, including how HA links together and its molecular weight, which changes the swelling rate and drug diffusion. This property allows medical researchers to develop drugs with enzyme-responsive delivery that increases drug release when hyaluronidase appears in targeted tissue areas. HA hydrogel systems have an exceptional ability to reach specific areas in the body. HA interacts with cell surface receptors such as CD44 to carry drugs precisely where they are required in target cell populations, especially in cancer cells that display higher than normal CD44 levels. By delivering drugs precisely to their intended targets, these systems boost the treatment power and reduce the impact of medicine on unrelated cells

ROLE OF CARRAGEENAN IN HYDROGEL FABRICATION FOR DRUG DELIVERY

Carrageenan, a natural carbohydrate substance, shows potential for various drug delivery uses through hydrogel production. Carrageenan helps in the design of targeted drug delivery systems because its natural composition allows it to hold different drugs and control their release efficiently while responding to changes in its environment. Carrageenan hydrogels are effective drug delivery systems that encapsulate various substances, including drugs, proteins, and cells, with release controlled by polymer concentration and environmental factors. They enable temperature- and ion-responsive drug delivery. In wound healing, they maintain moisture, promote cell migration, and prevent infection. For oral delivery, carrageenan hydrogels create slow-release matrices, improving drug absorption and protecting drugs through the digestive tract, reducing dosing frequency (Li et al., 2013).

AGAROSE IN HYDROGEL FABRICATION FOR DRUG DELIVERY

Drugs released from agarose polymer hydrogels take advantage of swelling control to maintain constant drug delivery over extended periods. Once drugs are placed in hydrogels, they form strong bonds within their network and slowly emit drug particles at specific rates. Huston the structure of a hydrogel allows us to craft its drug-delivery characteristics exactly for medical needs. The extended drug release capability of agarose hydrogels helps to maintain therapeutic effects over time with fewer medical treatments and makes treatment plans easier to follow. Researchers have developed agarose hydrogels that react to selected environmental triggers, such as acidity levels, heat changes, or light beams (Lin et al., 2019).

Agarose Hydrogels: Versatile Biomaterials for Controlled Drug Delivery

The field of drug delivery studies agarose from red seaweed because this natural polysaccharide features both good physical characteristics and safe body compatibility. This research looks at why agarose stands out as a complete drug delivery biomaterial owing to its important traits. Agarose-based hydrogels offer thermoreversible gelation, allowing them to remain liquid at high temperatures and solidify upon cooling, ideal for injectable drug delivery (Chai et al., 2017). They exhibit excellent biocompatibility and biodegradability, triggering minimal immune responses and degrading into harmless byproducts. Agarose's porous network enables controlled diffusion-based drug release, with porosity adjustable by polymer concentration. Its high hydrophilicity ensures strong water retention, supporting sustained release while maintaining gel shape. Drug release rates can be precisely tuned by altering agarose concentration and network structure, making agarose hydrogels highly adaptable for controlled and targeted therapeutic delivery applications.

Role of Agarose in Hydrogel Drug Delivery Systems

Agarose-based hydrogels are promising platforms for drug delivery and tissue engineering due to their 3D water-retentive networks, enabling sustained and targeted drug release (Raeisi & Farjadian, 2024). Their thermoreversible nature allows for injectable applications, with gels forming at target sites such as tumors or infections. Drug release rates can be tuned by modifying agarose concentration or additives. Functionalization with biological ligands like antibodies or peptides enhances specificity for cancer or infection therapy. Agarose hydrogels also support wound healing by delivering therapeutic agents and antibiotics through controlled release, accelerating tissue regeneration and improving treatment outcomes in biomedical applications.

SYNTHETIC POLYMERS IN HYDROGEL FABRICATION FOR DRUG DELIVERY

Modern drug delivery applications use synthetic polymers to build hydrogels, because these materials allow researchers to precisely design their chemical makeup. These customized hydrogels show multiple useful attributes for medical use, such as responding to signals in specific ways while remaining sufficiently strong and delivering drugs consistently over time (Slaughter et al., 2009).

The Versatile Role of Polyethylene Glycol in Hydrogel Drug Delivery

PEG-based hydrogels are essential for medical delivery because they provide many benefits to the development of controlled drug therapies. PEG hydrogels exhibit strong compatibility with many therapeutic agents, including small drugs and complex nucleic acids. The water-friendly chemical structure of PEG enables this material to manage water-based medicines in controlled release patterns during treatment. Doctors control drug release from PEG-based hydrogels by changing the crosslinking level and PEG size, which allows drugs to diffuse or swell as the matrix breaks down. Special drug release settings are required to treat diseases that require ongoing treatment over extended periods. PEGylation of medications and nanoparticles improves the shelf-life of treatments, improves drug absorption by the body, and is more effective in the circulation. The attachment of PEG chains helps drugs escape immune system detection, allowing them to remain in the bloodstream longer while directing treatment to specific targets for better patient outcomes. Researchers and pharmaceutical scientists are increasingly selecting polyethylene glycol (PEG) because of its versatility as a material for hydrogel-based drug delivery systems owing to its multiple benefits.

Improved Drug Bioavailability

PEG makes weakly dissolvable drugs more available and easier to release, because it creates a durable water-loving environment for drug storage. This technology helps drugs that might otherwise have absorption and distribution problems in the body because PEG-based hydrogels work to enhance the movement of these drugs through the system.

Customizable Release Profiles

The delivery of drugs from PEG networks depends on the choices we make for crosslinking density, PEG size, and drug selection. PEG hydrogels are useful when their drug release profiles match the specific therapeutic needs.

Reduced Toxicity and Immunogenicity

PEG does not harm patients during drug delivery because it neither irritates nor triggers immune responses during extended medical treatment. Understanding PEG's low toxic effects of PEG helps doctors to use it in various medical situations.

Injectability and Minimally Invasive Delivery

PEG hydrogel formulations permit injection as a painless method for delivering drugs to specific sites. Through localized delivery methods, PEG-based hydrogels help drugs reach places that are difficult to access, while targeting exact treatment sites. The many benefits of PEG-based hydrogels make them popular for medical drug delivery studies, as scientists search for better ways to use this material effectively (Suk et al., 2015). PEG hydrogels have high potential to help deliver injectable drugs into the body. The hydrogels easily transform from liquid to gel form during injection; therefore, therapies can reach their target destinations using minimal incisions. PEG-based hydrogels simplify drug delivery because they change their state during injection, without requiring surgical intervention.

PEG-Based Hydrogels in Drug Delivery

PEG hydrogels show great promise as drug delivery solutions because they can form gels that are injectable like liquids and turn solid when used. These properties make delivery methods possible without cutting into the body. These hydrogels provide many benefits through drug time-release management and protection for sensitive therapies with better medicinal absorption.

Cancer Therapy

Researchers have widely tested PEG hydrogels for delivering chemotherapeutic treatments in cancer medicine. Scientists have combined PEG with other materials and nanoparticles to allow drugs to reach tumor locations safely and effectively.

Wound Healing

PEG-based hydrogels improve wound healing by creating a moist environment that fits well with medication delivery systems that use antibiotics or growth factors. Hydrogels accelerate healing and block infection by maintaining moisture while slowly delivering their constituent medicines.

Gene Delivery

Scientists have used PEG hydrogels to protect DNA and RNA molecules while managing their timed release. This approach provides new options for the successful delivery of gene therapy.

Ophthalmic Drug Delivery

Ophthalmologists use PEG hydrogels to slow the release of medical treatments for glaucoma and eye infections. Their ability to bond with biological surfaces and hold moisture makes them an expert choice in drug supply systems. The many benefits of PEG-based hydrogels drive their frequent use in medicine since they help drugs work better and make treatment easier for patients (Knop et al., 2010).

POLYVINYL ALCOHOL (PVA) IN HYDROGEL FABRICATION FOR DRUG DELIVERY

Polyvinyl Alcohol (PVA) functions as a synthetic polymer in the development of hydrogel drug delivery systems. Owing to its outstanding water solubility and flexibility, PVA has important applications in the design of drugs that remain active over an extended period. Drug developers favor PVA because they create durable hydrated networks that steadily deliver medications over a long treatment duration.

The Role of Polyvinyl Alcohol in Hydrogel Drug Delivery Systems

Polyvinyl alcohol represents a strong material choice for medical drug carriers because it supports many options to produce hydrogel systems that deliver therapeutic substances effectively. Through network formation, PVA hydrogels accept both water-soluble and fat-soluble drugs, with drug release shaping mainly from the diffusion motion.PVA hydrogels gain mechanical stability through physical freeze-thaw crosslinking, which creates crystal domains that control drug release patterns over prolonged periods. The use of glutaraldehyde and borate as crosslinking agents creates stronger hydrogels that help control how drugs exit the system (Ali et al., 2023). PVA hydrogels can be changed with temperature via thermal triggers by placing thermosensitive ingredients or other polymers, such as PNIPAM. The system sends drugs from the system when temperature changes occur inside or outside of the body.

Advantages of Polyvinyl Alcohol in Hydrogel Drug Delivery Systems

PVA-based hydrogels are widely used in drug delivery due to their excellent biocompatibility, material stability, and customizable release profiles. Drug release rates can be controlled by adjusting polymer concentration, crosslinking levels, and drug-loading techniques, allowing flexibility to meet varied therapeutic needs. PVA hydrogels are safe for human use, showing low toxicity and compatibility with living tissues. Their thermally stable and pH-resistant properties make them suitable for both localized and systemic drug delivery. Additionally, their injectability enables minimally invasive treatments, making them particularly effective for targeting chronic diseases and cancer with site-specific, sustained drug delivery.

PVA-Based Hydrogels in Drug Delivery

Polyvinyl alcohol hydrogels show great potential as versatile drug delivery materials owing to their supportive traits such as high-water content and adjustable swelling behavior. Polyvinyl alcohol hydrogels serve different drug delivery methods that help treat specific medical requirements in targeted administration areas. Scientists have used PVA hydrogels to create injectable medicine delivery systems that directly apply treatment to targeted body parts. Doctors can inject these gels straight into body areas where treatment is necessary to slowly release drugs that fight cancer or ease pain. Researchers have used PVA hydrogels to help wounds heal and damaged tissue grows back. They have added antimicrobial medicines and growth-promoting factors to control infections during recovery. The water retention and biocompatible nature of PVA hydrogels support precise drug delivery in eye treatment of glaucoma and other eye conditions. The diverse uses of PVA hydrogels include passing genes into target cells because they safeguard nucleic acids before they are delivered over time (Gulfam et al., 2022).

PAM-BASED HYDROGELS IN DRUG DELIVERY

Polyacrylamide hydrogels show promise as drug delivery platforms because they provide benefits that overcome standard drug administration problems. Researchers have designed hydrogel materials to supply drugs steadily, which helps patients take their medications less often. PAM hydrogels enable precise drug release programming through crosslinking and polymer concentration regulation (Xu et al., 2020). Our targeted control techniques allow the pace of medication release to remain constant over various treatment times. Injectable PAM hydrogels can help doctors deliver drugs directly to hard-to-reach body parts without major medical procedures. PAM hydrogels effectively hold high drug concentrations that make it easier to transfer pharmaceutical products through their structure. Hydrogels transform easily to hold many types of medicines, including medicines and biological particles. Researchers have widely explored PAM hydrogels because they can adapt well and can be customized. The future development of this technology promises to enhance both the performance and patient safety of medical treatments.

THERMORESPONSIVE POLY HYDROGELS FOR CONTROLLED DRUG DELIVERY

Poly based hydrogels exhibit advanced thermoresponsive features, making them an ideal choice for drug delivery. Poly based hydrogels exhibit temperature-sensitive changes, which provide an effective method for managing drug release. The thermoresponsive drug-release properties make poly-based hydrogels a successful choice for pharmaceutical delivery systems. Through temperature-sensitive drug delivery systems based on poly-based hydrogels, scientists can release drugs at specific temperatures, making them valuable for targeted therapies, particularly in cancer treatment. The drug delivery system of poly-based hydrogels controls drug release from the material. By injecting or applying drug-loaded poly-based hydrogels to the body, the system releases medicine when the skin or body temperature crosses the lower critical solution temperature threshold of the polymer. Through temperature-sensitive release, this approach effectively controls drug delivery and helps patients follow their treatment plans better (Matanović et al., 2014). The heat sensitivity of poly-based hydrogels helps deliver drugs to targeted areas. Heat from a disease site triggers the hydrogel to release medicine precisely, which is necessary to help patients benefit from therapy more effective.

THE VERSATILITY OF PNIPAM-BASED HYDROGELS IN DRUG DELIVERY

Research has shown that PNIPAM-based hydrogels can deliver medication effectively through their many useful features beyond standard drug delivery technology. Thermoresponsive PNIPAM hydrogels deliver drugs precisely to cancer sites through their targeted drug delivery mechanisms. The targeted release strategy helps PNIPAM hydrogels to deliver cancer drugs better while preventing harmful effects throughout the body.

Transdermal Drug Delivery

PNIPAM-based hydrogels can serve as effective drug delivery patches through the skin. Doctors apply hydrogels to skin areas that require pain treatment, and these products deliver drugs at specific temperatures for localized relief.

Gene Therapy and RNA Delivery

PNIPAM hydrogels show flexibility by facilitating gene therapy delivery. Their temperature response enables the exact delivery of DNAs or RNAs from these polymers to targeted areas.

Implantable Drug Delivery Systems

The temperature-sensitive nature of PNIPAM makes these hydrogels effective for implanting extended-release drugs into patients. Health professionals design these hydrogels to discharge drugs when the body temperature changes to treat chronic conditions.

Wound Healing and Tissue Regeneration

The PNIPAM-based hydrogel system is important for wound healing because it transports antibiotics and bioactive agents. They work through temperature-triggered volume changes to maintain proper hydration while gradually releasing the healing medicine to prevent infections (Wang et al., 2020).

PLGA-Based Hydrogels in Drug Delivery

PLGA has attracted the attention of researchers in drug delivery because it is biodegradable and biocompatible. PLGA's combination of lactic acid and glycolic acid in PLGA allows scientists to customize the breakdown rate for drug delivery applications. Maryland University researchers established PLGA hydrogels as the preferred choice for drug delivery when needed because of their slow breakdown into safe human body chemicals, lactic acid, and glycolic acid. PLGA is a smart polymer because it responds to external signals to create different types of disease treatment and diagnosis systems. PLGA is extensively used in pharmaceutical and biomedical fields because it can degrade safely, work well with body tissues, and deliver drugs at planned rates.

POLY (LACTIC-CO-GLYCOLIC ACID) IN HYDROGEL DRUG DELIVERY SYSTEMS

Poly(lactic-co-glycolic acid) is widely used as a drug delivery polymer, specifically in hydrogel system development. PLGA-based hydrogels excel at delivering medicine steadily, which helps patients avoid taking medication more often. Hydrogels made from PLGA serve as adaptable platforms to transport drugs, including small molecules, proteins, and peptides. PLGA hydrogels can be used in biodegradable implants and injection systems to eliminate manual removal after drug release. The polymer breakdown process and medicine release timeline control depended on the molecular weight adjustment points and glycolic acid portion mix proportions. PLGA hydrogels are excellent at delivering proteins and peptides because their matrix shields sensitive biologics from degradation and releases them slowly over time. PLGA hydrogels show endless use, including in advanced medicine delivery systems. Our hydrogel formulations enhance targeted drug delivery by hosting specific molecules, such as antibodies or peptides, which boost drug effects while reducing side effects throughout the body. A delivery system made from PLGA hydrogels permits the deployment of various therapeutic agents for enhanced therapeutic benefits. The material PLGA earns its name as a 'Smart Polymer' due to its reactions to stimuli. Experts have thoroughly explored its basic properties and breakdown behavior while documenting numerous drug delivery applications. Scientists can adjust PLGA's physical and mechanical features of PLGA to produce better connections with life science materials and direct drug delivery particles toward particular tissue areas (Agrawal et al., 2019). PLGA has shown great success as a drug delivery system, making it a good choice for future medical tools.

The Role of Polyacrylic Acid in Drug Delivery

Polyacrylic acid works well as a polymer for delivering medicines across different pharmaceutical applications owing to its many useful properties. pH-Responsive Drug Release: Hydrogel made from PAA shows a unique property of releasing drugs based on changes in environmental pH. The hydrogels expand or contract because of pH changes that release the drug at precise points in the digestive system to improve the effects and absorption rates of the medication. Bioadhesive Drug Delivery: Modifying Polyacrylic acid-based hydrogel materials makes them stick to biological surfaces for longer periods. The system can deliver medications to specific tissue areas, such as injured skin and eye surface areas. Drug loading and sustained release: Because PAA easily binds to water molecules, it can hold and release drugs for an extended time. The drug delivery system allows drugs to pass through the swollen hydrogel to provide a steady drug release and reduce the need for frequent doses (Chen et al., 2018).

HYBRID POLYMERS IN HYDROGEL DRUG DELIVERY SYSTEMS

Drug delivery system development is a leading industrial priority because precise therapeutic agent control through release and targeting mechanisms drives optimal treatment results. Hybrid polymers act as a strategic strategy for hydrogel drug delivery design by combining natural, synthetic, and semi-synthetic polymers to improve delivery systems through their individual attributes. Hydrogels serve as flexible drug delivery platforms because they create water-swollen three-dimensional polymer networks that shield therapeutic agents and control their release profiles either through continued deposition or remote activation. Hybrid hydrogels merge distinctive polymer materials to create synergistic combinations that develop tailored, managed drug delivery protocols to address precise therapeutic requirements, including programmed drug release, focused delivery, and external stimulus responsiveness.

TYPES OF HYBRID POLYMERS USED IN HYDROGEL DRUG DELIVERY SYSTEMS

Hybrid hydrogel systems, combining natural and synthetic polymers, offer advanced drug delivery capabilities with controlled release and biocompatibility. Examples include chitosan-PEG blends, which merge biofamiliarity and water retention, and alginate-PVA systems, combining encapsulation and mechanical strength for wound care and sustained delivery. PLGA-PAA hybrids exhibit pH-sensitive, location-specific release ideal for gastrointestinal applications. Chitosan-PAM hydrogels provide mechanical integrity and bioadhesion for protein delivery and wound healing. Stimuli-responsive hybrids like PNIPAM-HA respond to environmental changes, such as temperature, to enable smart, targeted drug release. These systems enhance therapeutic effectiveness and are suitable for diverse biomedical applications.

HYBRID POLYMER HYDROGELS IN DRUG DELIVERY

Hybrid hydrogels are versatile drug delivery platforms used in cancer therapy, wound healing, oral delivery, and tissue engineering. pH-sensitive and thermoresponsive hydrogels enable targeted drug release at tumor sites, minimizing side effects. Cationic polymers like polyethylene imine facilitate DNA/RNA encapsulation, stabilized by polyethylene glycol. Combining natural polymers (chitosan, alginate) with synthetics (PLGA, PVA) enhances biocompatibility, moisture retention, and precise delivery of growth factors and antimicrobials for wound healing. Polyacrylic acid-PLGA hydrogels enable pH-triggered oral drug release. These hybrids also support tissue regeneration by providing scaffolds that promote cell growth while delivering therapeutic agents effectively.

CHITOSAN-PEG HYBRID HYDROGELS: ENHANCING DRUG DELIVERY AND TISSUE ENGINEERING APPLICATIONS

The combined properties of chitosan and polyethylene have attracted significant attention from biomedical engineering researchers. The natural polysaccharide chitosan acts as a biodegradable material with both biocompatible and mucoadhesive properties, and PEG enhances its hydrophilic behavior and water retention properties. Hybrid chitosan-PEG hydrogels demonstrate expanded swelling properties and superior mechanical capabilities and higher biocompatibility which aligns them with numerous applications. This hybrid hydrogel framework serves as an essential drug delivery platform for sustained and controlled drug release applications. The extended drug release capabilities of chitosan-PEG hydrogels make them suitable for lengthy chronic disease care treatments as well as both wound therapy and transdermal medication delivery. Owing to its biodegradability and biocompatibility, the hybrid material provides a basis for tissue engineering scaffold production, particularly for bone and cartilage regeneration (Magli et al., 2021). Researchers have intensively studied the production and testing of chitosan-PEG hybrid hydrogels. Researchers have employed click reactions combined with photopatterning techniques and 3D printing methods to produce complex hydrogel tissues through which they can introduce biological molecules and motifs to guide cellular responses.

ALGINATE-PVA HYBRID HYDROGELS: SYNERGISTIC PROPERTIES AND BIOMEDICAL APPLICATIONS

Brown algae-derived alginate stands out for biomedical applications because it combines the properties of biocompatibility and divalent-cation-sensitive hydrogel formation with hydrated calcium ions. The natural physical weaknesses of alginate-based hydrogels restrict their practical use in specific applications. Studies have focused on the development of alginate-based hybrid materials by combining alginate with the synthetic polymer polyvinyl alcohol, because of its mechanical robustness and water stability properties (Rijal, 2023). The alginate-PVA hybrid system integrates the advantageous features of both materials to create hydrogels that display better structural stability, enhanced swelling behavior, and mechanical flexibility.

The hybrid hydrogel exhibits improved stability owing to the calcium-dependent gelation mechanism of alginate, which combines with PVA's contribution of PVA to provide structural flexibility and mechanical resilience. Given their amalgamation of superior properties, these hybrid alginate-PVA hydrogels have extensive medical applications, specifically in drug transport. The primary function of alginate-PVA hybrid creation is to develop prolonged drug delivery hydrogels that serve as oral drug delivery systems. Many therapeutic regimens require drug delivery systems that provide extended-release functionality, and alginate-PVA hybrids meet these requirements. Because of its compatibility with biological systems and its stable hydrogel-making ability, the hybrid is a preferred material for pharmaceutical drug delivery through the mouth. The alginate-PVA hybrid solution is effective as an injectable drug delivery system, in addition to its established oral drug delivery applications.

GELATIN-PNIPAM HYBRID HYDROGELS FOR BIOMEDICAL APPLICATIONS

Hybrid hydrogels have emerged through biomaterial research, which integrate attractive properties from separate polymeric materials into single devices for advanced biomedical purposes. A hybrid system consisting of gelatin and poly exhibits distinctive thermoresponsive properties while retaining its biocompatible nature. The natural substance gelatin extracted from collagen produces both material compatibility with biological systems and cell-positive interactions. The symbiotic combination of poly together with the well-established thermoresponsive poly creates a hydrogel platform showing biocompatibility and biodegradability while supporting temperature-dependent drug unloading. The temperature-responsive nature of gelatin-PNIPAM hybrid hydrogels provides exceptional opportunities for targeted drug delivery systems, especially in cancer therapy (Siqueira et al., 2018).

HYALURONIC ACID-POLYACRYLAMIDE HYBRID HYDROGELS

Biomaterials characterized as hydrogels demonstrate promising potential in tissue engineering applications, along with drug delivery techniques (Lee & Kim, 2018). Among the different hydrogel formulations, the Hyaluronic–polyacrylamide hybrid hydrogel demonstrated extraordinary properties. The hybrid structure links the natural benefits of Hyaluronic Acid to the water-retention properties of polyacrylamide, resulting in a versatile biocompatible formulation. The combination of Hyaluronic and polyacrylamide produces a hybrid hydrogel that presents promising potential for various biomedical applications. The natural biopolymer Hyaluronic Acid facilitates cell connection and cellular bonding, and polyacrylamide maintains the gel form that works well in biological solutions through its structural strength and water control function. This synergistic relationship between materials leads to hydrogels that demonstrate powerful water absorbency while maintaining excellent biocompatibility, which results in tissue-mimicking behavior. Soft tissue repair using injectable Hyaluronic Acid-Polyacrylamide hybrid hydrogels represents the most prominent real-world application. These hydrogels demonstrate characteristics that make them suitable for cartilage healing and drug transport applications because they both benefit from injection into the body and are biocompatible and explandable. These hydrogels excel at encapsulating water-soluble drugs because of their extensive ability to absorb solutions while providing sustained drug release, thus transforming them into important devices for medication delivery.

COLLAGEN-PLGA HYBRID: SYNERGISTIC SCAFFOLD FOR TISSUE REGENERATION AND CONTROLLED DRUG DELIVERY

Hybrid materials formed by combining collagen with PLGA are promising technological components suitable for multiple biomedical applications in regenerative medicine and drug delivery systems. The naturally occurring extracellular matrix component collagen offers biopsyologically compatible properties that enable cell attachment, proliferation, and differentiation. The synthetic biodegradable polymer PLGA possesses both controlled degradation pathways and sustained drug-release abilities. A combined scaffold that copies natural extracellular matrix functions alongside controlled drug delivery has emerged from this two-component system. Regenerative Medicine Applications: The collagen-PLGA hybrid hydrogel has been extensively used in tissue engineering applications for bone tissue development, cartilage repair, and skin tissue regeneration. As the scaffold ages, it degrades naturally through the PLGA material; however, the collagen portion creates an environment that aids biological cell behavior during growth and development. Drug-Eluting Implants: The collagen-PLGA hybrid provides a suitable foundation for creating therapeutic implants that deliver drugs gradually over time. The structure of PLGA serves exclusively as a support scaffold for drug release combined with tissue healing promotion properties of collagen (Ding & Zhu, 2018).

HYDROGELS AND THEIR APPLICATIONS IN TARGETED DRUG DELIVERY

The delivery of both monoclonal antibodies and hormones using injectable hydrogels provides a localized treatment for rheumatoid arthritis and osteoporosis. The delivery of both local anesthetics and anti-inflammatory drugs enables hydrogel application to manage long-term pain. Hydrogels serve as carriers for genetic materials used in medical treatments for genetic conditions and cancers. PNIPAM-based hydrogels serve as mainstream materials in temperature-responsive injectable delivery systems because they transform into gels at human body temperature to provide sustained drug release.

IMPLANTABLE HYDROGEL SYSTEMS FOR LONG-TERM CANCER AND HORMONAL THERAPIES

The management of medical conditions, including cancer and hormonal imbalances, requires drugs to operate fully through extended timeline delivery. Current methods of oral medication and repeated injection therapy do not reach their goals because patients develop drug resistance and fail to follow medical instructions (Rossi et al., 2018). Hydrogel implant technologies represent an advanced approach for delivering drugs in controlled amounts that directly act on target locations for extended periods of time. The hydrogel network underwent a time-dependent but unfavorable degradation process, creating stable drug delivery rates according to hydrogel analytical studies. Through their direct delivery method, hydrogels transport chemotherapeutic agents to cancer tumors and produce fewer adverse effects in the body. Functioning as a controlled drug delivery method, hydrogels supply hormonal medications at easily controlled release levels for periods ranging from monthly to multi-month durations. PLGA hydrogels have become renowned delivery platforms as pharmaceutical workers utilize them to sustain anticancer drug delivery, together with hormonal drug system regulation. The minimally invasive delivery method for these injectable hydrogels surpasses the surgical procedures traditionally required for the standard hydrogels. Hydrogels feature a distinctive porous network structure along with a high swelling capacity, together with solid mechanical properties that enable controlled drug delivery from their matrix. The hydrogel system design plays a crucial role because optimizing drug release kinetics and therapeutic efficacy depends on the selection of polymer composition, crosslinking mechanisms, and degradation rates.

HYDROGELS AS SUSTAINED ANTIMICROBIAL WOUND HEALING SCAFFOLDS

Wound healing involves tissue regeneration, infection control, and inflammation management, but chronic wounds like diabetic ulcers and burns often face delayed recovery due to persistent infections and poor nutrient supply. Hydrogels are promising in wound care as moist dressings that keep wounds hydrated and protected from contaminants. They act as scaffolds supporting tissue regeneration while enabling sustained release of antimicrobial agents such as silver nanoparticles, antibiotics, and antiseptics to prevent infections. Gelatin-based hydrogels incorporating agents like silver sulfadiazine or honey enhance moisture retention, antimicrobial activity, and accelerate healing, especially in chronic wounds. Hydrogels also aid burn treatment through moisturizing and temperature regulation. However, their limited mechanical strength and rapid degradation restrict applications. To address this, researchers have developed novel nanohybrid hydrogels, such as laponite combined with PVA-alginate, which offer improved mechanical integrity and effectively support natural healing, showing great potential as advanced wound dressings.

CHALLENGES AND FUTURE PERSPECTIVES IN HYDROGEL-BASED DRUG DELIVERY

Hydrogels demonstrate remarkable potential as sustained drug delivery platforms because they offer several distinct qualities, including compatibility with biological systems, degradation abilities, and localized delivery control of therapeutic payloads. Many obstacles hinder the enhancement of the effectiveness and commercial potential of hydrogel-based drug delivery systems, despite their existing drug delivery benefits. The main difficulty in drug delivery using hydrogels involves hydrophobic drug compounds because their water-loving polymeric matrix lacks suitable conditions for storing hydrophobic therapeutics. Scientists have investigated two approaches to overcome this limitation: creating hydrogels with specifically placed hydrophobic domains or integrating different drug delivery techniques.

ADVANCING DRUG DELIVERY THROUGH ENGINEERED HYDROGELS: OVERCOMING LIMITATIONS IN LOADING AND RELEASE PRECISION

Hydrogel-based drug delivery systems face challenges in precisely controlling the release kinetics of therapeutic agents and efficiently loading diverse drugs, especially large molecules and hydrophobic compounds. Their natural water affinity limits the incorporation of lipophilic drugs, but strategies like embedding nanoparticles, micelles, and liposomes have improved this. Despite their excellent biocompatibility and tunability, hydrogels struggle with unpredictable drug release due to variable swelling, diffusion, and polymer degradation rates influenced by environmental factors like temperature, pH, and ionic strength. To address these issues, "smart" hydrogels are being developed with responsive crosslinkers that react to stimuli such as pH, temperature, and light, enabling controlled, on-demand drug release. These multifunctional hydrogels aim to deliver a broad range of therapeutics, including gene vectors and small molecules, while maintaining precise kinetics. Although significant progress has been made, further innovation is necessary to enhance drug loading capacity and refine release control for wider clinical applications.

Hydrogel Formulations for Drug Delivery

Hydrogel-based drug delivery systems require solving the stability problem faced by hydrogel formulations during storage, in addition to in vivo delivery. Premature drug release and undesirable mechanical property changes among hydrogels result from dehydration, oxidative degradation, or loss of crosslinking agents (Li et al., 2021). Scientific research has aimed to develop pioneering crosslinking agents and polymer modifications that enhance hydrogel stability, combined with better biodegradability and reduced side effect potential. The encapsulation of hydrogels within protective matrices offers enhanced stability and improved drug release properties in abnormal storage environments. Hydrogel formulations face industrial scalability obstacles which make maintaining quality standards and achieving consistent production alongside affordability difficult. Product performance variability becomes inconsistent because of variations that occur between different batches of hydrogels at both crosslinking density and drug-loading levels. The development of automated standardized manufacturing equipment for hydrogel production is essential to guarantee quality assurance and reproducible outcomes. Research on 3D printing combined with microfluidic systems allows the development of scalable methods to build customized drug delivery devices that regulate drug distribution uniformly while surpassing traditional production limits. Currently, the drug delivery field utilizes hydrogels as promising platforms because their ability to manipulate swelling patterns enables precise control over release, which extends drug effectiveness over time. The synthesis of hydrogels requires material selection choices, ranging from fully synthetic polymers to natural constituents, such as polysaccharides and proteins, based on the target application.

Regulatory Challenges in Hydrogel-Based Products

Hydrogel-based drug delivery systems face significant regulatory challenges during development and commercialization, requiring strict compliance with complex FDA rules for safety and effectiveness. Thorough testing is essential to validate biocompatibility, drug release dynamics, and safe degradation products. The combination nature of these products adds complexity, demanding adherence to both medical device and pharmaceutical regulations. Collaboration between researchers, manufacturers, and regulatory agencies is crucial to establish standardized evaluation protocols and reliable in vitro and in vivo testing methods. Streamlined regulatory processes and ongoing cooperation are necessary to safely advance hydrogel technologies from research to clinical use.

Emerging Trends in Bioinspired and Smart Hydrogels

Recent hydrogel research has advanced significantly with bioinspired and smart hydrogels showing vast biomedical potential. Biomimetic hydrogels, made from natural polymers like collagen and hyaluronic acid, mimic body tissues and improve cellular interactions for effective drug delivery and tissue regeneration. These hydrogels incorporate enzyme-sensitive bonds and cell-binding sites to guide medication transport and healing. Smart hydrogels respond to stimuli such as pH, temperature, electricity, light, and magnetic fields to control drug release. Future research focuses on developing multifunctional hydrogels responsive to multiple triggers and integrating nanotechnology for precise targeting of drugs to tumors or cells, enhancing therapeutic efficacy.

CONCLUSION

The drug delivery field has adopted polymer-based hydrogels as a leading platform because these materials offer biocompatibility and biodegradability, alongside their ability to control drug release through swelling, crosslinking, and external stimuli responsiveness. The diverse applications of these polymers extend across oral and injectable formulations, and implantable and transdermal systems, leading to better therapeutic results and fewer adverse effects. The potential of precision medicine using smart hydrogels, bioinspired designs, and innovative fabrication methods, including 3D printing and microfluidics, continues to expand. The development of polymer-based hydrogels faces ongoing obstacles in maximizing drug loading potential while ensuring manufacturing scalability and regulatory compliance. Future research needs to improve hydrogel stability and develop targeted delivery systems using nanotechnology, while creating patient-specific formulations through AI-driven design methods. The future of sustained drug delivery can be transformed by polymer-based hydrogels, which will provide more efficient personalized and safer therapeutic solutions for medical conditions.

REFERENCES

Agrawal, A., Rellegadla, S., & Jain, S. (2019). Biomedical applications of PLGA particles. In *Elsevier eBooks* (p. 87). Elsevier BV., DOI: 10.1016/B978-0-12-816913-1.00004-0

Ahmed, T. A., & Aljaeid, B. M. (2016). Preparation, characterization, and potential application of chitosan, chitosan derivatives, and chitosan metal nanoparticles in pharmaceutical drug delivery. *Drug Design, Development and Therapy*, *483*, 483. Advance online publication. DOI: 10.2147/DDDT.S99651 PMID: 26869768

Ali, S., Ranjha, N. M., Ahmad, B., Khan, A. A., Hassan, F. U., Aziz, T., Alharbi, M., Alshammari, A., Alasmari, A. F., & Alharbi, M. E. (2023). Sustained Release of Drug Facilitated Through Chemically Crosslinked Polyvinyl Alcohol-Gelatin (PVA-GE) Hydrogels. A sustainable biomedical approach. *Polish Journal of Chemical Technology*, *25*(2), 56–65. DOI: 10.2478/pjct-2023-0017

Bupphathong, , Quiroz, C., Huang, W., Chung, P.-F., Tao, H.-Y., & Lin, C.-H. (2022). Gelatin Methacrylate Hydrogel for Tissue Engineering Applications—A Review on Material Modifications. *Pharmaceuticals (Basel, Switzerland)*, *15*(2), 171. DOI: 10.3390/ph15020171 PMID: 35215284

Chai, Q., Jiao, Y., & Yu, X. (2017). Hydrogels for Biomedical Applications: Their Characteristics and the Mechanisms behind Them. *Gels (Basel, Switzerland)*, *3*(1), 6. DOI: 10.3390/gels3010006 PMID: 30920503

Chen, Y., Gao, Y., da Silva, L. P., Pirraco, R. P., Ma, M., Yang, L., Reis, R. L., & Chen, J. (2018). A thermo-/pH-responsive hydrogel (PNIPAM-PDMA-PAA) with diverse nanostructures and gel behaviors as a general drug carrier for drug release. *Polymer Chemistry*, *9*(29), 4063–4072. DOI: 10.1039/C8PY00838H

D'Amora, U., Ronca, A., Raucci, M. G., Dozio, S. M., Lin, H., Fan, Y., Zhang, X., & Ambrosio, L. (2019). In situ sol-gel synthesis of hyaluronan derivatives bio-nanocomposite hydrogels. *Regenerative Biomaterials*, *6*(5), 249–258. DOI: 10.1093/rb/rbz029 PMID: 31620307

Dicker, K. T., Gurski, L. A., Pradhan-Bhatt, S., Witt, R. L., Farach-Carson, M. C., & Jia, X. (2013). Hyaluronan: A simple polysaccharide with diverse biological functions. *Acta Biomaterialia*, *10*(4), 1558–1570. DOI: 10.1016/j.actbio.2013.12.019 PMID: 24361428

Ding, D., & Zhu, Q. (2018). Recent advances of PLGA micro/nanoparticles for the delivery of biomacromolecular therapeutics. *Materials Science and Engineering C*, *92*, 1041–1060. DOI: 10.1016/j.msec.2017.12.036 PMID: 30184728

Duceac, I. A., Lobiuc, A., Coseri, S., & Vere tiuc, L. (2019). *Tunable Hydrogels based on Chitosan, Collagen and Poly(Acrylic Acid) for Regenerative Medicine. In 2022 E-Health and Bioengineering Conference*. EHB., DOI: 10.1109/EHB47216.2019.8969886

Foox, M., & Zilberman, M. (2015). Drug delivery from gelatin-based systems. *Expert Opinion on Drug Delivery*, *12*(9), 1547–1563. DOI: 10.1517/17425247.2015.1037272 PMID: 25943722

Gulfam, M., Jo, S., Jo, S., Vu, T. T., Park, S.-H., & Lim, K. T. (2022). Highly porous and injectable hydrogels derived from cartilage acellularized matrix exhibit reduction and NIR light dual-responsive drug release properties for application in antitumor therapy. *NPG Asia Materials*, *14*(1), 8. Advance online publication. DOI: 10.1038/s41427-021-00354-4

Huang, C.-C. (2021). Characteristics and Preparation of Designed Alginate-Based Composite Scaffold Membranes with Decellularized Fibrous Micro-Scaffold Structures from Porcine Skin. *Polymers*, *13*(20), 3464. DOI: 10.3390/polym13203464 PMID: 34685220

Huang, H., Qi, X., Chen, Y., & Wu, Z. (2019). Thermo-sensitive hydrogels for delivering biotherapeutic molecules: A review. *Saudi Pharmaceutical Journal : SPJ : the Official Publication of the Saudi Pharmaceutical Society*, *27*(7), 990–999. DOI: 10.1016/j.jsps.2019.08.001 PMID: 31997906

Jacob, S., Nair, A. B., Shah, J., Sreeharsha, N., Gupta, S., & Shinu, P. (2021). Emerging Role of Hydrogels in Drug Delivery Systems, Tissue Engineering and Wound Management. *Pharmaceutics*, *13*(3), 357. DOI: 10.3390/pharmaceutics13030357 PMID: 33800402

Knop, K., Hoogenboom, R., Fischer, D., & Schubert, U. S. (2010). Poly(ethylene glycol) in Drug Delivery: Pros and Cons as Well as Potential Alternatives. *Angewandte Chemie International Edition*, *49*(36), 6288–6308. DOI: 10.1002/anie.200902672 PMID: 20648499

Lee, J., & Kim, H. (2018). Emerging properties of hydrogels in tissue engineering. *Journal of Tissue Engineering*, *9*, 2041731418768285. Advance online publication. DOI: 10.1177/2041731418768285 PMID: 29623184

. Li, L., He, Y., Zheng, X., Yi, L., & Nian, W. (2021). Progress on Preparation of pH/Temperature-Sensitive Intelligent Hydrogels and Applications in Target Transport and Controlled Release of Drugs. International Journal of Polymer Science. 2021(1). https://doi.org/DOI: 10.1155/2021/1340538

Li, L., Ni, R., Shao, Y., & Mao, S. (2013). Carrageenan and its applications in drug delivery. *Carbohydrate Polymers*, *103*(1), 1–11. Advance online publication. DOI: 10.1016/j.carbpol.2013.12.008 PMID: 24528694

Magli, S., Rossi, L., Consentino, C., Bertini, S., Nicotra, F., & Russo, L. (2021). Combined Analytical Approaches to Standardize and Characterize Biomaterials Formulations: Application to Chitosan-Gelatin Cross-Linked Hydrogels. *Biomolecules*, *11*(5), 683. DOI: 10.3390/biom11050683 PMID: 34062918

Matanović, M. R., Kristl, J., & Grabnar, P. A. (2014). Thermoresponsive polymers: Insights into decisive hydrogel characteristics, mechanisms of gelation, and promising biomedical applications. *International Journal of Pharmaceutics*, *472*(1), 262–275. DOI: 10.1016/j.ijpharm.2014.06.029 PMID: 24950367

. Mathews, S. (2017). Microencapsulation of Probiotics by Calcium Alginate and Gelatin and Evaluation of its Survival in Simulated Human Gastro- Intestinal Condition. International Journal of Current Microbiology and Applied Sciences. 6(4): 2080). https://doi.org/DOI: 10.20546/ijcmas.2017.604.245

Nordin, A. H., Ahmad, Z., Husna, S., Ilyas, R., Azemi, A., Ismail, N., Nordin, M., Ngadi, N., Siti, N., Nabgan, W., Norfarhana, A., & Azami, M. (2023). The State of the Art of Natural Polymer Functionalized Fe3O4 Magnetic Nanoparticle Composites for Drug Delivery Applications: A Review. *Gels (Basel, Switzerland)*, *9*(2), 121. DOI: 10.3390/gels9020121 PMID: 36826291

Raeisi, A., & Farjadian, F. (2024). Commercial hydrogel product for drug elivery based on route of administration. *Frontiers in Chemistry*, *12*, 1336717. Advance online publication. DOI: 10.3389/fchem.2024.1336717 PMID: 38476651

Rijal, G. (2023). Bioinks of Natural Biomaterials for Printing Tissues. In Bioengineering. 10(6): 705. DOI: 10.3390/bioengineering10060705

. Rossi, S. M., Murray, T., Cassidy, J., Lee, M. J., & Kelly, H. (2018). A Custom Radiopaque Thermoresponsive Chemotherapy-Loaded Hydrogel for Intratumoural Injection: An In Vitro and Ex Vivo Assessment of Imaging Characteristics and Material Properties. CardioVascular and Interventional Radiology. 42(2): 289).https://doi.org/DOI: 10.1007/s00270-018-2103-0

Siqueira, N. M., Cirne, M. F. R., Immich, M. F., & Poletto, F. (2018). Stimuli-responsive polymeric hydrogels and nanogels for drug delivery applications. Elsevier eBooks (p. 343). https://doi.org/DOI: 10.1016/B978-0-08-101997-9.00017-5

. Slaughter, B. V., Khurshid, S. S., Fisher, O. Z., Khademhosseini, A., & Peppas, N. A. (2009). Hydrogels in Regenerative Medicine. Advanced Materials. 21(32): 3307). https://doi.org/DOI: 10.1002/adma.200802106

Suk, J. S., Xu, Q., Kim, N., Hanes, J., & Ensign, L. M. (2015). PEGylation as a strategy for improving nanoparticle-based drug and gene delivery. *Advanced Drug Delivery Reviews*, *99*, 28–51. DOI: 10.1016/j.addr.2015.09.012 PMID: 26456916

Thu, H. E., Zulfakar, M. H., & Ng, S.-F. (2012). Alginate based bilayer hydrocolloid films as potential slow-release modern wound dressing. In International Journal of Pharmaceutics. 434(1): 375). https://doi.org/DOI: 10.1016/j.ijpharm.2012.05.044

. Wang, Y., Li, Z., Ouyang, J., & Karniadakis, G. E. (2020). Controlled release of entrapped nanoparticles from thermoresponsive hydrogels with tunable network characteristics. Soft Matter. 16(20): 4756). DOI: 10.1039/D0SM00207K

Xu, , Liu, Y., Fu, W., Yao, M., Ding, Z., Xuan, J., Li, D., Wang, S., Xia, Y., & Cao, M. (2020). Poly(N-isopropylacrylamide)-Based Thermoresponsive Composite Hydrogels for Biomedical Applications. *Polymers*, *12*(3), 580. DOI: 10.3390/polym12030580 PMID: 32150904

Younes, I., & Rinaudo, M. (2015). Chitin and Chitosan Preparation from Marine Sources. Structure, Properties and Applications. *Marine Drugs*, *13*(3), 1133–1174. DOI: 10.3390/md13031133 PMID: 25738328

Zou, Z., Zhang, B., Nie, X., Cheng, Y., Zhang, H., Liao, M., & Li, S. (2020). A sodium alginate-based sustained-release IPN hydrogel and its applications. *RSC Advances*, *10*(65), 39722–39730. DOI: 10.1039/D0RA04316H PMID: 35515393

Chapter 5
Integrating Physical Entrapment, Chemical Conjugation, Biocompatibility, and Mechanical–Diffusive Balance in Advanced Hydrogel Systems for Controlled:
Advanced Hydrogel Systems for Controlled Drug Delivery

Abul Kalam Azad
https://orcid.org/0000-0003-4874-1169
University MAIWP International, Malaysia

Wan Mohd Azizi Wan Sulaiman
University MAIWP International, Malaysia

Sajid Raza
IBADAT International University Islamabad, Pakistan

ABSTRACT

This chapter offers a detailed overview of hydrogel-based drug delivery systems,

DOI: 10.4018/979-8-3373-1047-3.ch005

emphasizing their potential as precision-engineered platforms for controlled therapeutic release. It examines two main drug incorporation strategies—physical entrapment and chemical conjugation—highlighting their mechanistic and kinetic differences and suitability in various physiological contexts. A key focus is the relationship between drug diffusion kinetics and the mechanical stability of hydrogel matrices. The discussion underscores how modifying hydrogel composition and structure can balance robustness with predictable release profiles. Additionally, the chapter addresses essential material properties such as biocompatibility and biodegradability, which are vital for safe and effective in vivo use. By integrating recent advancements, it provides insights into the rational design of next-generation hydrogels that enable targeted, sustained, and stimuli-responsive drug delivery, aligning with the goals of precision medicine.

INTRODUCTION

Medical drugs can enter hydrogel matrices through two dedicated approaches to build controlled-release systems. These approaches involve physical entrapment and chemical conjugation. Under physical entrapment conditions, drugs remain trapped inside the hydrophilic hydrogel network yet maintain their chemical independence from the polymer framework. The simple nature of physical entrapment methods enables drug maintenance, while maintaining chemical stability (Ishihara & Mizushima, 2010). Hydrogels release drugs with unclear kinetics, which sometimes depend on both the swelling patterns and the structural elements of the hydrogels. Drug molecules are bound through covalent linkages to polymer chains during the chemical conjugation of hydrogel systems. Because the drug binds directly to the polymer structure, it enables a precise control mechanism for regulating drug delivery times. Drug liberation from chemical bonds requires specific degradation mechanisms to achieve controlled release. The covalent attachment of drugs to polymers through chemical methods provides better release control; however, the drug modification process can alter biological activity outcomes. The utilization of polymer-based hydrogels for controlled drug delivery provides users with an adaptable platform because of their distinctive structural characteristics and capability to react with different triggering factors. Drugs should be integrated into polymers through physical trapping or chemical bonds according to the precise demands of the delivery systems, which include controlling drug release timing, drug chemical behavior, and treatment objectives. Polymer-based hydrogels have brought about a major paradigm shift in drug delivery systems by providing excellent advancements in controlled drug release. Owing to their 3D infinite network structure, these hydrogels can absorb large amounts of water and swell without degradation, making them suitable for drug entrapment and drug delivery. They allow for the slow and long-term delivery of active biological substances, enhancing healing effects without having to endure toxic effects linked to conventional delivery systems. This novel strategy utilizes the favorable characteristics of polymers, such as teamwork, degradability, and adjustability, enabling the plausibility of designing drug release patterns suitable for different patients and therapy regimens. With the advancement of studies in this field, the functionalities of polymer-based hydrogels in this context continue to broaden, which will be helpful in improving the patient environment and reforming the treatment modalities in health care (Nguyen et al., 2023).

KEY BENEFITS OF SUSTAINED DRUG RELEASE INCLUDE

Controlled-release systems have several main advantages, which have had a profound impact on patients' treatment protocols and the efficient usage of active pharmaceutical ingredients. First, such systems ensure that patients strictly abide by doctors' prescriptions by extending the period between dosages, mainly for diseases such as diabetes, hypertension, and cancer (Murugesan et al., 2019). Sustained release forms are therefore advantageous to patients own, hence have better patients' compliance leading to increased therapeutic efficiency in the body system (Bala et al., 2021). Additionally, sustained-release approaches achieve first-pass constant concentration levels, which are expected to have a therapeutic effect because drug levels do not go up and down, making toxicity and therapeutic failure potential risks. Thus, the depot maintains a slow release and minimizes drug over-administration to sensitive tissues, decreasing the systematic side effects while simultaneously increasing the safety aspects due to its ability to deliver the drug to only the specific area of interest. Apart from enhanced patient health, sustained-release systems also provide better use of active pharmaceutical ingredients. These systems lower drug wastage because the bioavailability of the applied APIs is optimized; hence, the use of diminished concentration to achieve comparable therapeutic effects. Sustained-release drug delivery systems have various uses in different therapeutic areas, such as cardiovascular diseases, neurological disorders, infections, cancer, regenerative medicine, and tissue engineering.

OVERVIEW OF HYDROGELS AS VERSATILE CARRIERS IN DRUG DELIVERY SYSTEMS

Hydrogels are three-dimensional polymeric structures that swell with a large quantity of water and remain intact. They have been approved as excellent carriers for drug delivery, as they provide fairly distinct benefits. Another main feature of hydrogels is that they are biocompatible and biodegradable. Many hydrogels are synthesized from biocompatible polymers that can be chemically adapted to enhance their biological compatibility. In addition, the majority of hydrogels are biodegradable, breaking down into harmless products once their drug delivery task is performed. It also has a large drug-carrying capacity because of its high porosity and water content, and can encapsulate a large number of water-soluble drugs, proteins, peptides, and even cells. The rate of drug release from hydrogels can be controlled and altered by adjusting parameters such as crosslinking density, polymer type, and degree of swelling, which implies both diffusional and kinetic release mechanisms (Pal et al., 2009). Hydrogels can also be sensitive to certain environmental signals, such as pH, temperature, light, or particular biomolecules, allowing controlled drug delivery in certain areas. The use of hydrogels is also convenient because they can be administered through the mouth or intravenous, topical, and biodegradable systems, making them widely applicable in many medical disciplines (Dreiss, 2020). In a related sense, hydrogels can provide a protective setting that prevents drug degradation by enzymes or light or, at extreme pH, enhance drug stability and solubility.

COMPARISON WITH OTHER DRUG DELIVERY SYSTEMS

Hydrogels offer distinct advantages over alternative drug delivery systems while also presenting unique challenges (Prasanth et al., 2019. Below is a comparative analysis:

Table 1. Comparison of drug delivery systems: Hydrogels, liposomes, microparticles/nanoparticles, and matrix tablets

Feature	Hydrogels	Liposomes	Microparticles/Nanoparticles	Matrix Tablets
Drug Loading Capacity	High for hydrophilic drugs; customizable for others	Moderate, primarily for lipophilic drugs	High for both hydrophilic and lipophilic drugs	Limited by formulation constraints
Release Mechanism	Sustained release via swelling or degradation	Burst release followed by sustained release	Highly tunable; burst or controlled release	Gradual release, less adaptable
Biocompatibility	High, especially with natural polymers	High, depending on lipid composition	Variable; depends on polymer and surfactant	Moderate; excipients may impact compatibility
Stimuli Responsiveness	Easily engineered for pH, temperature, etc.	Limited responsiveness	Can be designed for specific triggers	Minimal; generally fixed release profiles
Ease of Manufacture	Moderate; depends on polymer choice and method	High for simple systems, low for complex ones	Moderate to high, requires specialized methods	High, with well-established techniques
Drug Protection	Strong for hydrophilic and sensitive drugs	Good for lipophilic drugs	Moderate to high	Minimal; drugs are exposed to the environment

PROPERTIES OF HYDROGELS RELEVANT TO SUSTAINED DRUG RELEASE

Hydrogels are an interesting class of biomaterials that have recently attracted considerable interest for drug delivery. Moreover, these water-swollen polymeric networks can accommodate and entrap a large amount of water or biological fluids, making these materials more suitable for sustained release systems.

Swelling Behavior and Water Absorption Capacity

According to the current definition, hydrogels can be described as highly hydrophilic and crosslinked polymer systems capable of swelling and retaining high water content. The main phenomenon occurs when water penetrates the polymeric matrix of the hydrogel without dissolving the hydrogel itself owing to swelling.

Relevance to Sustained Drug Release

The response to swelling is equally important for controlling the diffusion and release of the incorporated drug. The swelling could increase the amount of aqueous drug allowed to diffuse out of the matrix, thereby increasing the diffusion rate of highly water-soluble drugs. In addition, hydrogels can expand because of changes in conditions such as the pH or ionic strength of the environment, thus proposing the idea of controlled release of drugs under specific conditions (Sosnik & Seremeta, 2017).

Factors Influencing Swelling

It is noteworthy that the nature of the polymer matrix greatly determines the swelling behavior of the hydrogels. For instance, polyacrylamide and alginate are considered to be more hydrophilic swelling polymers than hydrophobic swelling polymers. The porosity of the polymer network is also an important factor, and a higher crosslinking density decreases swelling and controls drug release. Iadic hydrogels can also vary the swelling characteristics in response to certain stimuli, such as temperature and pH, making a wide range of possible uses of these hydrogels for the controlled release of drugs.

Crosslinking Density and Its Impact on Drug Release Rates

Many reactions, as part of polymer chemistry, include crosslinking, that is, the covalent or physical attachment of polymer chains, creating a 3D structure. Such a complex arrangement of networks greatly affects the physicochemical characteristics of polymeric substrates, especially with regard to their drug delivery applications. The crosslinking density, which quantifies the degree of connectivity of the polymer chains, significantly influences drug release patterns. Increased crosslinking leads to a smaller pore size in the hydrogel matrix, which stereo-typically affects the diffusion of the confined drug, thus prolonging the release time. Because of this property, highly crosslinked hydrogels are suitable for sustained drug delivery systems, where an optimum and protracted release rate is needed. On the other hand, low cross linking density results in the enlargement of pore size through which the rh has a faster diffusion rate the kind of system that is appropriate for fast release of drugs (Caccavo et al., 2016). Chemical crosslinking methods can generate fully thermally and mechanically stable hydrogel structures, whereas physical crosslinking allows for reversible hydrogel formation, which aids in improved control over drug delivery. These dynamic crosslinking approaches allow the fabrication of hydrogel-forming drug delivery systems that release incorporated medicines at desired rates, thereby addressing a wide range of therapeutic requirements.

Biocompatibility and Biodegradability of Hydrogels for Sustained Drug Delivery

Hydrogels aremuft polymers which have received a great deal of attention in drug delivery systems because of the properties they possess, including biocompatibility and biodegradability. One essential consideration in hydrogel design is biocompatibility, which has the potential for interaction with biological tissue systems without toxicity, immune response, or inflammation. Hyaluronic acid, chitosan, and polyethylene glycol are natural or synthetic biocompatible polymers commonly used in medicine (Ho et al., 2022). Other properties include biodegradability of hydrogels, because the material degrades under certain conditions to form non-hazardous byproducts that make surgical removal after drug delivery unnecessary. Based on polymer chemistry and the degree of crosslinking, the degradation rate of a hydrogel can be altered to offer a suitably controlled and long-term drug delivery system. Another advantage of hydrogels is that they are biocompatible and biodegradable, which makes them more suitable for sustained drug delivery. Biocompatibility prevents any adverse reactions towards the hydrogels and allows for lengthy periods without negative effects, whereas biodegradability allows for slow release of the drug as the hydrogel slowly degrades, thereby achieving consistent results (Far et al., 2022).

OPTIMIZATION OF HYDROGEL-BASED DRUG DELIVERY SYSTEMS: BALANCING DIFFUSION AND MECHANICAL STABILITY

They have been used as multifunctional drug delivery systems because they can accommodate any type of therapeutic agent and can be designed for controlled and/or sustained drug release (Pérez-Luna & González-Reynoso, 2018). An important factor in designing hydrogel-based drug delivery systems is the control of drug diffusion and mechanical strength of the hydrogel network.

Diffusion-Controlled Drug Release

The rate-controlling mechanism of drug release from hydrogels involves the diffusion of drug molecules into the hydrogel matrix. The porosity, particle size, and molecular weight of the drug molecules, interconnecting channels in the hydrogel, and mesh size all appear to play a major role in influencing the diffusion rates and hence the time course of drug release. Long-term and steady delivery of the therapeutic agent is advantageous because general diffusion-controlled release is beneficial for many therapeutic applications.

Mechanical Stability of Hydrogels

Ideally, hydrogels need to be stable at the places where they are located, and the environment has to be physiological, which includes mechanical stress due to movement of the tissue or pressure. The mechanical properties of hydrogels can be improved either by increasing the crosslinking density or by including fillers or secondary polymers, such as nanomaterials.

Balancing Diffusion and Mechanical Stability

A major drawback in the fabrication of hydrogel-based drug delivery systems is the difficulty in achieving a high mechanical strength and at the same time rapid drug release. Long-term stability is important in mechanical applications because the hydrogel matrix must not degrade and release the drug before the appropriate time. However, these advancements also have drawbacks: crosslinking membranes can also slow diffusion if done to a high degree, or if they incorporate reinforcing agent (Narayanaswamy & Torchilin, 2019). Thus, the physico-mechanical characteristics of hydrogel-based drug delivery systems must be meticulously fine-tuned to achieve the required mechanical strength and porosity to allow optimal release of the drug.

DIFFUSION-CONTROLLED RELEASE MECHANISMS IN HYDROGEL DRUG DELIVERY SYSTEMS

Drugs are normally released through hydrogel systems via a mechanism called diffusion-controlled release, which is a typical drug delivery approach. The drug escapes the hydrogel matrix through diffusion of drug molecules through the gel network structure. Following drug encapsulation, the hydrogel creates a barrier that controls drug movement. Drug molecules move freely through the gel network, causing drug release that results from the establishment of drug concentration gradients (Mauri et al., 2018). The release timing depends on three main hydrogel characteristics: drug molecular size, gel swelling patterns, and hydrophilic properties. The porosity of hydrogels plays a major role in diffusion-controlled release because increased porosity enhances the molecular diffusion speed. The atomic size of drug molecules impacts diffusion because large molecules penetrate the material slower than smaller ones. The swelling behavior of hydrogels alters the release rate by modifying the diffusion pathway and lbrack;9] Drug delivery applications increasingly rely on hydrogels as the main platform because these materials serve to retain drugs inside their porous structure and distribute them at controlled release rates [1& rbrack;. The precise modulation of hydrogel compositions and structures enables healthcare professionals to tailor both release speeds and durations for therapeutic purposes and lbrack;2& rbrack;. This delivers several benefits, including extended treatment periods, fewer prescribed doses, increased patient comfort and effectiveness, and lbrack;2&rbrack.

SWELLING-CONTROLLED DRUG DELIVERY: MECHANISMS, FACTORS, AND APPLICATIONS

Controlled drug delivery methods have become popular in pharmaceutical development owing to their extended drug concentration stability, diminished side effects, and enhanced patient adherence to treatment. The swelling characteristics of the hydrogel matrices function as a mechanism for controlling drug release (Hardenia et al., 2019). After hydrogel-based drug delivery systems come into contact with water or physiological fluids, they absorb a hydrophilic polymer network liquid to induce swelling. During this swelling process, large pores develop within the polymer matrix, and these newly formed pores enhance drug diffusion paths or improve drug transport through hydrogel networks. The control of the release rate depends on the swelling degree, which can be affected by the polymer composition, crosslinking density, and environmental conditions such as pH and ionic strength. Hydrogel swelling depends heavily on the polymer's water affinity and its water-absorption capacity. Hydrophilic polymers have expanded markedly alongside significant drug release acceleration. An increased density of crosslinking agents in hydrogel structures minimizes swelling while slowing down drug release kinetics. The swelling characteristics of hydrogels and drug release kinetics are influenced by environmental conditions, including pH and ionic strength measurements. Hydrogels with environmental sensitivity can function as drug delivery systems through programmed substance release, triggered by specific physiological stimuli. Example designs use pH-triggered hydrogels keyed to swell upon diverse pH levels observed throughout the gastrointestinal tract or additional body locations. The swelling response of these hydrogel systems extends beyond oral medicine by enabling controlled drug delivery throughout the gastrointestinal tract to treat wounds through the delivery of antimicrobial agents at a steady rate.

BIODEGRADATION-TRIGGERED DRUG RELEASE: A REVIEW OF MECHANISMS AND APPLICATIONS

This innovative method of biodegradation-triggered drug release utilizes a polymer matrix degradation process to control drug delivery systems. This review examines fundamental operating principles while exploring important procedural variables and describes the different applications of this technology. The degradation of polysaccharide matrices leads to sustained drug release through biodegradation triggers. The drug release rate comes from the biodegradation kinetic behavior, which researchers can modify to align with therapeutic needs. The pharmaceutical field adopts two main polymers, poly(lactic-co-glycolic acid) and polycaprolactone, because they break down naturally using enzymatic hydrolysis and biochemical mechanisms (Hines & Kaplan, 2013). Various variables determine both the biodegradation speed and drug release characteristics. Selecting the right polymer for the system is essential because diverse polymers release drugs at different speeds that match the target medication delivery profiles. Esterase enzymes found in bodily tissues work alongside the polymer concentration to accelerate the degradation processes (Narančić et al., 2020). Hydrogel formulation affects drug release kinetics through its chemical structure and crosslinking density, which directly controls the rate of degradation. Biodegradation-triggered drug delivery systems have broad potential for biomedical applications in various fields. These devices utilize these structures for long-term therapies as part of implants, drug-delivery systems, and tissue regeneration scaffolds because drug release occurs during scaffold or matrix destruction.

DUAL OR MULTI-MECHANISTIC RELEASE SYSTEMS: OPTIMIZING DRUG DELIVERY THROUGH SYNERGISTIC APPROACHES

Researchers are focusing on diverse strategies in modern pharmaceutical research to control drug delivery mechanisms, because controlled drug delivery remains a fundamental pharmaceutical concern. Dual-release systems and multi-mechanistic forms represent pharmaceutical developments that unite distinct release techniques to boost drug release control. Dual- or multi-mechanistic release systems enable sequential delivery and synergistic action by making different mechanisms function in concert to release drugs according to set release rates and time conditions or environmental triggers. The hybrid release system integrates the drug delivery functions of diffusion, swelling, and biodegradation through a singular hydrogel matrix. This system combines elements of diffusion-controlled release with swelling-driven release, followed by drug delivery based on hydrogel biodegradation during a specified timeline. This method enables optimized delivery profiles to release drugs at distinct times for drug combination therapies (Modi et al., 2013). The diverse range of dual/multi-mechanistic delivery systems include cancer therapy and biphasic drug delivery to maintain stable drug levels in patients with diabetes and cardiovascular diseases.

Stimuli-Responsive Hydrogel Systems for Targeted Drug Delivery

The "smart hydrogel" designation describes hydrogel materials capable of conducting structural modifications under definite external or internal stimulus triggers which include pH effects or temperature fluctuations as well as illumination sources as well as biological enzymatic activity. Hydrogel materials exhibit control over swelling, shrinking behavior, and degradation dynamics, thus allowing precise drug delivery schedules. The unique ability of hydrogels to release therapeutic agents based on specific stimuli makes them highly valuable for drug delivery systems that require targeted delivery (Rizzo & Kehr, 2020). The main types of stimuli-responsive hydrogel systems and their applications in tailored drug delivery include hydrogels with pH-triggered behavior that function as drug delivery systems for targeted compound delivery to particular body areas, including the intestines and cancerous tissue environments. This research explored temperature-sensitive hydrogels with volume phase transition capabilities at different temperatures for injectable drug delivery systems and in situ-forming systems. Additionally, light- and enzyme-responsive hydrogels have been studied to advance specific drug delivery techniques (Shi et al., 2019).

PH-Sensitive Hydrogels for Gastrointestinal Targeting

Targeted drug delivery techniques with pH-sensitive hydrogels present themselves as advanced approaches that safeguard delicate pharmaceutical compounds against stomach acidity while directing their release exclusively to the neutral pH solution of the small intestine. Hydrogel networks employ ionic functional groups such as carboxyl and amine moieties that display protonation/deprotonation behavior according to environmental pH changes, causing hydrogel expansion-contraction, which controls drug delivery rates (Zhao et al., 2019). Hydrogels show pH-triggered versatility in their capability to produce customizable delivery platforms that target specific areas within the GI tract system. Hydrogels made from chitosan, alginate, or acrylic acid display engineered properties that allow them to release their contents when the stomach pH increases to the intestinal pH. These pH-sensitive materials enhance the absorption potential of oral medications through targeted protection against drug degradation and the precise delivery of medication. The pH-responsive nature of these hydrogels allows precise control of drug release, targeting physiological modifications across different disease states. Modern healthcare depends on these hydrogels because they enable precise drug delivery during pathological events such as cancer and diabetes. These events cause pH changes, which could otherwise lead to adverse medical outcomes.

Thermo-Sensitive Hydrogels for Localized Drug Delivery

Thermosensitive hydrogels serve as future-based solutions for delivering therapeutic substances locally through a temperature-dependent release mechanism. These hydrogel systems incorporate polymeric chains that transform from a hydrophilic state below 37°C to a hydrophobic state as the body temperature reaches this critical threshold. The thermoreversible natural behavior of these hydrogels transforms them from a liquid to a gel state during administration, while gently releasing drugs from their encapsulating systems at specific delivery sites.The extensive application of thermoresponsive hydrogels in pharmaceutical delivery includes their use in precise localized delivery systems that support chronic disease management strategies. Hydrogel technology permits the delivery of anticancer agents, anti-inflammatory drugs, and local anesthetics to precise target areas, including tumor wounds and joints, while maintaining extended drug release durations. Temperature-sensitive hydrogels can deliver drugs through precise control, while matching the therapeutic needs of chronic illnesses, including arthritis. Among the thermosensitive polymers that have been extensively studied for drug delivery applications, researchers have focused on poly. This substance forms a gel upon body temperature exposure while releasing drugs, including anticancer agents, anti-inflammatory medications, and local anesthetics. The medication has a liquid form that changes into a targeted gel state when it reaches its location, which reduces the requirement for invasive operations. Through their formulation, these hydrogels represent a versatile drug delivery system because they can encompass both hydrophobic and hydrophilic pharmaceutical agents. Controlled sustained drug release from thermo-sensitive hydrogel systems enables improved therapeutic efficacy and reduces systemic side effects because drugs go directly to their target site (Tian et al., 2021).

Light-Responsive and Enzyme-Sensitive Systems for Precision Therapy

Hydrogels show considerable potential for targeted delivery applications because they respond to precise specific wavelengths of light wavelengths with structural transformations. Hydrogels routinely include photosensitive functionalities, such as azobenzenes or nitrobenzyl groups, to enable controlled drug release by means of light-induced reversible changes such as photoisomerization and photocleavage (Rwei et al., 2015). Light-responsive hydrogels demonstrate unique potential because they enable controlled drug delivery through precise spatial and temporal control, which reduces side effects across the body. Shear-sensitive hydrogels show exceptional potential for precision therapy because they provide controlled techniques to send medical agents to tumors and target disease strands through exposure to illumination. The release of drugs can be precisely triggered by exposing azobenzene-based hydrogels to UV light, as part of their design for controlled therapeutic drug delivery (Costa et al., 2018). Targeted drug delivery elements exist in enzyme-sensitive hydrogels that form part of the stimuli-responsive system design. The hydrogel contains answerable functional groups that break down specific enzymes in bodily fluids or disease locations, enabling selective drug release mechanisms. By leveraging enzyme-sensitive hydrogels for targeted drug delivery, physicians can achieve precision therapy through drug release activation by specific enzymes found within the target disease sites. Drug delivery can utilize polymeric hydrogels containing peptide sequences that are released upon matrix metalloproteinases activity for targeted tumor delivery. Hydrogels activated by external stimuli allow controlled drug release into selected target areas to enhance therapy, while reducing damage to normal tissues. Precision therapy is an important advancement in hydrogel development, which responds to light inputs together with enzyme-based sensitivity for drug delivery applications. These systems represent a promising method that demonstrates both increased therapeutic intervention effectiveness and a decreased chance of off-target effect

Multi-Stimuli-Responsive Hydrogels for Tailored Drug Delivery

Hydrogels designed as multi-stimuli-responsive systems show promise as next-generation drug delivery systems that control therapeutic agent release according to both natural body changes and external triggers. Hydrogels engineered for dual- or multiple-stimulus detection allow better controlled drug delivery through mechanisms that respond to pH, temperature, light and magnetic fields, and enzymatic conditions. These hydrogels enhance therapeutic delivery through sequential drug release because they detect multiple environmental signals, leading to optimized spatiotemporal control. Through the combined use of pH-sensitive and temperature-sensitive hydrogels, drug delivery occurs first in gastric acid conditions before tumor-specific temperature release, enabling precise targeted drug delivery within the site. Refined control of drug delivery is essential for chronic disease management because it requires consistent optimized drug release methods. Hydrogel systems can be developed as multi-stimuli-responsive platforms that release anti-inflammatory drugs together with insulin for rheumatoid arthritis and diabetes patients, while delivering medications in corresponding patient-specific requirements. Evolving chemical systems for drug delivery stem from an enhanced awareness of the complex relationship between the designed hydrogel properties and physiological signaling cues (Thirupathi et al., 2022) Research teams achieved controlled interactive hydrogels through polymer layout optimization to create systems which monitor local environmental shifts thereby optimizing drug delivery techniques and output while lessening negative effects.

Hydrogel Fabrication Techniques for Drug Delivery Applications

The creation of hydrogel-based drug delivery systems requires careful attention to three fundamental parameters, starting with precise control over polymer network structures and ending with consideration of both swelling properties and drug-loading efficiency. The fabrication approach establishes fundamental parameters that determine the drug release behavior, hydrogel strength properties, and biological compatibility capabilities. The drug delivery capabilities of polymer-based hydrogels make them attractive delivery systems because their swelling behavior enables precise control of drug release, leading to extended duration of drug delivery (Chyzy et al., 2020). The porous structure of hydrogels enables them to retain drugs during drug delivery, which allows controlled substance leakage. Precise tailoring of therapeutic requirements is possible through hydrogel structure management and material composition control. Hydrogels exhibit continuous drug release ability, which delivers several benefits including extended therapeutic outcomes, fewer medication administrations, and higher treatment compliance for patients.

BULK AND SOLUTION POLYMERIZATION: MECHANISMS, ADVANTAGES, AND APPLICATIONS IN HYDROGEL FORMULATIONS

Polymer-based hydrogels have become popular in biomedical applications because they replicate the characteristics of natural tissues exceptionally well. The selection of polymerization procedures controls both the functional properties and the application possibilities of these hydrogel systems.

Emulsion and Suspension Polymerization: Strategies for Controlled Drug Delivery

Scientists across the drug delivery field have displayed growing interest in emulsion and suspension polymerization because these methods enable the controlled production of micro- and nanosized hydrogel particles with specific properties. Emulsion polymerization enables monomer solubilization in oil-phase droplets that exist within an aqueous bulk, while an emulsifier activates polymerization throughout the dispersed hydrophobic droplets. This production technique enables researchers to achieve precise particle size control, which may be valuable when designing drug delivery systems. Emulsion polymerization enables the efficient production of drug-loaded microgels that exhibit controlled release capabilities suitable for specific drug delivery applications, including tumor-specific targeting and vaccine delivery (Sapra et al., 2013). The basic difference between suspension polymerizations lies in the use of monomer dispersals in non-solvent media containing aqueous droplets. Diamond-shaped hydrogel particles along with uniform beads become possible using this technique, and both elements demonstrate excellent stability for controlled-release applications. Hydrogel-based microspheres and microcapsules obtained through suspension polymerization allow controlled drug release (Aguirre et al., 2023). Polymer-based hydrogels are widely used in drug delivery applications because their swelling behavior allows them to control drug release to provide extended drug delivery. Therapeutic needs guide framework parameter modification of hydrogel structures, which enables exact control of both release profiles and durations. The consistent drug release capability of hydrogels provides multiple advantages, including prolonged therapeutic effects, reduced dosing frequency, and enhanced patient treatment adherence. Hydrogels have become more effective in drug delivery after engineers design them to activate with particular stimuli, including pH and temperature changes.

Microfabrication and 3D Printing Techniques for Advanced Hydrogel-Based Drug Delivery

Hydrogels serve as versatile biomaterials for controlled drug delivery applications because of their biocompatibility, high water content, and adjustable physicochemical characteristics (Lin & Metters, 2006). The micropatterning of hydrogels for microscale applications using photolithography, soft lithography, and molding provides a precise manufacturing process for these materials. Through microfabrication, researchers have created devices with complex microscale features from hydrogels that enable drug delivery through targeted pathways. Microfabrication provides precise geometric control of hydrogels, which helps develop implantable systems, on-demand drug delivery devices, and targeted biosensors for medication control. Additive manufacturing technology, referred to as 3D printing, is a strong method to construct complicated hydrogel-based drug delivery systems. Hydrogel material deposition occurs through sequential layers with printing equipment, such as inkjet machines, fused deposition modeling, and stereolithography. Engineers can formulate hydrogel inks into custom-made 3D structures by adding drugs for targeted local deliveries. 3D printing enables the creation of patient-specific drug delivery implants and wound dressings with controlled release kinetics, which has attracted its use in cancer treatment and wound care. Both microfabrication and 3D printing offer unique advantages in the development of advanced hydrogel-based drug delivery systems. Far smaller hardware devices with very precise elements can be made possible through microfabrication, whereas 3D printing supports the creation of detailed drug delivery scaffold structures according to individual needs.

Enhancing Drug Stability through Encapsulation Techniques in Hydrogels

The delivery of pharmaceutical agents requires stable formulations and effective administration methods to optimize drug performance. Hydrogel matrices provide a flexible drug delivery approach through drug entrapment or inclusion to enhance both drug stability and availability, along with controlled release capabilities. Coacervation or solvent evaporation is a widespread method in pharmaceutical materials science used to encapsulate drugs within forming hydrogels, while the polymer precipitates from solutions or evaporates the solvent. Nanoparticle-based encapsulation researchers embed drugs within nanostructures such as liposomes and micelles alongside nanocapsules before dispersing them within hydrogel matrices. Hydrogel matrix analysis with sensitive pharmacological agents delivers double benefits by reinforcing drug stability while enhancing delivery precision. Hydrophilic properties provide either hydrophobic or hydrophilic drugs with better encapsulation outcomes owing to their adaptable characteristics (Abaee et al., 2017). Hydrogels provide multiple advantages when drugs are integrated within their structures. Hydrogels stabilize sensitive active pharmaceutical ingredients while simultaneously shielding drugs that break down during biological processes. Hydrogels provide extended drug-release properties that create therapeutic persistence at lower dosage rates, thereby resulting in better therapeutic outcomes and increased patient adherence to treatment. These encapsulation techniques enable neglectable therapeutic agents such as proteins and peptides and nucleic acids to be delivered through protected carriers. Sustained stability of hydrogel-based systems is essential for maintaining the efficacy of encapsulated vaccines throughout storage as well as during application.

Hydrogel-Based Sustained Drug Delivery: Advancements and Applications

Hydrogel-based drug delivery systems show great promise as drug delivery solutions because of their high-water content retention, biodegradability, and biocompatibility (Pham et al., 2020). These materials allow researchers to engineer specific responses through stimuli, such as pH variation, temperature changes, and light activation, for therapeutic agent movement control. The sustained and controlled drug release capabilities of hydrogels represent their fundamental application in sustained drug delivery systems. The swelling behavior of polymer-based hydrogels can be used to control drug delivery because it leads to gradual controlled drug release rates. The sustained-release feature has beneficial effects, including an extended therapeutic duration along with diminished dosing requirements, which enhances treatment adherence.

OVERCOMING ENZYMATIC DEGRADATION IN ORAL DRUG DELIVERY USING PH-SENSITIVE HYDROGELS

Oral drug delivery is the most widely chosen route for medication delivery because of its ability to enhance both convenience of use and patient cooperation. Oral drug delivery methods typically encounter two major problems: gastrointestinal tract absorption rate and enzyme-mediated drug breakdown. Enzymes present in the stomach and intestine break down sensitive drugs, which reduces both drug availability and treatment reliability. The delivery of drugs through the mouth faces multiple obstacles, yet pH-sensitive hydrogels demonstrate great potential as resolving solutions. Such hydrogels shield drugs during their transit through the acidic environment of the stomach before liberating them into the neutral pH region of the small intestine. Biologics and proteins, together with peptides, can be successfully encapsulated within hydrogels to protect them from enzymatic destruction during their passage through the digestive system. Constructive applications of pH-responsive hydrogels consist of protein peptide vaccine protection from enzymatic degradation, while also enabling drug delivery systems with predictable release profiles and targeted medical therapy methods to reach designated gastrointestinal sections, including the colon (Gupta et al., 2002). Chitosan-based hydrogels are an effective option for oral drug delivery applications. Structurally engineered chitosan-based hydrogels shield medications during stomach passage, while releasing content continuously through the small intestine. These hydrogels act biodegradably so that all compounds are released without producing toxic substances during or after treatment. Hydrogels designed for dual-stimulus behavior to integrate pH and temperature sensitivity have been investigated as precise drug delivery systems. Hydrogel delivery systems have the potential to trigger drug release through pH and temperature variations that affect specific areas of the body, including the gastrointestinal tract.

Injectable Hydrogels: Localized and Controlled Release for Chronic Diseases

Treatment of chronic diseases requires ongoing drug delivery systems that sustain drug release for long durations. Drug delivery systems that use injectable methods enable minimal procedures for direct drug transport to action sites, while providing prolonged sustained drug release. The therapeutic effectiveness of drug delivery depends on achieving precisely controlled release over prolonged periods. When hydrogels are delivered as liquid injections, they transform into in-situ gels inside the target area. Hydrogels function as drug depots that accept and release small molecules, proteins, and gene medicines, and deliver controlled drug quantities over extended durations. Through their injectable approach, hydrogels enable minimally invasive procedures where the hydrogel network transforms into a gel to continuously disperse medications. Hydrogels have become a notable focus in the development of local drug delivery solutions for multiple medical applications, particularly chronic disease treatment. Skilled management of hydrogel swelling enables drug release control, which results in prolonged drug delivery. Different drug molecules integrate within hydrogel structures to permit their diffusion from interstitial spaces toward biological environments for controlled-release applications. The constructed polymer hydrogels have built-in mechanisms to observe particular stimuli from their environment, including temperature and pH conditions, thus boosting their precise drug-release functionality (Pang et al., 2020).

CONCLUSION

Hydrogel-based drug delivery systems represent a versatile and promising platform for achieving controlled, targeted, and patient-specific therapeutic outcomes. This chapter has explored the dual strategies of physical entrapment and chemical conjugation for incorporating drugs into hydrogels, outlining their distinct advantages, limitations, and performance across diverse physiological conditions. A key insight is that no single approach is universally optimal; instead, the choice between entrapment and conjugation must be guided by the drug's physicochemical properties, the intended site of action, and the therapeutic goals. Equally important is the balance between critical material characteristics—biocompatibility, biodegradability, diffusion kinetics, and mechanical integrity. Achieving this balance requires thoughtful structural and compositional design, where polymer type, crosslinking method, and environmental responsiveness are carefully tuned. Mechanical robustness must be sufficient to maintain functional integrity, while the hydrogel's network must permit predictable and controlled drug diffusion. Overall, this chapter underscores the importance of an integrated, multidisciplinary approach to hydrogel design. By aligning drug loading strategies with the material and biological requirements of specific clinical applications, researchers and developers can create next-generation delivery vehicles that are not only effective and reliable but also adaptable to the evolving demands of precision medicine.

REFERENCES

Abaee, A., Mohammadian, M., & Jafari, S. M. (2017). Whey and soy protein-based hydrogels and nano-hydrogels as bioactive delivery systems. *Trends in Food Science & Technology*, *70*, 69–81. DOI: 10.1016/j.tifs.2017.10.011

Aguirre, M., Ballard, N., Gonzalez, E., Hamzehlou, S., Sardon, H., Calderon, M., Paulis, M., Tomovska, R., Dupin, D., Bean, R. H., Long, T. E., Leiza, J. R., & Asua, J. M. (2023). Polymer Colloids: Current Challenges, Emerging Applications, and New. *Macromolecules*, *56*(7), 2579–2607. DOI: 10.1021/acs.macromol.3c00108 PMID: 37066026

Bala, R., Rana, R. K., & Madaan, R. (2021). Formulation Design and Optimization of Sustained Released Matrix Tablets of Propranolol HCl Using Natural and Synthetic Polymers. *ACTA Pharmaceutica Sciencia*, *59*(2), 321–341. DOI: 10.23893/1307-2080.APS.05920

Caccavo, D., Cascone, S., Lamberti, G., Barba, A. A., & Larsson, A. (2016). Swellable Hydrogel-based Systems for Controlled Drug Delivery. IntechOpen eBooks. https://doi.org/DOI: 10.5772/61792

Chyzy, A., Tomczykowa, M., & Płońska-Brzezińska, M. E. (2020). Hydrogels as Potential Nano-, Micro- and Macro-Scale Systems for Controlled Drug Delivery Materials, 13(1): 188. https://doi.org/DOI: 10.3390/ma13010188

Costa, D. F., Mendes, L. P., & Torchilin, V. P. (2018). The effect of low- and high-penetration light on localized cancer therapy. *Advanced Drug Delivery Reviews*, *138*, 105–116. DOI: 10.1016/j.addr.2018.09.004 PMID: 30217518

Dreiss, C. A. (2020). Hydrogel design strategies for drug delivery. *Current Opinion in Colloid & Interface Science*, *48*, 1–17. DOI: 10.1016/j.cocis.2020.02.001

Far, B. F.. (2022). A Review on Biomedical Application of Polysaccharide-Based Hydrogels with a Focus on Drug Delivery System. *Polymers*, *14*(24), 5432. DOI: 10.3390/polym14245432 PMID: 36559799

Gupta, P., Vermani, K., & Garg, S. (2002). Hydrogels: From controlled release to pH-responsive drug delivery. *Drug Discovery Today*, *7*(10), 569–579. DOI: 10.1016/S1359-6446(02)02255-9 PMID: 12047857

Hardenia, A., Maheshwari, N., Hardenia, S. S., Dwivedi, S., Maheshwari, R., & Tekade, R. K. (2019). *Scientific Rationale for Designing Controlled Drug Delivery Systems. eBooks*. Elsevier BV., DOI: 10.1016/B978-0-12-817909-3.00001-7

Hines, D. J., & Kaplan, D. L. (2013). Poly(lactic-co-glycolic) Acid-Controlled-Release Systems: Experimental and Modeling Insights. *Critical Reviews in Therapeutic Drug Carrier Systems, 30*(3), 257–276. DOI: 10.1615/CritRevTherDrugCarrierSyst.2013006475 PMID: 23614648

Ho, T., Chang, C.-C., Chan, H.-P., Chung, T., Shu, C., Chuang, K., Duh, T., Yang, M., & Tyan, Y. (2022). Hydrogels: Properties and Applications in Biomedicine. *Molecules (Basel, Switzerland), 27*(9), 2902. DOI: 10.3390/molecules27092902 PMID: 35566251

Ishihara, T., & Mizushima, T. (2010). Techniques for efficient entrapment of pharmaceuticals in biodegradable solid micro/nanoparticles. *Expert Opinion on Drug Delivery, 7*(5), 565–575. DOI: 10.1517/17425241003713486 PMID: 20230305

Lin, C., & Metters, A. T. (2006). Hydrogels in controlled release formulations: Network design and mathematical modelling. *Advanced Drug Delivery Reviews, 58*(12), 1379–1408. DOI: 10.1016/j.addr.2006.09.004 PMID: 17081649

Mauri, E., Negri, A., Rebellato, E., Masi, M., Perale, G., & Rossi, F. (2018). Hydrogel-Nanoparticles Composite System for Controlled Drug Delivery. *Gels (Basel, Switzerland), 4*(3), 74. DOI: 10.3390/gels4030074 PMID: 30674850

Modi, K., Modi, M., Mishra, D., Mittal, P., Sorathiya, U., & Shelat, P. (2013). Oral controlled release drug delivery system: An overview. *International Research Journal of Pharmacy, 4*(3), 70–76. DOI: 10.7897/2230-8407.04312

Murugesan, S., Byran, G., Lakshmanan, K., Karri, V. V. S. R., & Radhakrishnan, A. (2019). Oral Modified Drug Release Solid Dosage Form with Special Reference to Design; An Overview. *Current Drug Research Reviews, 12*(1), 16–25. DOI: 10.2174/2589977511666191121094520 PMID: 31755398

Narančić, T., Cerrone, F., Beagan, N., & O'Connor, K. E. (2020). Recent Advances in Bioplastics: Application and Biodegradation. *Polymers, 12*(4), 920. DOI: 10.3390/polym12040920 PMID: 32326661

Narayanaswamy, R., & Torchilin, V. P. (2019). Hydrogels and Their Applications in Targeted Drug Delivery. *Molecules (Basel, Switzerland), 24*(3), 603. DOI: 10.3390/molecules24030603 PMID: 30744011

Nguyen, H. T., Chien, T. B., & Cư ng, Đ. X. (2023). Polymer-Based Hydrogels Applied in Drug Delivery: An Overview. *Gels (Basel, Switzerland), 9*(7), 523. DOI: 10.3390/gels9070523 PMID: 37504402

Pal, K., Banthia, A. K., & Majumdar, D. K. (2009). Polymeric Hydrogels: Characterization and Biomedical Applications. *Designed Monomers and Polymers*, *12*(3), 197–220. DOI: 10.1163/156855509X436030

Pang, X., ... (2020). Engineering Thermo-pH Dual Responsive Hydrogel for Enhanced Tumor Accumulation, Penetration, and Chemo-Protein Combination Therapy. https://doaj.org/article/dbd573f72d7c451ab5abeb5271ab8eba

Pérez-Luna, V. H., & González-Reynoso, O. (2018). Encapsulation of Biological Agents in Hydrogels for Therapeutic Applications. *Gels (Basel, Switzerland)*, *4*(3), 61. DOI: 10.3390/gels4030061 PMID: 30674837

Pham, T.-N., Su, C.-F., Huang, C.-C., & Jan, J. (2020). Biomimetic hydrogels based on L-Dopa conjugated gelatin as pH-responsive drug carriers and antimicrobial agents. *Colloids and Surfaces. B, Biointerfaces*, *196*, 111316. DOI: 10.1016/j.colsurfb.2020.111316 PMID: 32827950

Prasanth, A., Kumar, A., Shruthi, B., & Sindhu, S. (2019). Kinetic study and in vitro drug release studies of nitrendipine loaded arylamide grafted chitosan blend microspheres. *Materials Research Express*, *6*(12), 125427. DOI: 10.1088/2053-1591/ab5811

Rizzo, F., & Kehr, S. (2020). Recent Advances in Injectable Hydrogels for Controlled and Local Drug Delivery. *Advanced Healthcare Materials*, *10*(1), 2001341. Advance online publication. DOI: 10.1002/adhm.202001341 PMID: 33073515

Rwei, A. Y., Wang, W., & Kohane, D. S. (2015). Photoresponsive nanoparticles for drug delivery. *Nano Today*, *10*(4), 45. DOI: 10.1016/j.nantod.2015.06.004 PMID: 26644797

Sapra, B., Thatai, P., Bhandari, S., Sood, J., Jindal, M. K., & Tiwary, A. K. (2013). A Critical Appraisal of Microemulsions for Drug Delivery: Part II. *Therapeutic Delivery*, *5*(1), 83–94. DOI: 10.4155/tde.13.125 PMID: 24341819

Shi, Q., Liu, H., Tang, D., Li, Y., Li, X., & Xu, F. (2019). Bioactuators based on stimulus-responsive hydrogels and their emerging biomedical applications. *NPG Asia Materials*, *11*(1), 64. Advance online publication. DOI: 10.1038/s41427-019-0165-3

Sosnik, A., & Seremeta, K. P. (2017). Polymeric Hydrogels as Technology Platform for Drug Delivery Applications. *Gels (Basel, Switzerland)*, *3*(3), 25. DOI: 10.3390/gels3030025 PMID: 30920522

Thirupathi, K., Phan, T. T. V., Santhamoorthy, M., Ramkumar, V., & Kim, S. (2022). pH and Thermoresponsive PNIPAm-co-Polyacrylamide Hydrogel for Dual Stimuli-Responsive Controlled Drug Delivery. *Polymers*, *15*(1), 167. DOI: 10.3390/polym15010167 PMID: 36616517

Tian, M. L., Zhou, J., Qi, X., & Shen, R. (2021). Thermo-sensitive hydrogel and their biomedical applications. *IOP Conference Series. Earth and Environmental Science*, *714*(3), 32062. DOI: 10.1088/1755-1315/714/3/032062

Zhao, S., Zhu, H., Chen, Z., Shuai, S., Zhang, N., Liu, Y., Rao, Z., Li, Y., Zhao, C., Zhou, K., Ge, W., & Hao, J. (2019). Preparation and properties of a temperature- and pH- responsive polypeptide hydrogel. *Materials Research Express*, *6*(8), 85711. DOI: 10.1088/2053-1591/ab253e

Chapter 6
Alginate-Based Microbeads:
A Promising Platform for Targeted Drug Delivery Systems

Fatema Tuz Zohera
https://orcid.org/0009-0006-4718-5390
International Islamic University Malaysia, Malaysia

Abul Kalam Azad
University MAIWP International, Malaysia

ABSTRACT

Alginate-based microbeads are promising carriers for targeted drug delivery due to their biocompatibility, biodegradability, and gel-forming capacity. Formed via ionotropic gelation with divalent cations, these microbeads protect labile drugs and enable sustained, controlled, and site-specific release. Tuning parameters like the M/G ratio and crosslinking density allows precise control of drug release kinetics and mechanical properties. Recent advances—including co-polymers, nanoparticles, and stimuli-responsive systems—have improved targeting, drug retention, and stability. Applications span gastrointestinal, colonic, and tumor-targeted therapies. With low toxicity and regulatory acceptability, alginate microbeads are suitable for oral, injectable, and localized administration. This review outlines recent developments, challenges, and future directions, highlighting alginate microbeads as a versatile platform in precision drug delivery.

DOI: 10.4018/979-8-3373-1047-3.ch006

INTRODUCTION

Targeted drug delivery has revolutionized pharmaceutical science by enhancing therapeutic efficacy while minimizing adverse effects associated with conventional drug administration. Unlike systemic drug distribution, targeted delivery systems aim to concentrate therapeutic agents specifically at the site of action, reducing toxicity to healthy tissues and improving patient compliance. Among the various carriers explored for such delivery strategies, biodegradable and biocompatible polymers have demonstrated significant potential due to their ability to encapsulate a wide array of therapeutic agents and control their release kinetics. Alginate, a naturally occurring anionic polysaccharide extracted primarily from brown seaweed (Phaeophyceae), has emerged as a versatile biopolymer for developing drug delivery systems. It consists of linear copolymers of β-D-mannuronic acid (M) and α-L-guluronic acid (G) residues, arranged in homopolymeric (M or G) and heteropolymeric (MG) blocks. The physicochemical characteristics of alginate, such as viscosity, gelation ability, and porosity, are influenced by the M/G ratio, molecular weight, and degree of polymerization, making it highly adaptable for pharmaceutical applications. Alginate-based microbeads are typically fabricated through ionotropic gelation using divalent cations like calcium (Ca^{2+}), which crosslink the G-blocks to form hydrogel matrices capable of encapsulating drugs in a protective network. These microbeads offer several advantages: high biocompatibility, mild preparation conditions suitable for sensitive bioactives, and modifiable release profiles based on crosslinking density and bead structure. Furthermore, their ability to respond to physiological stimuli such as pH, enzymes, or ionic strength makes them suitable for site-specific delivery, particularly in gastrointestinal and tumor-targeting applications (George & Abraham, 2006; Lee & Mooney, 2012). The present work aims to explore the formulation, functionalization, and therapeutic potential of alginate-based microbeads as a promising platform for targeted drug delivery, while also addressing current challenges and future perspectives in their clinical translation.

ALGINATE: STRUCTURE AND PROPERTIES

Source and Composition

Alginate is a naturally derived linear polysaccharide predominantly extracted from the cell walls of brown seaweeds such as *Laminaria*, *Ascophyllum*, and *Macrocystis* species, and also produced by certain bacteria, including *Azotobacter vinelandii* and *Pseudomonas* species (Lee & Mooney, 2012). Chemically, alginate consists of two monomeric units: β-D-mannuronic acid (M) and α-L-guluronic acid (G), arranged in homopolymeric M-blocks, G-blocks, and alternating MG-blocks. The sequence and distribution of these blocks determine the physicochemical behavior of the polymer.

Gelation Mechanism

One of the most notable features of alginate is its ability to form hydrogels through ionotropic gelation. This process occurs when the G-blocks of the alginate chain interact with divalent cations such as Ca^{2+}, forming junction zones in a cooperative manner, often referred to as the "egg-box" model (Draget et al., 2006). The efficiency of gel formation and mechanical properties of the hydrogel depend on the G-content and concentration of crosslinking ions.

Physicochemical Properties

The physical and mechanical properties of alginate are largely determined by its molecular weight and M/G ratio. A higher proportion of G-blocks yields stronger and more brittle gels, while M-rich alginates form more elastic and flexible gels. These properties are crucial for pharmaceutical applications where mechanical integrity, swelling behavior, and permeability influence drug encapsulation and release kinetics (George & Abraham, 2006).

Key properties of alginate include:

- **Biodegradability**: Naturally broken down by environmental or microbial enzymes.
- **Biocompatibility**: Non-toxic and well-tolerated in biological systems.
- **Mucoadhesion**: Facilitates localized delivery in mucosal tissues.
- **Ionotropic Gelation**: Forms hydrogels in the presence of divalent cations like Ca^{2+} or Ba^{2+}.

Biocompatibility and Pharmaceutical Relevance

Alginate is inherently biocompatible, hydrophilic, and biodegradable, and is classified as Generally Recognized as Safe (GRAS) by the U.S. FDA. These characteristics make it suitable for biomedical and pharmaceutical applications. Its mild gelation conditions allow encapsulation of sensitive molecules like proteins, peptides, and probiotics without compromising their activity, making alginate an attractive matrix for drug delivery systems (Lee & Mooney, 2012).

RATIONALE FOR USING ALGINATE IN DRUG DELIVERY

The selection of an appropriate polymeric material is critical in the development of advanced drug delivery systems. Alginate, a naturally derived, non-toxic, and biodegradable polysaccharide, has emerged as a highly promising candidate for pharmaceutical applications due to its unique physicochemical and biological characteristics.

Biocompatibility and Safety

Alginate is widely regarded for its excellent biocompatibility, making it suitable for both systemic and localized drug delivery. It does not elicit significant immunogenic or toxic responses in vivo, and it is recognized as Generally Recognized as Safe (GRAS) by the U.S. Food and Drug Administration (George & Abraham, 2006). This ensures patient safety, particularly in long-term or sustained-release therapies.

Mild Gelation Conditions

One of alginate's most favorable features is its ability to form hydrogels under mild, aqueous conditions through ionotropic gelation with divalent cations such as Ca^{2+}. This process does not require high temperatures or organic solvents, which is particularly advantageous for encapsulating heat-sensitive or biologically labile compounds such as proteins, enzymes, probiotics, and nucleic acids (Lee & Mooney, 2012).

Controlled and Targeted Release Capabilities

Alginate-based matrices can be engineered to provide controlled, sustained, or targeted release of therapeutic agents. Modulating parameters such as the molecular weight, M/G ratio, crosslinking density, and microbead size allows for precise control over drug diffusion and degradation kinetics. Furthermore, alginate hydrogels exhibit responsiveness to environmental stimuli such as pH and ionic strength, making them suitable for site-specific delivery, particularly in the gastrointestinal tract or tumor microenvironments.

Versatility in Formulation Design

Alginate can be formulated into various delivery vehicles, including microbeads, nanoparticles, films, and scaffolds, providing flexibility for different administration routes—oral, topical, injectable, or implantable. Its compatibility with other natural or synthetic polymers also enables the development of hybrid systems to overcome challenges such as burst release or mechanical weakness.

In summary, alginate's multifunctional properties—ranging from safe biological interaction to customizable release profiles—justify its extensive use as a drug carrier in modern therapeutics and drug delivery which is driven by several advantages mentioned below:

- **Mild Processing Conditions**: Ideal for heat- or pH-sensitive drugs like peptides and proteins.
- **Targeted Release**: pH-sensitive dissolution enables site-specific delivery (e.g., colon-targeting).
- **Customizable Properties**: Gel strength and porosity can be modulated by the M:G ratio, crosslinking ions, and processing conditions.
- **High Encapsulation Efficiency**: Suitable for a wide range of drugs, including small molecules, proteins, and nucleic acids.

FABRICATION TECHNIQUES FOR ALGINATE MICROBEADS

Alginate microbeads are prominent drug delivery vehicles owing to their biocompatibility, mild gelation conditions, and ability to encapsulate diverse therapeutic agents. Various fabrication methods have been optimized to control bead size, morphology, and drug release profiles. This section discusses four common techniques used in alginate microbead production.

Ionic Gelation

Ionic gelation is the most frequently employed technique for microbead fabrication due to its simplicity and gentle processing. A sodium alginate solution containing the drug is dropped into a solution of divalent cations, most commonly calcium chloride. The calcium ions crosslink the guluronic acid segments of alginate chains, leading to immediate hydrogel bead formation based on the egg-box model of gelation (Gombotz & Wee, 1998). Process parameters including alginate and calcium concentrations, droplet size, and stirring rate influence bead characteristics such as size distribution and mechanical strength.

Emulsion Gelation

Emulsion gelation involves dispersing the aqueous alginate solution into a hydrophobic oil phase (e.g., light mineral oil) to form a water-in-oil emulsion. Upon addition of calcium ions, gelation occurs within the dispersed aqueous droplets, generating microbeads (Lee et al., 2000). This method yields smaller and more uniform beads relative to ionic gelation. Control over emulsification conditions, such as agitation speed and surfactant concentration, permits fine-tuning of bead size and encapsulation efficiency, which is particularly useful for hydrophilic drug delivery.

Jet-Cutting

In the jet-cutting technique, the alginate solution is extruded through a nozzle, creating a continuous jet that is segmented into uniform droplets by a high-frequency vibrating cutter. These droplets fall into a calcium chloride bath where ionic crosslinking forms stable microbeads (Turgeon et al., 2007). Jet-cutting enables precise control over bead diameter and shape by adjusting vibration frequency, flow rate, and nozzle size. The scalability and reproducibility of this technique make it favorable for industrial-scale production.

Electrostatic Droplet Generation

Electrostatic droplet generation employs an electric field to control droplet detachment from a nozzle tip. The alginate solution is electrically charged, producing monodisperse droplets that are collected in a calcium chloride solution to form crosslinked microbeads (Vehring, 2008). This method provides excellent control over particle size and uniformity, essential for consistent drug release profiles. It is also gentle enough to encapsulate sensitive biological agents such as proteins and live cells.

CROSSLINKING METHODS AND STABILIZATION

Crosslinking is a critical step in the fabrication of alginate microbeads, determining their mechanical strength, stability, and drug release behavior. Stabilization techniques further improve bead integrity under physiological conditions and allow for controlled delivery properties.

Ionic Crosslinking

Ionic crosslinking remains the predominant method for alginate gelation due to its simplicity and biocompatibility. Calcium ions (Ca^{2+}) are the most commonly used divalent cations because of their non-toxic nature and physiological relevance. The gelation mechanism is described by the "egg-box" model, wherein the G-blocks of alginate chains chelate calcium ions to form a three-dimensional network of junction zones that stabilize the hydrogel matrix (Smidsrød & Skjåk-Bræk, 1990). Although calcium provides adequate gel strength for many applications, other divalent cations such as barium (Ba^{2+}), strontium (Sr^{2+}), and magnesium (Mg^{2+}) have been investigated. Barium ions form stronger and more stable gels due to higher affinity binding, but their potential toxicity limits clinical use (Grant et al., 1973). Strontium offers intermediate gel strength and enhanced biocompatibility, whereas magnesium forms weaker gels but can influence degradation rates.

Covalent Crosslinking

To overcome the relatively weak stability of ionically crosslinked alginate under physiological ionic conditions, covalent crosslinking methods have been employed. This approach forms permanent chemical bonds between alginate polymer chains, enhancing mechanical integrity and reducing premature degradation. Carbodiimide chemistry, typically using 1-ethyl-3-(3-dimethylaminopropyl) carbodiimide (EDC), enables the formation of amide bonds between alginate carboxyl groups and primary amines, improving gel stability without introducing significant toxicity (Kim et al., 2014). Similarly, glutaraldehyde can crosslink alginate by reacting with hydroxyl or amine groups, although its potential cytotoxicity requires careful control of residual reagent removal (Martins et al., 2009). Covalent crosslinking enhances bead durability in biological fluids, making it suitable for long-term drug release applications.

Coatings and Layer-by-Layer Assembly

Surface modification through coatings or layer-by-layer (LbL) assembly offers additional stabilization and functionalization of alginate microbeads. Coating microbeads with cationic polymers such as chitosan or gelatin creates a polyelectrolyte complex shell that improves mechanical strength, reduces swelling, and modulates drug diffusion (Mohanty et al., 2016). LbL assembly allows alternate deposition of oppositely charged polymers, enabling fine-tuning of permeability and responsiveness to environmental stimuli such as pH or ionic strength, which is advantageous for targeted or triggered drug release (De Geest et al., 2007). These coatings also provide biocompatible interfaces that can enhance cell interaction or protect encapsulated bioactives.

DRUG LOADING STRATEGIES

Passive Encapsulation:

Drugs are incorporated by mixing directly with the alginate solution before gelation, a straightforward method particularly effective for hydrophilic compounds (Feng et al., 2013).

Adsorption:

Post-fabrication loading involves soaking pre-formed microbeads in drug solutions, suitable for thermolabile or pH-sensitive drugs to avoid degradation during bead formation (Huang et al., 2010).

Complexation:

Charged drugs can be loaded through electrostatic interactions with oppositely charged alginate chains, enhancing loading efficiency and enabling controlled release (Luo & Wang, 2014).

Layered Loading:

Multilayered microbeads allow spatial drug placement in distinct layers, facilitating sequential or multi-phase drug release profiles (Park et al., 2017).

CONTROLLED AND TARGETED DRUG RELEASE MECHANISMS

Diffusion-Controlled Release:

Drug molecules diffuse through the alginate gel matrix, with release kinetics governed by bead size, porosity, and crosslinking density (Park et al., 2017).

Swelling-Controlled Release:

Water uptake causes the hydrogel matrix to swell, enlarging pores and facilitating drug diffusion (Peppas et al., 2000).

7.3 pH-Responsive Release:

Alginate remains intact in acidic gastric conditions but dissolves in neutral to basic intestinal pH, enabling site-specific drug delivery (Yang et al., 2019).

Enzyme-Responsive Systems:

Incorporation of enzyme-degradable polymers allows drug release to be triggered selectively by enzymes overexpressed in pathological sites (Li & Mooney, 2016).

APPLICATIONS IN TARGETED DRUG DELIVERY

Gastrointestinal Tract

Alginate microbeads are widely used for oral drug delivery due to their ability to protect bioactives from gastric degradation and enable targeted release in the intestines. The gel matrix remains stable in acidic pH, but dissolves or swells in the neutral-to-alkaline pH of the intestine, making them ideal for site-specific release (Yang et al., 2019). This is particularly beneficial for the oral delivery of sensitive molecules such as **insulin**, **probiotics**, and **colon-targeted drugs**, where premature degradation in the stomach must be avoided (Krasaekoopt et al., 2004). For example, encapsulated probiotics show improved viability during gastric transit and colon colonization.

Cancer Therapy

Alginate-based microbeads offer a promising strategy for localized and targeted delivery of chemotherapeutics, minimizing systemic toxicity. The porous structure allows sustained drug release, and functionalization of the bead surface enables ligand-mediated targeting. For instance, folic acid-conjugated alginate beads can selectively bind to folate receptors overexpressed in certain tumor cells, thereby improving drug accumulation at the tumor site (Luo et al., 2016). Additionally, co-encapsulation with imaging agents can facilitate theranostic applications in cancer treatment.

Wound Healing

In wound management, alginate beads can serve as carriers for **antibiotics**, **antioxidants**, or **growth factors** to enhance localized tissue regeneration. Their high water absorption capacity maintains a moist wound environment, essential for epithelial repair, while enabling controlled release of therapeutic agents (Boateng et al., 2008). Furthermore, alginate's intrinsic hemostatic properties add value in acute wound care.

Pulmonary and Nasal Delivery

Due to their biocompatibility and mucoadhesive nature, alginate microspheres are being developed for **inhalation** and **nasal delivery** routes. These delivery modes offer direct access to the systemic circulation or local action in the respiratory tract, bypassing first-pass metabolism. Alginate beads loaded with anti-inflammatory agents or vaccines have shown effective mucosal adhesion and prolonged residence time in the nasal cavity or lungs (Illum, 2007), enhancing therapeutic outcomes.

IN VIVO PERFORMANCE AND BIOCOMPATIBILITY

In vivo evaluations have consistently affirmed the biocompatibility of alginate-based microbeads. Upon administration, alginate undergoes gradual degradation into non-toxic, absorbable byproducts with minimal immunogenicity (Andersen et al., 2015). Pharmacokinetic studies indicate that encapsulated drugs exhibit **prolonged systemic circulation**, **sustained release profiles**, and **enhanced site-specific accumulation**, especially when surface-functionalized with targeting ligands (Liu et al., 2020).

Examples:

- **Oral Insulin Delivery**: Alginate beads enhanced insulin bioavailability in diabetic rats, reducing blood glucose levels significantly and protecting insulin from gastric degradation (Shojaei et al., 2021).
- **Tumor-Targeted Doxorubicin Delivery**: Folic acid-modified alginate beads loaded with doxorubicin showed selective tumor targeting, reduced tumor mass, and minimal cardiotoxicity in mouse models (Khalil et al., 2016).
- **Probiotic Delivery**: Alginate-encapsulated *Lactobacillus* strains showed improved survival through the gastrointestinal tract and effective colonization in the gut, supporting intestinal health (Krasaekoopt et al., 2004).

- **Ocular Drug Delivery**: Alginate microspheres containing timolol maleate demonstrated prolonged intraocular pressure reduction in rabbit models, indicating sustained ocular drug release (Al Khateb et al., 2016).
- **Pulmonary Delivery**: Alginate microparticles containing anti-tubercular drugs (e.g., rifampicin) delivered via inhalation showed enhanced drug retention in the lungs and reduced bacterial load in TB-infected rats (Pandey et al., 2003).
- **Intraperitoneal Chemotherapy**: Alginate microbeads loaded with cisplatin and administered intraperitoneally achieved localized drug retention and reduced systemic toxicity in ovarian cancer models (Vossen et al., 2017).
- **Wound Healing**: Alginate microbeads loaded with vascular endothelial growth factor (VEGF) enhanced tissue regeneration and re-epithelialization in full-thickness wound models in rats (Lee et al., 2010).

CHALLENGES AND LIMITATIONS

Poor Mechanical Strength

Alginate microbeads often lack sufficient mechanical strength, particularly under physiological conditions. This can result in structural breakdown and premature drug release, especially in dynamic environments like the gastrointestinal tract (George & Abraham, 2006).

Uncontrolled Swelling and Erosion

The highly hydrophilic nature of alginate can lead to uncontrolled swelling and subsequent erosion of the microbeads in aqueous media, which may alter the release kinetics and reduce therapeutic efficiency (Augst et al., 2006).

Burst Release of Drug

Drugs encapsulated through passive mixing may exhibit an initial burst release due to poor entrapment within the polymer matrix, especially for low molecular weight or water-soluble compounds (Lee & Mooney, 2012).

Limited Stability In Vivo

Alginate gels crosslinked with calcium or other divalent cations may undergo ion exchange with sodium ions in bodily fluids. This compromises structural integrity and reduces the efficacy of the delivery system (Hariyadi & Islam, 2020).

Poor Encapsulation Efficiency for Hydrophobic Drugs

The inherent hydrophilicity of alginate hinders the encapsulation of poorly water-soluble drugs unless modified with hydrophobic carriers or blended with other polymers to enhance compatibility (Liu et al., 2020).

Batch-to-Batch Variability

Natural variability in alginate composition, particularly in the mannuronic acid to guluronic acid (M/G) ratio, affects gelation properties and results in inconsistencies across different production batches (Andersen et al., 2015).

Immunogenic Impurities

Unpurified commercial alginate may contain residual proteins, endotoxins, or other impurities that can provoke an immune response. Thus, pharmaceutical-grade purification is necessary for biomedical use (Lee & Mooney, 2012).

Difficulty in Large-Scale Production

Achieving precise control over bead size, drug loading efficiency, and morphological uniformity at industrial scale remains technically challenging due to limitations in droplet generation and crosslinking control (Jain & Jain, 2016).

EMERGING TRENDS AND FUTURE PERSPECTIVES

Smart and Stimuli-Responsive Microbeads

Recent advancements focus on engineering alginate microbeads that respond to environmental stimuli such as pH, temperature, or redox gradients. These "smart" systems enhance site-specific and controlled drug release. For instance, pH-responsive systems remain intact in the acidic gastric environment but dissolve in the neutral pH of the intestine, enabling colon-targeted drug delivery (Zhao et al., 2019). Similarly, redox-sensitive microbeads leverage intracellular glutathione concentrations for cancer-targeted drug release (Mandal et al., 2021).

3D-Printed Alginate Beads

The integration of 3D bioprinting enables the design of alginate-based microbeads with highly controlled geometry, porosity, and multi-layer drug loading. This approach offers customization for patient-specific drug release profiles and the fabrication of complex microbead architectures for sequential or gradient release (Goyanes et al., 2015). 3D printing also supports on-demand drug manufacturing, crucial for emergency and remote healthcare applications.

Gene and Protein Delivery

Alginate microbeads are being tailored for the encapsulation and delivery of biomacromolecules such as DNA, siRNA, and therapeutic proteins. Surface modifications and co-encapsulation with stabilizing agents like polycations or nanoparticles help protect labile molecules and enhance cellular uptake (Yin et al., 2020). These developments have shown promise in gene therapy and regenerative medicine.

Dual-Function Beads (Theranostics)

There is growing interest in dual-function or "theranostic" alginate microbeads, which combine therapeutic delivery with diagnostic imaging. For example, beads embedded with iron oxide nanoparticles allow magnetic resonance imaging (MRI) tracking while simultaneously delivering chemotherapeutic agents (Wang et al., 2022). Such systems facilitate real-time monitoring of treatment efficacy, particularly in cancer.

Personalized Medicine

Emerging research focuses on customizing alginate microbeads to individual patient physiology through AI-driven drug release modeling and pharmacogenomic data. Personalized microbeads could adapt drug release kinetics based on a patient's metabolic profile, disease state, or treatment response (Liang et al., 2021). This trend aligns with the future of precision medicine, especially for chronic or genetically variable conditions.

HYBRID AND COMPOSITE MICROBEADS

Combining alginate with other polymers enhances the physicochemical and functional properties of microbeads. Chitosan–alginate complexes improve mucoadhesion and provide pH-responsive drug release due to the polyelectrolyte interaction. Similarly, blending with synthetic polymers like PLGA or PCL can strengthen mechanical stability and extend release profiles, making them suitable for long-acting injectables or implants (Tønnesen & Karlsen, 2002).

ALGINATE MICROBEADS IN VACCINE DELIVERY

Alginate microbeads are being explored as delivery vehicles for oral and mucosal vaccines. Their ability to shield antigens from harsh gastric environments and release them in the intestine supports immune system interaction. Co-encapsulation of adjuvants like CpG or MPLA can enhance immunogenicity. This platform has shown promise for hepatitis B, tetanus toxoid, and DNA vaccines (de Vos et al., 2010).

ENVIRONMENTAL AND ETHICAL ASPECTS

Alginate is sourced sustainably from brown algae, making it an environmentally friendly polymer. However, overharvesting and marine biodiversity disruption are growing concerns. Efforts toward sustainable seaweed farming and ethical bioprospecting practices are increasingly emphasized in pharmaceutical applications (Chopin et al., 2012).

ALGINATE BEADS IN CELL AND TISSUE ENGINEERING

Alginate beads are used for encapsulating living cells such as pancreatic islets, stem cells, and probiotics. Their gentle gelation conditions and porous nature support cell viability and metabolic activity. In tissue regeneration, alginate scaffolds have been applied in cartilage repair, nerve regeneration, and wound healing (Augst et al., 2006).

ALGINATE-BASED MICROBEADS IN ANTIMICROBIAL AND ANTIVIRAL THERAPY

Alginate microbeads have shown significant promise in the delivery of antimicrobial and antiviral agents, particularly for localized infections and resistant microbial strains. The controlled release of antibiotics like ciprofloxacin or ampicillin from alginate matrices enhances efficacy while reducing systemic side effects and resistance development. In viral infections, such as those caused by HSV or influenza, encapsulating antiviral drugs (e.g., acyclovir or oseltamivir) in alginate beads may improve mucosal retention and bioavailability. Moreover, incorporating silver nanoparticles or essential oils into alginate carriers has been explored to further enhance antimicrobial potency (Liakos et al., 2016).

INDUSTRIAL AND COMMERCIAL TRANSLATION OF ALGINATE MICROBEADS

Bridging the gap between laboratory research and industrial-scale production remains a major focus. Alginate microbeads have already been commercialized for controlled-release fertilizers, probiotics, and cosmetic actives, paving the way for pharmaceutical-scale applications. Optimization of spray-drying, fluidic extrusion, and sterile encapsulation techniques under cGMP conditions is crucial for pharmaceutical manufacturing. Moreover, intellectual property protection, regulatory pathways (e.g., FDA's GRAS status for alginate), and market analysis play a vital role in translating alginate-based systems into market-ready products (Grant et al., 2022).

CONCLUSION

Alginate-based microbeads represent a promising and adaptable platform in the field of targeted drug delivery, owing to their inherent biocompatibility, gentle gelation process, and ability to encapsulate diverse therapeutic agents. Their physicochemical tunability supports tailored release profiles, while recent innovations—such as hybrid composites, stimuli-responsive systems, and 3D-printed architectures—have significantly expanded their clinical potential. Applications across gastrointestinal, oncological, pulmonary, and vaccine delivery domains underscore their versatility. Moreover, the integration of gene delivery, theranostics, and personalized medicine is steering these systems toward next-generation therapeutics. Despite persisting limitations related to scale-up, regulatory compliance, and long-term stability, continued progress in polymer chemistry, nanotechnology, and translational research is steadily overcoming these hurdles. With sustainability and biocompatibility at the forefront, alginate microbeads are poised to play a pivotal role in the future of precision and regenerative medicine.

REFERENCES

Al Khateb, K., Ozhmukhametova, E. K., Mussin, M. N., Seilkhanov, T. M., Rakhypbekov, T. K., & Khutoryanskiy, V. V. (2016). In situ gelling systems based on Pluronic F127/alginate: Development and ocular delivery of timolol maleate. *International Journal of Pharmaceutics*, *502*(1-2), 70–79. DOI: 10.1016/j.ijpharm.2016.02.027 PMID: 26899977

Andersen, T., Auk-Emblem, P., & Dornish, M. (2015). 3D cell culture in alginate hydrogels. *Microarrays (Basel, Switzerland)*, *4*(2), 133–161. DOI: 10.3390/microarrays4020133 PMID: 27600217

Augst, A. D., Kong, H. J., & Mooney, D. J. (2006). Alginate hydrogels as biomaterials. *Macromolecular Bioscience*, *6*(8), 623–633. DOI: 10.1002/mabi.200600069 PMID: 16881042

Boateng, J. S., Matthews, K. H., Stevens, H. N. E., & Eccleston, G. M. (2008). Wound healing dressings and drug delivery systems: A review. *Journal of Pharmaceutical Sciences*, *97*(8), 2892–2923. DOI: 10.1002/jps.21210 PMID: 17963217

Chopin, T.. (2012). Seaweed aquaculture for integrated multi-trophic aquaculture (IMTA): A key towards sustainability. [ander]. *Journal of Phycology*, *48*(5), 983–992.

De Geest, B. G., De Koker, S., Sukhorukov, G. B., Kreft, O., Parak, W. J., & Sanders, N. N. (2007). Polyelectrolyte microcapsules for biomedical applications. *Soft Matter*, *3*(6), 711–719. DOI: 10.1039/B617074H

de Vos, P., Faas, M. M., Spasojevic, M., & Sikkema, J. (2010). Encapsulation for preservation of functionality and targeted delivery of bioactive food components. *International Dairy Journal*, *20*(4), 292–302. DOI: 10.1016/j.idairyj.2009.11.008

Draget, K. I., Skjåk-Bræk, G., & Smidsrød, O. (2006). Alginates from algae. In Phillips, G. O., & Williams, P. A. (Eds.), *Handbook of Hydrocolloids* (pp. 379–395). Woodhead Publishing., DOI: 10.1533/9781845693149.3.379

Feng, J., Zhao, L., Huang, Z., & Du, Y. (2013). Alginate-based microcapsules for drug delivery: Effect of preparation method on drug release. *International Journal of Pharmaceutics*, *453*(2), 497–503. DOI: 10.1016/j.ijpharm.2013.06.034

George, M., & Abraham, T. E. (2006). Polyionic hydrocolloids for the intestinal delivery of protein drugs: Alginate and chitosan—a review. *Journal of Controlled Release*, *114*(1), 1–14. DOI: 10.1016/j.jconrel.2006.04.017 PMID: 16828914

Gombotz, W. R., & Wee, S. F. (1998). Protein release from alginate matrices. *Advanced Drug Delivery Reviews*, *31*(3), 267–285. DOI: 10.1016/S0169-409X(97)00124-5 PMID: 10837629

Goyanes, A., Buanz, A. B., Basit, A. W., & Gaisford, S. (2015). Fused-filament 3D printing (3DP) for fabrication of tablets. *International Journal of Pharmaceutics*, *476*(1–2), 88–92. DOI: 10.1016/j.ijpharm.2014.09.044 PMID: 25275937

Grant, G. T., Morris, E. R., Rees, D. A., Smith, P. J. C., & Thom, D. (1973). Biological interactions between polysaccharides and divalent cations: The egg-box model. *FEBS Letters*, *32*(1), 195–198. DOI: 10.1016/0014-5793(73)80770-7

Grant, S. M., & Maleki, H., & GhavamiNejad, A. (2022). Translation of alginate-based delivery systems: From bench to market. *Journal of Drug Delivery Science and Technology*, *69*, 103095.

Hariyadi, D. M., & Islam, N. (2020). Current strategies to improve the bioavailability of poorly water-soluble drugs. *Journal of Drug Delivery Science and Technology*, *55*, 101458. DOI: 10.1016/j.jddst.2019.101458

Huang, Y., Leobandung, W., Foss, A., & Peppas, N. A. (2010). Molecular aspects of muco- and bioadhesion: Tethered structures and site-specific drug delivery. *Journal of Controlled Release*, *65*(1-2), 63–71. DOI: 10.1016/S0168-3659(99)00233-3 PMID: 10699271

Illum, L. (2007). Nanoparticulate systems for nasal delivery of drugs: A real improvement over simple systems? *Journal of Pharmaceutical Sciences*, *96*(3), 473–483. DOI: 10.1002/jps.20718 PMID: 17117404

Jain, D., & Jain, V. (2016). Novel microencapsulation techniques for controlled drug delivery—A review. *International Journal of Pharmaceutical Sciences and Research*, *7*(11), 4373–4380.

Kim, H., Park, H., & Kim, S. H. (2014). Enhanced stability and controlled release of alginate hydrogel beads through carbodiimide-mediated covalent crosslinking. *Carbohydrate Polymers*, *101*, 98–105. DOI: 10.1016/j.carbpol.2013.08.076

Krasaekoopt, W., Bhandari, B., & Deeth, H. (2004). The influence of coating materials on some properties of alginate beads and survivability of microencapsulated probiotic bacteria. *International Dairy Journal*, *14*(8), 737–743. DOI: 10.1016/j.idairyj.2004.01.004

Lee, K. Y., & Mooney, D. J. (2010). Alginate: Properties and biomedical applications. *Progress in Polymer Science*, *37*(1), 106–126. DOI: 10.1016/j.progpolymsci.2011.06.003 PMID: 22125349

Li, J., & Mooney, D. J. (2016). Designing hydrogels for controlled drug delivery. *Nature Reviews. Materials*, *1*(12), 16071. DOI: 10.1038/natrevmats.2016.71 PMID: 29657852

Liakos, I., Rizzello, L., Scurr, D. J., Pompa, P. P., & Bayer, I. S. (2016). Alginate-based antimicrobial films and fibers: Current advances and future perspectives. *Advances in Colloid and Interface Science*, *237*, 55–64.

Liang, Y., Zhao, X., Hu, T., Chen, B., Yin, Z., Ma, P. X., & Guo, B. (2021). Flexible and Highly Sensitive Wearable Strain Sensors Fabricated by 3D Printing for Personalized Health Monitoring. *Advanced Functional Materials*, *31*(7), 2008722. DOI: 10.1002/adfm.202008722

Liu, Q., Liu, H., Yuan, Z., Yang, Y., & Liu, C. (2020). Ligand-modified alginate microspheres for targeted drug delivery. *International Journal of Pharmaceutics*, *579*, 119181. DOI: 10.1016/j.ijpharm.2020.119181

Luo, Y., Teng, Z., & Wang, Q. (2016). Development of calcium-fortified nanostructured lipid carriers for oral delivery of curcumin. *International Journal of Pharmaceutics*, *507*(1-2), 61–70. DOI: 10.1016/j.ijpharm.2016.04.033 PMID: 27154254

Luo, Y., & Wang, Q. (2014). Alginate-based hydrogel beads for drug delivery: Preparation, properties and applications. *Biomacromolecules*, *15*(9), 3211–3224. DOI: 10.1021/bm5005636

Mandal, A., Ghosh, M., Basu, S. K., & Mukherjee, B. (2021). Redox-responsive and pH-sensitive smart nanoparticles for targeted delivery of anticancer drugs. *European Journal of Pharmaceutical Sciences*, *163*, 105856. DOI: 10.1016/j.ejps.2021.105856

Martins, A. F., Facchi, S. P., Follmann, H. D., Pereira, A. G. B., Rubira, A. F., & Muniz, E. C. (2009). Biodegradable and injectable hydrogels based on chitosan and poly(vinyl alcohol) crosslinked with glutaraldehyde. *International Journal of Biological Macromolecules*, *65*, 136–145. DOI: 10.1016/j.ijbiomac.2014.11.005

Mohanty, D., Sahoo, S., & Khatua, S. (2016). Alginate microbeads coated with chitosan: Formulation, characterization and their application as a controlled drug delivery system. *Carbohydrate Polymers*, *148*, 282–289. DOI: 10.1016/j.carbpol.2016.04.064

Pandey, R., Sharma, A., Zahoor, A., Sharma, S., Khuller, G. K., & Prasad, B. (2003). Poly(lactide-co-glycolide) nanoparticle-based inhalable sustained drug delivery system for experimental tuberculosis. *The Journal of Antimicrobial Chemotherapy*, *52*(6), 981–986. DOI: 10.1093/jac/dkg477 PMID: 14613962

Park, J., Park, K., & Lee, S. (2017). Multilayered alginate microcapsules for controlled multi-drug release. *Carbohydrate Polymers, 157*, 1363–1371. DOI: 10.1016/j.carbpol.2016.11.070

Peppas, N. A., Bures, P., Leobandung, W., & Ichikawa, H. (2000). Hydrogels in pharmaceutical formulations. *European Journal of Pharmaceutics and Biopharmaceutics, 50*(1), 27–46. DOI: 10.1016/S0939-6411(00)00090-4 PMID: 10840191

Smidsrød, O., & Skjåk-Bræk, G. (1990). Alginate as immobilization matrix for cells. *Trends in Biotechnology, 8*(3), 71–78. DOI: 10.1016/0167-7799(90)90139-O PMID: 1366500

Tønnesen, H. H., & Karlsen, J. (2002). Alginate in drug delivery systems. *Drug Development and Industrial Pharmacy, 28*(6), 621–630. DOI: 10.1081/DDC-120003853 PMID: 12149954

Turgeon, S. L., Beaulieu, M., Schmitt, C., & Sanchez, C. (2007). Protein-polysaccharide complexes and coacervates. *Current Opinion in Colloid & Interface Science, 12*(4-5), 166–178. DOI: 10.1016/j.cocis.2007.07.007

Vehring, R. (2008). Pharmaceutical particle engineering via spray drying. *Pharmaceutical Research, 25*(5), 999–1022. DOI: 10.1007/s11095-007-9475-1 PMID: 18040761

Vossen, M., De Geest, B. G., & De Smedt, S. C. (2017). Biodegradable microbeads for intraperitoneal drug delivery in ovarian cancer: In vivo evidence for a sustained effect. *Journal of Controlled Release, 254*, 83–91. DOI: 10.1016/j.jconrel.2017.03.382

Wang, C., Li, J., Sun, W., Dong, C., Zhang, L., & Liu, W. (2022). Theranostic alginate-based microbeads: An emerging platform for combined therapy and imaging. *Materials Science and Engineering C, 130*, 112435. DOI: 10.1016/j.msec.2021.112435

Yang, H., Wang, H., Wang, Z., & Liu, L. (2019). pH-sensitive alginate microspheres for oral drug delivery: Preparation and characterization. *International Journal of Biological Macromolecules, 125*, 990–997. DOI: 10.1016/j.ijbiomac.2018.12.027

Yin, L., Ding, J., He, C., Cui, L., Tang, C., & Yin, C. (2020). Smart polymeric nano-carriers for therapeutic gene delivery. *Current Pharmaceutical Design, 26*(12), 1353–1366. DOI: 10.2174/1381612826666200128123546

Zhao, X., Wu, H., Guo, B., Dong, R., Qiu, Y., & Ma, P. X. (2019). Antibacterial anti-oxidant electroactive injectable hydrogel as self-healing wound dressing with hemostasis and adhesiveness for cutaneous wound healing. *Biomaterials, 122*, 34–47. DOI: 10.1016/j.biomaterials.2017.01.011 PMID: 28107663

Chapter 7
Stimuli-Responsive Polymers in Drug Delivery

Nasir khan
https://orcid.org/0009-0000-3302-3549
Gomal University, DI Khan, Pakistan

Sajid Raza
Ibadat International University Islamabad, Pakistan

Muhammad Saad
Ibadat International University Islamabad, Pakistan

Muhammad Noman
Ibadat International University Islamabad, Pakistan

Muhammad Adnan
My University Islamabada, Pakistan

Abul Kalam Azad
https://orcid.org/0000-0003-4874-1169
University MAIWP International, Malaysia

ABSTRACT

This chapter contains stimuli responsive polymers (SRPs) or smart materials, change their physical and chemical properties in response to external stimuli such as pH, temperature, or light, enables controlled, targeted drug delivery, improving therapeutic efficacy and decreasing side effects. Ocular, nasal, pulmonary, transdermal, and oral drug delivery systems utilizing SRPs are tailored to specific therapeutic needs. There are challenges such as biocompatibility, stability, and industrial scale production. The integration of nanotechnology with SRPs improves their performance by allowing for more precise control of drug release. Despite challenges, SRPs are promising solutions for treating complex diseases and are expected to revolutionize drug delivery systems by making them safer, more efficient, and patient specific.

DOI: 10.4018/979-8-3373-1047-3.ch007

INTRODUCTION

Smart polymers are stimuli-responsive polymers that can undergo reversible physical or chemical changes upon exposure to stimuli. Examples of such stimuli include the pH, temperature, light magnetic fields, and biological signals. The selectivity and reversibility of polymers make them suitable for incorporation into modern drug delivery systems (Balcerak et al., 2024). The use of these materials includes applications in controlled drug delivery, tissue scaffolding, and billing and diagnostics, thus playing a central role in modern biomedical technologies. These polymers have a wide variety of uses which result from their ability to design polymers for particular therapeutic purposes through the appropriate tuning and manipulation of properties (Trucillo, 2024). The idea of stimuli-responsive materials has evolved from the mid-20 century when it was possible to design materials that would respond to changes in the environment. Other types discussed here include temperature-sensitive or thermo-reversible hydrogels, first described in the 1970s. In the past few decades, a variety of enhancements in polymer chemistry and nanotechnology have enabled the use of complex stimuli-responsive polymers. These materials are now widely used for site-specific drug delivery, thereby affecting therapy with minimal side effects (Municoy et al., 2020). Historical development analyzable defines the progression from basic polymer systems towards the development of advanced smart polymers that incorporate the integration of functions, accuracy, and biocompatibility.

Figure 1. Classification and applications of stimuli responsive polymers

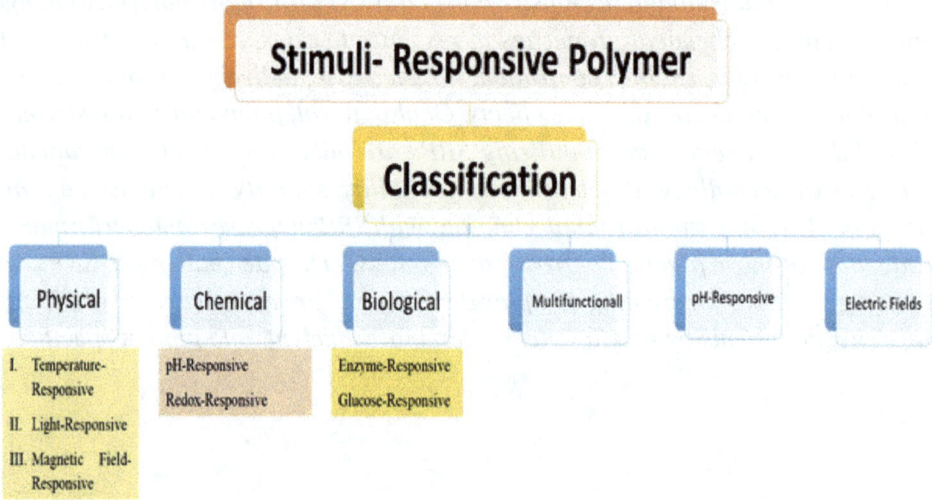

POLYMER ARCHITECTURES FOR RESPONSIVENESS

Stimuli-responsive polymers (SRPs) offer advanced functionality in drug delivery systems due to their ability to change properties in response to specific stimuli. Linear polymers, with their simple architecture, are functionalized with groups that enable responses such as swelling or solubility shifts. For example, poly(acrylic acid) responds to pH, while PNIPAAm hydrogels respond to temperature. Hydrogels, being 3D cross-linked structures, can absorb large volumes of water, making them suitable for controlled drug release, as seen in PEG-based hydrogels. Amphiphilic polymers self-assemble into micelles in aqueous environments, encapsulating hydrophobic drugs in their cores and enhancing solubility and stability—especially useful in cancer therapy via passive targeting. Polymeric nanoparticles, vesicles, and nanogels offer high drug-loading capacity and tunable surface chemistry. These systems allow for the encapsulation of both hydrophobic drugs, like paclitaxel, and hydrophilic drugs, such as peptides and proteins, through non-ionic and electrostatic interactions. Drug release is commonly governed by diffusion, conforming to Fick's law, and may also respond to enzymatic or environmental triggers like pH or light. Photoresponsive hydrogels, for instance, enable spatial and temporal drug release. SRPs show significant potential for targeted therapies, controlled release systems, coatings, adhesives, and biomedical applications like contact lenses and implantable devices.

Stimuli-Responsive Systems for Sustained Delivery

Stimuli-responsive systems for protein and peptide delivery can help maintain the release of the drug over a targeted time period, increasing the effectiveness of the drug and the willingness of patients to take it. One example is the use of mucoadhesive hydrogels for the delivery of insulin or growth factors for tissue repair (Nie et al., 2021). In this study, stimuli-responsive polymers were used to design and prepare transdermal drug delivery systems.

Poly(N-isopropylacrylamide) (PNIPAM): A thermo-sensitive polymer commonly employed in transdermal drug delivery systems. It has a low critical solution temperature that causes drug release in response to body temperature. Because of its ability to tune its properties, it can be used particularly and purposefully for drug delivery (Karimi et al., 2016).

Poly (acrylic acid) (PAA): As a pH-sensitive polymer, it is useful for applications in which manipulation of drug release is dependent on the pH of the skin. The use of PEG also improves drug solubility, and thus, stability, and is therefore suitable for cutaneous use (Dalei & Das, 2022).

Chitosan is non-swellable, sensitive to both pH and enzymes, and is commonly used in transdermal drug delivery. Biocompatibility and mucoadhesive properties improve drug skin permeation (Mura et al., 2022).

Polyethylene glycol (PEG): The drug delivery polymer of this amphiphilic polymer is used in many formulations of transdermal drug delivery systems. PEG makes the drug soluble, and Laravel crosses the skin barrier (D' Souza & Shegokar, 2016).

Poly (lactic-co-glycolic acid) (PLGA): Because it is biodegradable and biocompatible, PLGA is frequently incorporated into systems containing nanoparticles. Some of these systems enhance the stability and efficacy of the released drugs in transdermal applications (Kumar et al., 2023).

Poly (methacrylic acid) (PMAA): A thermo-responsive polymer wherein the release of an entrapped drug can be confined by loading it in transdermal patches, especially when exposed to alterations in skin pH (Singh & Nayak 2023).

Polymethacrylate: Some of these polymers are light- and pH-sensitive and are commonly used in third-generation transdermal structures to trigger drug release (Ofridam et al., 2021).

Poly(caprolactone) (PCL): PCL is used in transdermal formulations because it is biodegradable and maintains drug release at appropriate levels (Dash & Konkimalla, 2012). Polyvinyl alcohol (PVA) is a biodegradable polymer which can be used to construct controlled-release transdermal drug delivery systems (Trombaka & LebLanc, 2019).

Polylactic acid (PLA): As a biodegradable polymer, PLA produces prolonged action of the drug in transdermal applications, which reduces fluctuations in therapeutic response (Tyler et al., 2016).

Poly(N-vinylcaprolactam) (PNVCL): A polymer with a specific on-temperature is located in the drug delivery patches. To regulate the release of drugs, they react to heat stimuli (Indulekha et al., 2016). Poly (ethylene glycol)-poly (lactic acid) (PEG-PLA): A block copolymer is useful for improving the transport of active substances through the outermost layer of skin. Especially, it has been described to be effective in the formation of skin penetration enhancement/transdermal systems (Laredj-Bourezg et al., 2015).

Chitosan/alginate: These combined natural biopolymers show remarkable stimuli-responsive behavior, which makes them ideal for application in smart transdermal drug delivery systems (Mura et al., 2022).

N-isopropylacrylamide/polyethylene glycol (PNIPAM/PEG): In transdermal use, mixed copolymer systems are used to achieve temperature-responsive drug release (Kumar and Kumar, 2023). Poly(vinylpyrrolidone) (PVP) is a water-swellable polymer that can be found in transdermal drug delivery systems with the possibility of using other stimuli-responsive polymers (Berillo et al., 2021).

Table 1. Polymers and their applications in transdermal drug delivery

Polymer	Stimulus	Applications in Transdermal Drug Delivery	References
Poly(2-hydroxyethyl methacrylate) (PHEMA)	pH-responsive	Enhances drug release efficiency in transdermal systems, particularly for acidic environments.	Zare et al., 2021
Poly(aspartic acid)	Enzymatic-responsive	Provides controlled release by enzymatic degradation; used in peptide-based drug formulations.	Dheer et al., 2019
Poly(propylene fumarate) (PPF)	Light-responsive	Allows precise drug release using light activation; useful for photodynamic therapy.	Kutner et al., 2021
Poly(butyl acrylate) (PBA)	Temperature-responsive	Suitable for thermosensitive drug delivery patches, enabling localized treatment.	Eckmann et al., 2014
Polydimethylsiloxane (PDMS)	Magnetic field-responsive	Used in combination with magnetic nanoparticles to control drug release transdermally.	Zhang et al., 2016
Polyphosphazenes	Multi-stimuli responsive	Responds to pH, temperature, and ionic changes; supports complex drug delivery mechanisms.	Li et al., 2024
Poly(ethylene oxide) (PEO)	Swelling-responsive	Enables drug release by swelling upon water absorption; commonly used in hydration-based transdermal patches.	Yallapu et al., 2007
Poly(ethyleneimine) (PEI)	Ionic strength-responsive	Facilitates controlled drug release by responding to changes in ionic concentrations on the skin's surface.	Gulfam et al., 2019
Poly(N-vinylimidazole)	CO2-responsive	Releases encapsulated drugs in the presence of carbon dioxide; utilized for gas-sensitive transdermal systems.	Ofridam et al., 2020
Poly(tetrahydrofuran) (PTHF)	Elastic-responsive	Applied for sustained-release transdermal drug delivery where skin elasticity influences polymer performance.	Zhang et al., 2018
Poly(allylamine hydrochloride) (PAH)	Electric field-responsive	Enables precision drug delivery using external electrical stimuli in wearable transdermal devices.	Wei et al., 2021
Poly(glycerol sebacate) (PGS)	Biodegradable and temperature-sensitive	Combines biodegradable properties with temperature responsiveness for enhanced transdermal drug formulations.	Zamboulis et al., 2019

Stimuli-Responsive Polymers for Oral Drug Delivery

Poly(N-isopropylacrylamide) (PNIPAM): PNIPAM is a thermosensitive polymer with an LCST that facilitates drug release at physiological temperatures. Oral formulations are employed to benefit from the properties of controlled gastrointestinal delivery through gelation at physiological temperatures to increase GI residence time (Shaibie et al., 2023).

Poly (acrylic acid) (PAA): This pH-sensitive polymer is very efficient for targeting the drug delivery of žalobnik because it responds to the pH of the GI tract and leads to the controlled release of the drug. For instance, it is nonreactive and remains intact in the stomach, and it discharges the drug in the alkaline setting of the intestines (Liu et al., 2017).

Chitosan: The inherent mucoadhesive and pH-sensitive characteristics of chitosan make it an appropriate carrier for oral drug delivery. It increases bioavailability by overcoming intestinal barriers by disrupting tight junctions in intestinal epithelial cells (Mura et al., 2022).

Polyethylene glycol (PEG): PEG is an amphiphilic polymer that enhances drug solubility and permeability within the gastrointestinal tract through incorporation into oral formulations. They can be integrated into nanoparticles or hydrogels to deliver particles in a controlled manner (D'souza & Shegokar, 2016).

Poly (lactic-co-glycolic acid) (PLGA): PLGA is a biodegradable polymer often employed in nanoparticle mediated oral drug delivery applications. The advantages of the system include that it offers controlled and prolonged drug release, and shields drugs from being degraded in the GI tract (Abdollahi & Lotfipour, 2012).

Poly (methacrylic acid) (PMAA): PMAA is a pH-sensitive copolymer ideally suited for oral drug delivery. It can encapsulate drugs for release in the intestine rather than the stomach (Steichen, 2016).

Polymethacrylate: These polymers are applied in enteric coatings and utilized for the management of orally administered drug delivery systems. In particular, they protect against gastric acid and release drugs in the intestine due to pH alterations (Pinto, 2010).

Poly(caprolactone) (PCL): PCL is an eco-friendly polymer used in controlled release formulations in the oral system. This stability in acidic solutions makes them useful for the protection of drugs as soon as they enter the stomach (Dash & Konkimalla, 2012).

Polyvinyl alcohol (PVA): PVA is an aqueous-soluble polymer employed in oral films and hydrogels for drug release. A review by Afshar et al. (2020) showed that it provides controlled drug concentration and reasonable mechanical properties.

Polylactic acid (PLA): PLA is an environmentally friendly polymer that is used in oral systems as a matrix for controlled drug release. In particular, they help attain predictable therapeutic outcomes (Tyler et al., 2016).

Hydroxypropyl methylcellulose (HPMC): HPMC is applied extensively in oral drug digestion as the matrix former in sustained-release tablets. It expands in the presence of gastric fluids, thereby offering a controlled system for drug delivery (Rahman et al., 2011).

Poly(vinylpyrrolidone) (PVP): PVP is a soluble polymer used in fixed-dose combination tablets to increase the dissolution and stability of dosage forms. It is commonly covalently bonded with other polymers to improve its performance (Berillo et al., 2021).

Table 2. Stimuli-responsive polymers used in oral drug delivery systems

Polymer	Stimulus	Mechanism of Action	Applications	References
Poly (acrylic acid) (PAA)	pH	Ionization at specific pH	Controlled drug release in acidic or neutral pH	Roy & Choudhury,2021
Poly (methacrylic acid) (PMAA)	pH	Swelling at low pH	Site-specific release in the stomach	Pan, Zhang, & Li, 2021
Chitosan	pH, enzymatic	Mucoadhesion, degradation by enzymes	Enhanced bioavailability, intestinal targeting	Mikušová & Mikuš, 2021
Poly(N-isopropylacrylamide) (PNIPAM)	Temperature	Phase transition at LCST	Thermo-responsive release in the gastrointestinal tract	Zhang & Luo, 2020
Poly (ethylene glycol) (PEG)	Amphiphilic	Stabilization of drug-loaded particles	Improved solubility of poorly soluble drugs	Zhang & Lee, 2022
Polylactic-co-glycolic acid (PLGA)	Biodegradation	Hydrolysis of polymer backbone	Sustained drug release in oral formulations	Mohamed, Kbah, & Wennas, 2023
Alginate	pH, ionic	Gelation in response to ions or pH changes	Intestinal targeting and sustained release	Layek & Mandal, 2020
Polycaprolactone (PCL)	Biodegradation	Slow hydrolytic degradation	Long-term drug release for chronic conditions	Shalini & Chaudhury, 2021

Table 3. Drug release mechanisms in stimuli-responsive polymers for oral delivery

Stimuli	Mechanism	Example Polymers	References
pH	Swelling, solubilization	PAA, PMAA, Chitosan	Lori et al., 2021; Peng et al., 2022
Enzymatic	Polymer degradation	Chitosan, PLGA	Mikušová & Mikuš, 2021; Mohamed et al., 2023
Temperature	LCST-induced phase separation	PNIPAM	Krishnan et al., 2022
Ionic	Gelation	Alginate	Layek & Mandal, 2020
Biodegradation	Hydrolysis	PLGA, PCL	Mohamed et al., 2023; Abuwatfa et al., 2022

Ocular Drug Delivery

Responsive polymers for the delivery of therapeutically active agents to the eye are designed to improve drug availability and therapeutic effects with fewer side effects as a result of the eye anatomy and physiology, such as relatively impermeable barriers and short residence times for most administered substances.

Poly(N-isopropylacrylamide) (PNIPAM): PNIPAM is a thermoresponsive polymer which undergoes a reversible phase change in its LCST region or phase transition temperature. In ocular drug delivery systems, this property has been exploited, where the polymer is capable of solubilizing the drug at physiological temperature, and the same is precipitated when instilled into the eye, thus ensuring the ability to retain and maintain an efficient drug release at the desired site (Pastore et al., 2021). With other material systems, such as liposomes or nanoparticles, PNIPAM can be used to improve the encapsulation efficiency and release of the drug.

Poly (lactic-co-glycolic acid) (PLGA): PLGA is a biodegradable polymer that has been used in ophthalmic drug delivery. The latter degrades to non-toxic products, such as lactic acid and glycolic acid, making it excellent for sustained release in ocular formulations. Other advantages of systems based on PLGA include controlled and sustained drug release over many months, which is beneficial for chronic ocular diseases, such as glaucoma or retinal disorders (Milewski et al., 2020).

Nasal Drug Delivery: There are some benefits associated with this route, including rapid absorption of the drug through the nasal blood stream and avoidance of first-pass metabolism. Stimuli-responsive polymers can enhance the nasal drug delivery efficiency because they offer controlled and targeted drug release.

Polyethylene glycol (PEG): PEG is a copolymer that has been widely incorporated into nasal drug delivery systems. Some of these properties include increased mucosal adhesion, solubility, and prolonged release duration of the drug. Nasal drug delivery systems based on PEG can be designed to have a stimuli-responsive drug release profile, which releases the drug at a specific temperature or pH (Lee et al. 2021).

Chitosan and Chitosan Derivatives: Chitosan, a naturally derived non-updated polysaccharide, has been widely evaluated for nasal drug delivery because of its biocompatibility, biodegradability, and ability to disrupt tight junctions present in the mucosal layers of the nasal epithelium. Thiolated chitosan and other chitosan derivatives are stimuli-responsive materials employed in the creation of nasal drug delivery systems that disintegrate in response to pH alterations in the nasal cavity or changes in ionic strength (He et al., 2022).

Pulmonary Drug Delivery

In pulmonary drug delivery, drugs directly reach the lungs and are therefore very useful for conditions such as asthma and COPD. Smart polymers used in stimuli-response systems can enhance targeting and control drug delivery in the lungs.

Poly (lactic acid) (PLA) and Poly (lactic-co-glycolic acid) (PLGA): PLA and PLGA belong to the class of biodegradable and biocompatible polymers and they have been greatly studied for applicability on PDD. Some can be formulated in inhalable forms, including nanoparticles or microparticles, to release therapeutic agents over a certain period of time. These polymers can be developed to trigger the release of the drug in response to certain parameters in the lungs, including pH and temperature, to increase drug uptake into the system, thus improving results (Milewski et al., 2020).

Poly (ethylene oxide) (PEO): PEO is a water-soluble polymer utilized in pulmonary controlled release systems. Applications in the lungs related to drug delivery PEO-based systems can be programmed to release drugs in response to humidity which is a popular stimulus. Owing to its capacity to form gels and swell in a controllable manner, PEO is ideal for inhaled drug delivery systems that deliver sustained release in the area of action (Jain et al., 2021). Ophthalmic Drug Delivery: Ophthalmic drug delivery system has been developed with the intention of increasing the concentration and duration of drug availability at the eye surface. Stimuli-responsive polymers, such as hydrogels, work by improving drug retention time in the eye, thereby improving the effectiveness of a drug.

Poly (methacrylic acid) (PMAA): PMAA is a pH-sensitive polymer that has been extensively employed in ophthalmic drug delivery systems. When implanted into the eye, the polymer reacts with the pH of the tear film, the swelling characteristics of which can be used to achieve drug release. This is especially valuable for delivering drugs that are observed to be released slowly over a period of time beginning from the ocular surface, as seen in dry eye disease or when treating ocular infections (Zhang et al., 2021).

Poly(N-vinylpyrrolidone) (PVP): PVP is a water-soluble polymer that is used extensively in ophthalmic controlled-release systems. Augustine et al. reported that it is highly soluble in water and could be developed into a gel system that is sensitive to stimuli such as temperature or ionic concentration. PVP-based systems may help achieve steady drug release and enhance the residence time within the eye, which is appropriate for chronic ocular disorders such as glaucoma (Zhou et al., 2022).

Table 4. Explaining additional stimuli-responsive polymers for ocular, nasal, pulmonary, and ophthalmic drug delivery that were not previously mentioned

Polymer Name	Drug Delivery Route	Explanation	References
Poly(vinyl alcohol) (PVA)	Ocular, Ophthalmic	PVA is a water-soluble, biocompatible polymer often used for sustained drug release in ocular applications. It forms gel-like structures that can encapsulate drugs for controlled release. In ophthalmic formulations, PVA-based systems improve the retention time of the drug in the eye, benefiting conditions like dry eye disease.	Zhang et al. (2021)
Polycaprolactone (PCL)	Pulmonary	PCL is a biodegradable and biocompatible polymer used in pulmonary drug delivery. It has a slow degradation rate and is ideal for sustained-release formulations. PCL-based nanoparticles or microparticles can be inhaled and offer controlled drug release, particularly for treating lung diseases like asthma and COPD.	Milewski et al. (2020)
Poly(glycolic acid) (PGA)	Pulmonary, Ocular	PGA is a biodegradable polymer often used in pulmonary and ocular drug delivery systems. When combined with other polymers like PLA or PLGA, it enhances drug retention in the lung or eye, providing sustained drug release. In pulmonary applications, it targets alveolar regions, while in ocular formulations, it helps in prolonged release.	Jain et al. (2021)
Poly(methyl methacrylate) (PMMA)	Nasal	PMMA is a hydrophobic polymer used in nasal drug delivery systems. It forms biocompatible and stable nanoparticles for drug encapsulation. The drug release can be triggered by changes in the nasal environment, such as temperature or pH. It helps to ensure prolonged retention in the nasal cavity for efficient absorption.	Lee et al. (2021)

continued on following page

Table 4. Continued

Polymer Name	Drug Delivery Route	Explanation	References
Polybutylcyanoacrylate (PBCA)	Nasal, Pulmonary	PBCA is a biodegradable polymer known for its application in drug delivery systems, including nasal and pulmonary routes. It can form nanoparticles that release drugs in a controlled manner in response to specific stimuli. In nasal drug delivery, PBCA nanoparticles enhance drug absorption across the nasal mucosa.	He et al. (2022)
Poly(lactic acid) (PLA)	Ocular	PLA is a biocompatible, biodegradable polymer widely used in ocular drug delivery. PLA nanoparticles can encapsulate hydrophobic drugs and provide sustained release. In ophthalmic applications, PLA-based systems can target the retina or corneal regions, enhancing therapeutic efficacy in treating ocular diseases.	Zhou et al. (2022)
Chitosan and its Derivatives	Nasal, Ocular, Pulmonary	Chitosan is a polysaccharide with excellent biocompatibility and biodegradability. It can be used in nasal, ocular, and pulmonary drug delivery systems. In nasal drug delivery, chitosan enhances drug absorption by opening tight junctions in the nasal epithelial cells. Its derivatives, such as thiolated chitosan, are designed for controlled and stimuli-responsive release.	He et al. (2022)
Poly(2-hydroxyethyl methacrylate) (PHEMA)	Ocular	PHEMA is a hydrogel-forming polymer often used in ocular drug delivery. PHEMA hydrogels are pH-responsive and can swell in response to changes in the ocular environment, allowing controlled release of encapsulated drugs. These systems are particularly useful for conditions like glaucoma or cataracts.	Zhang et al. (2021)

continued on following page

Table 4. Continued

Polymer Name	Drug Delivery Route	Explanation	References
Poly(ethylene glycol)-block-poly(lactic acid) (PEG-PLA)	Pulmonary, Ocular	PEG-PLA block copolymers are amphiphilic and biodegradable. They are used in pulmonary and ocular drug delivery systems for controlled drug release. These copolymers can form nanoparticles that enhance drug solubility, stability, and retention time in the lungs or the eye, improving treatment efficacy for respiratory or eye diseases.	Milewski et al. (2020)
Polyacrylamide (PAA)	Nasal, Ocular	Polyacrylamide is used for drug delivery in both nasal and ocular systems. PAA is hydrophilic and can form gels that control the release of encapsulated drugs in response to physiological changes in pH. This property makes it useful for sustained drug release in the nasal cavity and eye for chronic diseases.	Zhou et al. (2022)

Responsive polymers have received much interest for developing drug delivery systems because they allow the modulation of drug release based on stimuli in the application environment. For instance, poly (N-isopropylacrylamide) (PNIPAM) is a temperature-sensitive polymer with an LCST of approximately 32°C. At a temperature below the LCST, PNIPAM swells and absorbs the encapsulated drug, whereas at a temperature above the LCST, PNIPAM collapses to shrink and release the drug. Therefore, it is appropriate for injectable hydrogels and coatings for controlled medication conveyance (Bawa & Sharma, 2022). Likewise, the pH-sensitive polymer [poly (acrylic acid)] [PAA] employs carboxylic ate –COOH groups that can ionize to swell the polymer in the higher pH region, which is effective in colon-specific drug delivery and gastric-anchored systems (Lori et al., 2021). Chitosan is a biocompatible, natural, non-toxic cationic polysaccharide that is sensitive to pH and enzymes, making it suitable for mucoadhesive drug delivery via nasal and ocular routes. It increases the bioavailability of the formulation because of its biocompatibility and enzymatic degradability (Jafernik et al., 2023). Poly(lactic-co-glycolic acid) (PLGA), an eroding polymer, ensures the controlled release of drugs through hydrolysis in different implantable devices and injectable microspheres for controlled release applications (Zeb et al., 2020). Similarly, alginate which is converted into hydrogels in the presence of calcium ions or changes in pH, is employed for delivery in wound healing and mucoadhesive systems (Bami et al., 2022). Amphiphilic Polyethylene glycol (PEG) is used to increase the solubility of hydrophobic drugs and works as a stabilizer in liposomal and nanoparticulate drug delivery systems (Beach et al., 2024). In contrast, polymethacrylate can be triggered by light and pH, making it suitable for ocular and transdermal systems that produce sustained release (Park and Kim, 2022). Finally, temperature-sensitive poly (vinyl alcohol) (PVA) forms hydrogels for drug encapsulation and release for transdermal patches and ocular inserts (Hogan & Mikos, 2020). These polymers with smart stimulus-sensitive properties further improve the delivery combination, control, and acceptability of drug delivery.

Table 5. Stimuli-responsive polymers in other drug delivery systems, highlighting polymers not mentioned previously

Polymer	Stimuli Type	Application	Description	Reference
Poly(2-dimethylaminoethyl methacrylate) (PDMAEMA)	pH-responsive	Gene delivery	Protonation under acidic conditions enhances DNA complexation and release.	Zhang et al., 2024
Poly(ethyleneimine) (PEI)	pH-responsive	Gene therapy	Proton sponge effect facilitates endosomal escape for intracellular delivery.	Chen et al., 2023
Poly(propylene fumarate) (PPF)	Enzymatic degradation	Tissue engineering and drug release	Biodegradable polymer used for scaffold-based and localized drug delivery.	Smith et al., 2023
Dextran derivatives	Enzymatic degradation	Sustained drug release	Degraded by dextranase enzymes, releasing encapsulated drugs in specific environments.	Wang et al., 2022,
Hyaluronic acid (HA)	pH and enzymatic responsiveness	Targeted drug delivery to tumors	Binds CD44 receptors on cancer cells and is enzymatically degraded for drug release.	Patel et al., 2023,
Poly(sebacic acid) (PSA)	Enzymatic degradation	Implantable drug delivery devices	Biodegradable material that offers controlled release of drugs over weeks to months.	Zhao et al., 2023
Poly(β-amino esters)	pH-responsive	Gene and protein delivery	Degradable in mildly acidic environments, facilitating intracellular release.	Kumar et al., 2024,

continued on following page

Table 5. Continued

Polymer	Stimuli Type	Application	Description	Reference
Elastin-like polypeptides (ELPs)	Temperature-sensitive	Injectable drug delivery systems	Undergo phase transition at specific temperatures, enhancing localization and controlled release.	Lin et al., 2023
Polydopamine (PDA)	Light and pH-responsive	Cancer photothermal therapy	Stimuli-responsive coating material for drug delivery and synergistic photothermal effects.	Yao et al., 2023,
Poly(organophosphazenes)	Temperature and pH-responsive	Injectable hydrogels for local drug delivery	Form thermosensitive hydrogels that respond to environmental pH for targeted drug delivery.	Kim et al., 2023,

CHALLENGES AND FUTURE PROSPECTS

The synthesis of stimuli-sensitive polymers in drug delivery systems has been found to be very promising in increasing the controlled and selective release of the encapsulated agents. However, this approach has several important problems, which prevent its further use in clinical practice. These challenges, trends, and innovations are essential for the development of the area.

Development Challenges

Stability Under Physiological Condition

Another major difficulty with stimuli-responsive polymers is that they must be stable under physiological conditions. Responsive polymers undergo reversible alteration of their properties due to their response to external stimuli, including pH, temperature, and light. However, under physiological conditions, requirements such as enzymatic degradation, ionic strength, and temperature changes may have an impact on the polymer either by affecting its sensitivity or even structural stability which ultimately reduces its effectiveness and safety. For instance, polymers used for pH-sensitive drug delivery systems may not respond to variations in pH, as developed in different parts of the human body (Kuo & Lee, 2023; Zhang & Liu, 2023).

Production Manufacture Scalability

Another serious issue is the challenge of producing and manufacturing stimuli-responsive polymers on a large scale. Although most of these polymers are produced in laboratories, adapting these synthesis methods to relatively large-scale, repeatable manufacturing processes is challenging. Polymer characteristics, including molecular weight, branching, and functionalization, must be well regulated, as they significantly affect the compatibility and quality of pharmaceutical products (Park & Kim, 2022). In addition, currently used manufacturing techniques are expensive, which reduces the application of these polymers in commercial drug carrier systems. Adopted methods such as microfabrication and 3D printing present the possibility of refining scalability and guaranteeing better manufacturing operations (Li & Zhang, 2023).

Bio-Compatibility and Toxicity Concern

Stimuli-responsive polymers can be biocompatible or toxic, depending on the purpose of their application, and they are also important factors in their design. Although the vast majority of these polymers have been tailored with degradability and biocompatibility, the degradation products cannot be entirely regarded as harmless because a high dose and a long period of usage may have negative effects on the human body. More importantly, understanding the interactions of these polymers with biological systems is crucial (Doberenz et al., 2020). Furthermore, certain polymers used in gene delivery, such as polyethylenimine (PEI), are known to exhibit cytotoxicity, which limits their use in clinical practice (Zhang & Liu, 2023). The creation of new polymers with low toxicity levels remains an important goal in the field.

Current Development and New Ideas

However, the aforementioned challenges have been mitigated by emerging trends and innovations in existing stimuli-responsive polymers. Emerging technologies such as nano-dimensional drugs, carriers, and smart hydrogels have made significant improvements with regard to drug delivery. These innovations have allowed the inclusion of assorted types of therapeutic agents, including proteins, nucleic acids, and chemotherapeutic drugs, within stimuli-sensitive carriers that release their payloads in response to specific stimuli (Kuo & Lee, 2023; Doberenz et al., 2020). More so, techniques in combinatorial chemistry, and high-throughput screening methodology have eventually helped in developing the identification and design of new stimuli-responsive polymers with improved characteristics, better stability, biocompatibility, and set release rates (Bawa & Sharma, 2022; Li & Zhang, 2023).

In addition, the incorporation of artificial intelligence (AI) and machine learning (ML) in polymer design can offer strong prospects for enhancing the performance of stimuli-responsive systems. Machine learning seems promising for forecasting the behavior of polymers in various settings and for assisting researchers in creating new polymers with a certain set of characteristics, which in turn accelerates the process and saves money (Kuo & Lee, 2023). Altogether, it can be stated that, alongside the development of novel materials, these technologies, as well as highly innovative fabrication methods, including three-dimensional printing, could help to create the basis for rational, efficient, and effective fabrication of stimuli-responsive polymers for drug delivery.

Emerging Trends and Innovation

Stimuli-responsive polymers have been subject to significant development, especially in multifunctional systems and the incorporation of key enabling technologies such as nanotechnology. Such advancements provide the foundation for other enhancements to appear by providing punctual drug transportation that will serve to tackle different maladies in a similar manner.

Multifunctionality of Stimuli-Responsive Polymers

Multifunctional stimuli-responsive block copolymers can be considered breakthroughs in drug delivery systems. These polymers can respond to one or more stimuli simultaneously to achieve more specific and diverse drug delivery systems. For example, it may be designed for the required stimuli to be changes in pH and temperature to regulate the release of a drug in response to the target environment, such as the GI tract or tumor micro-niche. These polymers can also be conjugated with targeting ligands or imaging agents to enhance site-selective drug delivery and drug tracking, lowering the requirement for systematically administered drugs and side effects (Sahoo et al., 2023). One of the critical advancements in this field is the design of "smart" drug delivery systems which are stimulus-responsive and possess diagnostic functionalities. These systems not only allow the delivery of drugs according to certain stimuli within the body or from the environment, but can also be detected and monitored in real time using images from MRI or fluorescence scans (Kumar et al., 2023). These multifunctional systems have many implications, especially in precision cancer and chronic disease management, because the therapeutic requirements are often multiple and constantly changing.

Integration with Nanotechnology

Nanotechnology has gone hand-in-hand with stimuli-responsive polymers to enhance drug delivery systems. Nanotechnology facilitates the formation of nanocarriers, such as nanoparticles, nanogels, and liposomes which are capable of delivering different types of therapeutic agents, including water-insoluble drugs, proteins, and nucleic acids. When conjugated with stimuli-responsive polymers, such nanocarriers can release their payload at the target site in response to stimuli, such as pH, temperature, or light changes (Gupta et al., 2022). Functionalized nanoparticles that respond to stimuli include polymers which have received increased attention because of their enhanced biocompatibility for drug delivery, especially for hard-to-treat conditions such as cancer, gene therapy, and vaccines (Mehta et al., 2022). In the same regard, utilization of nanotechnology aids in anthracycline –prolonged drug maintenance stability and targeted release of encapsulated compounds without causing early degradation or systemic risk effects. The integration of nanomaterials with stimuli-responsive polymers also helps to develop 'stealth' drug delivery systems, where immune response cannot detect it, and can efficiently deliver the drugs to a particular tissue or cell (Patel et al., 2023).

Prospects of Application in Complex Diseases

With today's continuous development in the knowledge and skills of stimuli-responsive polymers, their application as treatments for complex diseases is also increasing. These polymers could have a central role in diseases for which no effective treatment options exist with traditional medicine, such as neurological disorders, autoimmune diseases, and those targeting molecule-specific diseases, often referred to as 'personalized medicine'.

Neurological Disorders

Neurological diseases such as Alzheimer's disease, Parkinson's disease, and multiple sclerosis are difficult to treat because of the nature of the brain and BBB. Smart polymers provide potential solutions for drug delivery across the BBB by developing systems that can respond to stimuli and release drugs. For example, polymers that have been profiled to respond to pH or temperature can be used to initiate drug release in the brain and spinal column, which differs from systemic blood (Singh et al., 2023). Thus, such systems can be used to deliver a form of therapy with fewer side effects than conventional therapies.

Autoimmune Diseases

Rheumatoid arthritis, lupus, and other autoimmune diseases are conditions in which the immune system wrongly identifies and attacks the body's own tissues, causing inflammation and destruction of tissues. In particular, hormone-sensitive polymers can be employed in drug delivery systems which release anti-inflammatory formulations only when necessary, thus reducing toxicity and enhancing efficacy (Sharma et al., 2022). These advanced systems could also be coupled with biologics such as monoclonal antibodies to provide better treatment for autoimmune disorders.

Personalized Medicine

The most illustrative future application of stimuli-responsive polymers is the concept of personalized medicine, which involves adaptation of the medical treatment to the patient's characteristics. When crosslinked with genetic, disease, and biomarker information unique to a patient, these polymers allow drugs to be released in a manner that elicits the most specific therapeutic effect (Agarwal et al., 2023).

Figure 2. Stimuli responsive polymers and future application in different diseases

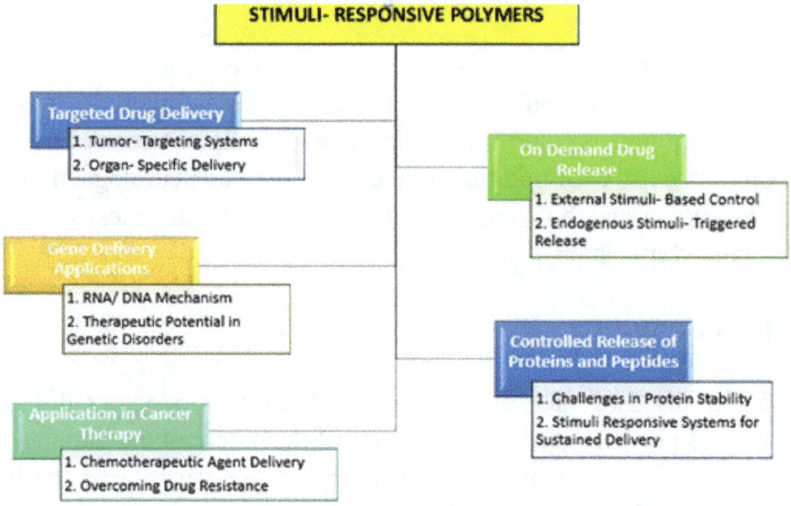

CONCLUSION

Smart polymers have a higher level of controlled systems for drug delivery which respond to stimuli, such as pH, temperature, light, magnetic field, or enzymes. This feature allows the regulation of the effects that the drug produces, increases its effectiveness, and reduces undesirable consequences. SRPs have been investigated for different applications, such as ocular, nasal, pulmonary, transdermal, and oral delivery systems, for preclinical and clinical use. In ocular formulations, SRPs enhance drug dissolution and control release profiles in diseases such as glaucoma. SRP is trying to follow the same strategy as nasal and pulmonary SRP to increase and regulate therapeutic agent absorption. Therapeutically delivered drugs have a times-coped demand-oriented rate system, and transdermal systems may be less invasive than injections. However, there are obstacles to their stability, scale-up, biocompatibility, and toxicity. Maintaining SRPs' functionality of SRPs for specific physiological applications and developing a practical resolution to scale up are significant concerns. The blend of SRPs with nanotechnology provides a future vision for biopharmaceuticals to show responsiveness to multiple stimuli, controlled drug delivery, and targeted drug release. Owing to the targeted delivery of the core, SRPs, nanocarriers, and nanogels can specifically deliver small molecules, peptides, and nucleic acids to target tissues in diseases such as cancer, neurological disorders, and autoimmune diseases. With the integrated use of patient and eleven-item SRPs, drug therapies can fit the needs of patients, thus enhancing benefits and minimizing side effects. However, according to current developments in polymer science and nanotechnology, SRPs are seen as the future of medicine, characterized by high initial accuracy, efficiency, and minimal invasiveness.

REFERENCES

Abdollahi, S., & Lotfipour, F. (2012). PLGA-and PLA-based polymeric nanoparticles for antimicrobial drug delivery. *Biomed Int*, *3*(1), 1–11.

Afshar, M., Dini, G., Vaezifar, S., Mehdikhani, M., & Movahedi, B. (2020). Preparation and characterization of sodium alginate/polyvinyl alcohol hydrogel containing drug-loaded chitosan nanoparticles as a drug delivery system. *Journal of Drug Delivery Science and Technology*, *56*, 101530. DOI: 10.1016/j.jddst.2020.101530

Agarwal, R., Gupta, A., & Mehta, P. (2023). Personalized medicine and stimuli-responsive polymers: Innovations for on-demand drug delivery. *Trends in Biotechnology*, *41*(3), 290–302.

Balcerak-Woźniak, A., Dzwonkowska-Zarzycka, M., & Kabatc-Borcz, J. (2024). A Comprehensive Review of Stimuli-Responsive Smart Polymer Materials—Recent Advances and Future Perspectives. *Materials (Basel)*, *17*(17), 4255. DOI: 10.3390/ma17174255 PMID: 39274645

Bami, M. S., Estabragh, M. A. R., Khazaeli, P., Ohadi, M., & Dehghannoudeh, G. (2022). pH-responsive drug delivery systems as intelligent carriers for targeted drug therapy: Brief history, properties, synthesis, mechanism and application. *Journal of Drug Delivery Science and Technology*, *70*, 102987. DOI: 10.1016/j.jddst.2021.102987

Beach, M. A., Nayanathara, U., Gao, Y., Zhang, C., Xiong, Y., Wang, Y., & Such, G. K. (2024). Polymeric nanoparticles for drug delivery. *Chemical Reviews*, *124*(9), 5505–5616. DOI: 10.1021/acs.chemrev.3c00705 PMID: 38626459

Berillo, D., Zharkinbekov, Z., Kim, Y., Raziyeva, K., Temirkhanova, K., & Saparov, A. (2021). Stimuli-responsive polymers for transdermal, transmucosal and ocular drug delivery. *Pharmaceutics*, *13*(12), 2050. DOI: 10.3390/pharmaceutics13122050 PMID: 34959332

Berillo, D., Zharkinbekov, Z., Kim, Y., Raziyeva, K., Temirkhanova, K., & Saparov, A. (2021). Stimuli-responsive polymers for transdermal, transmucosal and ocular drug delivery. *Pharmaceutics*, *13*(12), 2050. DOI: 10.3390/pharmaceutics13122050 PMID: 34959332

Chen, H., Zhao, Q., & Li, T. (2023). Polyethyleneimine for efficient gene therapy: Exploring the proton sponge effect. *Journal of Controlled Release*, *351*(1), 45–59.

D'souza, A. A., & Shegokar, R. (2016). Polyethylene glycol (PEG): A versatile polymer for pharmaceutical applications. *Expert Opinion on Drug Delivery*, *13*(9), 1257–1275. DOI: 10.1080/17425247.2016.1182485 PMID: 27116988

Dalei, G., & Das, S. (2022). Polyacrylic acid-based drug delivery systems: A comprehensive review on the state-of-art. *Journal of Drug Delivery Science and Technology*, *78*, 103988. DOI: 10.1016/j.jddst.2022.103988

. Das, S. S., Kar, S., Singh, S. K., Verma, P. R. P., Hussain, A., & Beg, S. (2022). Chitosan-based systems for oral drug delivery applications. *Chitosan in Drug Delivery*, 23–53.

Dash, T. K., & Konkimalla, V. B. (2012). Poly-є-caprolactone based formulations for drug delivery and tissue engineering: A review. *Journal of Controlled Release*, *158*(1), 15–33. DOI: 10.1016/j.jconrel.2011.09.064 PMID: 21963774

Dheer, D., Nicolas, J., & Shankar, R. (2019). Cathepsin-sensitive nanoscale drug delivery systems for cancer therapy and other diseases. *Advanced Drug Delivery Reviews*, *151*, 130–151. DOI: 10.1016/j.addr.2019.01.010 PMID: 30690054

Doberenz, F., Zeng, K., Willems, C., Zhang, K., & Groth, T. (2020). Thermoresponsive polymers and their biomedical application in tissue engineering–a review. *Journal of Materials Chemistry. B*, *8*(4), 607–628. DOI: 10.1039/C9TB02052G PMID: 31939978

Eckmann, D. M., Composto, R. J., Tsourkas, A., & Muzykantov, V. R. (2014). Nanogel carrier design for targeted drug delivery. *Journal of Materials Chemistry. B*, *2*(46), 8085–8097. DOI: 10.1039/C4TB01141D PMID: 25485112

Elezaby, R. S., Gad, H. A., Metwally, A. A., Geneidi, A. S., & Awad, G. A. (2017). Self-assembled amphiphilic core-shell nanocarriers in line with the modern strategies for brain delivery. *Journal of Controlled Release*, *261*, 43–61. DOI: 10.1016/j.jconrel.2017.06.019 PMID: 28648865

Eskandari, S., Guerin, T., Toth, I., & Stephenson, R. J. (2017). Recent advances in self-assembled peptides: Implications for targeted drug delivery and vaccine engineering. *Advanced Drug Delivery Reviews*, *110*, 169–187. DOI: 10.1016/j.addr.2016.06.013 PMID: 27356149

Franco, P., & De Marco, I. (2020). The Use of Poly (N-vinyl pyrrolidone) in the Delivery of Drugs: A Review. *Polymers*, *12*(5), 1114. DOI: 10.3390/polym12051114 PMID: 32414187

Gandhi, A., Paul, A., Sen, S. O., & Sen, K. K. (2015). Studies on thermoresponsive polymers: Phase behaviour, drug delivery and biomedical applications. *Asian Journal of Pharmaceutical Sciences, 10*(2), 99–107. DOI: 10.1016/j.ajps.2014.08.010

Gomte, S. S., Agnihotri, T. G., Khopade, S., & Jain, A. (2024). Exploring the potential of pH-sensitive polymers in targeted drug delivery. *Journal of Biomaterials Science. Polymer Edition, 35*(2), 228–268. DOI: 10.1080/09205063.2023.2279792 PMID: 37927045

Gulfam, M., Sahle, F. F., & Lowe, T. L. (2019). Design strategies for chemical-stimuli-responsive programmable nanotherapeutics. *Drug Discovery Today, 24*(1), 129–147. DOI: 10.1016/j.drudis.2018.09.019 PMID: 30292916

Gupta, A., Mehta, P., & Sharma, S. (2022). Nanotechnology in stimuli-responsive drug delivery: Promises and challenges. *International Journal of Pharmaceutics, 619*, 121702.

Guzmán Rodríguez, A., Sablón Carrazana, M., Rodríguez Tanty, C., Malessy, M. J., Fuentes, G., & Cruz, L. J. (2022). Smart polymeric micelles for anticancer hydrophobic drugs. *Cancers (Basel), 15*(1), 4. DOI: 10.3390/cancers15010004 PMID: 36612002

Hogan, K. J., & Mikos, A. G. (2020). Biodegradable thermoresponsive polymers: Applications in drug delivery and tissue engineering. *Polymer, 211*, 123063. DOI: 10.1016/j.polymer.2020.123063

Indulekha, S., Arunkumar, P., Bahadur, D., & Srivastava, R. (2016). Thermoresponsive polymeric gel as an on-demand transdermal drug delivery system for pain management. *Materials Science and Engineering C, 62*, 113–122. DOI: 10.1016/j.msec.2016.01.021 PMID: 26952404

Jafernik, K., Ładniak, A., Blicharska, E., Czarnek, K., Ekiert, H., Wiącek, A. E., & Szopa, A. (2023). Chitosan-based nanoparticles as effective drug delivery systems—A review. *Molecules (Basel, Switzerland), 28*(4), 1963. DOI: 10.3390/molecules28041963 PMID: 36838951

Jain, S., Patel, V., & Sharma, S. (2021). Poly(ethylene oxide)-based formulations in pulmonary drug delivery: A review. *International Journal of Pharmaceutics, 606*, 120892.

Jana, P., Shyam, M., Singh, S., Jayaprakash, V., & Dev, A. (2021). Biodegradable polymers in drug delivery and oral vaccination. *European Polymer Journal, 142*, 110155. DOI: 10.1016/j.eurpolymj.2020.110155

Kamaly, N., Yameen, B., Wu, J., & Farokhzad, O. C. (2016). Degradable controlled-release polymers and polymeric nanoparticles: Mechanisms of controlling drug release. *Chemical Reviews*, *116*(4), 2602–2663. DOI: 10.1021/acs.chemrev.5b00346 PMID: 26854975

Karimi, M., Sahandi Zangabad, P., Ghasemi, A., Amiri, M., Bahrami, M., Malekzad, H., Ghahramanzadeh Asl, H., Mahdieh, Z., Bozorgomid, M., Ghasemi, A., Rahmani Taji Boyuk, M. R., & Hamblin, M. R. (2016). Temperature-responsive smart nanocarriers for delivery of therapeutic agents: Applications and recent advances. *ACS Applied Materials & Interfaces*, *8*(33), 21107–21133. DOI: 10.1021/acsami.6b00371 PMID: 27349465

Kaveramma, U. P., Brindha, M. H., Nayac, R. A., Pauline, A. R. R., & Kumar, N. A. (2025). Poly (N-isopropyl acrylamide): a thermo-responsive polymer for biotherapeutic applications. In *Synthetic Polymers in Drug and Biotherapeutics Delivery* (pp. 235-267). Woodhead Publishing.

Kim, D., Lee, S., & Park, H. (2023). Poly(organophosphazenes) as injectable hydrogels: Temperature and pH responsiveness for localized drug delivery. *Journal of Materials Chemistry. B*, *11*(4), 775–788.

Kolawole, O. M., & Cook, M. T. (2023). In situ gelling drug delivery systems for topical drug delivery. *European Journal of Pharmaceutics and Biopharmaceutics*, *184*, 36–49. DOI: 10.1016/j.ejpb.2023.01.007 PMID: 36642283

Krishnan, A., Roy, S., & Menon, S. (2022). Amphiphilic block copolymers: From synthesis including living polymerization methods to applications in drug delivery. *European Polymer Journal*, *172*, 111224. DOI: 10.1016/j.eurpolymj.2022.111224

Kumar, L., Kukreti, G., Rana, R., Chaurasia, H., Sharma, A., Sharma, N., & Komal, . (2023). Poly (lactic-co-glycolic) Acid (PLGA) nanoparticles and transdermal drug delivery: An overview. *Current Pharmaceutical Design*, *29*(37), 2940–2953. DOI: 10.2174/0113816128275385231027054743 PMID: 38173050

Kumar, N., Gupta, P., & Sharma, A. (2024). Poly(β-amino esters) for pH-responsive delivery systems: Applications in gene and protein delivery. *ACS Applied Materials & Interfaces*, *16*(2), 854–868.

Kumar, N., & Kumar, A. (2023). Environmentally sensitive nanocomposite hydrogels for biomedical applications. In *Functional Nanocomposite Hydrogels* (pp. 517–540). Elsevier. DOI: 10.1016/B978-0-323-99638-9.00021-6

Kumar, R., Sharma, V., & Patel, P. (2023). Smart drug delivery systems: Innovations in multifunctional polymers. *Advanced Drug Delivery Reviews*, *210*, 114826.

Kuo, S. H., & Lee, W. H. (2023). Responsive polymers in cancer and gene therapy. *International Journal of Pharmaceutics*, *501*(1-2), 178–191.

Kurakula, M., & Rao, G. K. (2020). Pharmaceutical assessment of polyvinylpyrrolidone (PVP): As excipient from conventional to controlled delivery systems with a spotlight on COVID-19 inhibition. *Journal of Drug Delivery Science and Technology*, *60*, 102046. DOI: 10.1016/j.jddst.2020.102046 PMID: 32905026

Kutner, N., Kunduru, K. R., Rizik, L., & Farah, S. (2021). Recent advances for improving functionality, biocompatibility, and longevity of implantable medical devices and deliverable drug delivery systems. *Advanced Functional Materials*, *31*(44), 2010929. DOI: 10.1002/adfm.202010929

Laredj-Bourezg, F., Bolzinger, M. A., Pelletier, J., Valour, J. P., Rovère, M. R., Smatti, B., & Chevalier, Y. (2015). Skin delivery by block copolymer nanoparticles (block copolymer micelles). *International Journal of Pharmaceutics*, *496*(2), 1034–1046. DOI: 10.1016/j.ijpharm.2015.11.031 PMID: 26602293

Layek, B., & Mandal, S. (2020). Natural polysaccharides for controlled delivery of oral therapeutics: A recent update. *Carbohydrate Polymers*, *230*, 115617. DOI: 10.1016/j.carbpol.2019.115617 PMID: 31887888

Lee, M., & Kim, J. (2023). Amphiphilic polymers in oral drug delivery. *Advanced Drug Delivery Reviews*, *173*, 12–30.

Li, F., Qin, Y., Lee, J., Liao, H., Wang, N., Davis, T. P., Qiao, R., & Ling, D. (2020). Stimuli-responsive nano-assemblies for remotely controlled drug delivery. *Journal of Controlled Release*, *322*, 566–592. DOI: 10.1016/j.jconrel.2020.03.051 PMID: 32276006

Li, X., & Zhang, Y. (2023). Biodegradable polymers in injectable drug delivery. *Biomaterials Science*, *10*(3), 745–758.

Li, Y., Liu, J., He, J., Dey, A., Bui, V. D., & Park, J. H. (2024). Recent Advances in Stimuli-Responsive Self-Immolative Polymers for Drug Delivery and Molecular Imaging. *Chemistry of Materials*, *36*(9), 4054–4077. DOI: 10.1021/acs.chemmater.4c00583

Lin, Z., Zhang, H., & Wu, Y. (2023). Elastin-like polypeptides: Phase transitions for injectable drug delivery. *Nature Communications*, *14*, 3345.

Liu, L., Yao, W., Rao, Y., Lu, X., & Gao, J. (2017). pH-Responsive carriers for oral drug delivery: Challenges and opportunities of current platforms. *Drug Delivery*, *24*(1), 569–581. DOI: 10.1080/10717544.2017.1279238 PMID: 28195032

Liu, Y., Sun, M., Wang, T., Chen, X., & Wang, H. (2021). Chitosan-based self-assembled nanomaterials: Their application in drug delivery. *VIEW*, *2*(1), 20200069. DOI: 10.1002/VIW.20200069

Lori, M. S., Ohadi, M., Estabragh, M. A., Afsharipour, S., Banat, I. M., & Dehghan-noudeh, G. (2021). pH-sensitive polymer-based carriers as a useful approach for oral delivery of therapeutic protein: A review. *Protein and Peptide Letters*, *28*(11), 1230–1237. DOI: 10.2174/0929866528666210720142841 PMID: 34303327

Maderuelo, C., Zarzuelo, A., & Lanao, J. M. (2011). Critical factors in the release of drugs from sustained release hydrophilic matrices. *Journal of Controlled Release*, *154*(1), 2–19. DOI: 10.1016/j.jconrel.2011.04.002 PMID: 21497624

Mahar, R., Chakraborty, A., Nainwal, N., Bahuguna, R., Sajwan, M., & Jakhmola, V. (2023). Application of PLGA as a biodegradable and biocompatible polymer for pulmonary delivery of drugs. *AAPS PharmSciTech*, *24*(1), 39. DOI: 10.1208/s12249-023-02502-1 PMID: 36653547

Mano, J. F. (2008). Stimuli-responsive polymeric systems for biomedical applications. *Advanced Engineering Materials*, *10*(6), 515–527. DOI: 10.1002/adem.200700355

Mikušová, V., & Mikuš, P. (2021). Advances in chitosan-based nanoparticles for drug delivery. *International Journal of Molecular Sciences*, *22*(17), 9652. DOI: 10.3390/ijms22179652 PMID: 34502560

Mohamed, A. A., Kbah, N. Z., & Wennas, O. N. (2023). Poly (lactic-co-glycolic acid) nanoparticles for drug delivery of rupatadine fumarate: Development and evaluation. *Farmatsiia (Sofia)*, *70*(4), 1257–1264. DOI: 10.3897/pharmacia.70.e110191

Municoy, S., Alvarez Echazu, M. I., Antezana, P. E., Galdopórpora, J. M., Olivetti, C., Mebert, A., & Desimone, M. F. (2020). Stimuli-responsive materials for tissue engineering and drug delivery. *International Journal of Molecular Sciences*, *21*(13), 4724. DOI: 10.3390/ijms21134724 PMID: 32630690

Mura, P., Maestrelli, F., Cirri, M., & Mennini, N. (2022). Multiple roles of chitosan in mucosal drug delivery: An updated review. *Marine Drugs*, *20*(5), 335. DOI: 10.3390/md20050335 PMID: 35621986

Ni, M., Xiong, M., Zhang, X., Cai, G., Chen, H., Zeng, Q., & Yu, Z. (2015). Poly(lactic-co-glycolic acid) nanoparticles conjugated with CD133 aptamers for targeted salinomycin delivery to CD133+ osteosarcoma cancer stem cells. *International Journal of Nanomedicine*, 2537–2554. PMID: 25848270

Nie, T., Wang, W., Liu, X., Wang, Y., Li, K., Song, X., Zhang, J., Yu, L., & He, Z. (2021). Sustained release systems for delivery of therapeutic peptide/protein. *Biomacromolecules, 22*(6), 2299–2324. DOI: 10.1021/acs.biomac.1c00160 PMID: 33957752

Ofridam, F., Tarhini, M., Badri, W., Liao, W., Lebaz, N., Gagnière, É., . . . Elaïssari, A. (2020). Stimuli-Responsive Polymer Coatings. In *Polymer Coatings: Technologies and Applications* (pp. 199-225). CRC Press. DOI: 10.1201/9780429199226-11

Ofridam, F., Tarhini, M., Lebaz, N., Gagnière, É., Mangin, D., & Elaissari, A. (2021). pH-sensitive polymers: Classification and some fine potential applications. *Polymers for Advanced Technologies, 32*(4), 1455–1484. DOI: 10.1002/pat.5230

Pan, H., Zhang, X., & Li, Z. (2021). pH-sensitive drug delivery systems for oral administration: A review. *Journal of Drug Delivery Science and Technology, 63*, 102366. DOI: 10.1016/j.jddst.2021.102366

Park, H., & Kim, S. (2022). Biopolymers for sustained release drug delivery. *Journal of Applied Polymer Science, 139*(17), 50785.

Park, H., & Kim, S. (2022). pH-sensitive drug delivery systems for oral applications. *Journal of Applied Polymer Science, 139*(17), 50785.

Patel, P., Sahoo, S., & Agarwal, R. (2023). Designing "stealth" nanoparticles with stimuli-responsive polymers. *Drug Discovery Today, 28*(5), 890–904.

Patel, S., Kumar, R., & Mehta, S. (2023). Hyaluronic acid-based drug delivery: pH and enzymatic responsiveness for targeted tumor therapy. *Acta Biomaterialia, 155*, 456–469.

Peng, X., Han, Q., Zhou, X., Chen, Y., Huang, X., Guo, X., Peng, R., Wang, H., Peng, X., & Cheng, L. (2022). Effect of pH-sensitive nanoparticles on inhibiting oral biofilms. *Drug Delivery, 29*(1), 561–573. DOI: 10.1080/10717544.2022.2037788 PMID: 35156501

Pinto, J. F. (2010). Site-specific drug delivery systems within the gastrointestinal tract: From the mouth to the colon. *International Journal of Pharmaceutics, 395*(1-2), 44–52. DOI: 10.1016/j.ijpharm.2010.05.003 PMID: 20546856

Qu, M., Jiang, X., Zhou, X., Wang, C., Wu, Q., Ren, L., Zhu, J., Zhu, S., Tebon, P., Sun, W., & Khademhosseini, A. (2020). Stimuli-responsive delivery of growth factors for tissue engineering. *Advanced Healthcare Materials, 9*(7), 1901714. DOI: 10.1002/adhm.201901714 PMID: 32125786

Rahman, M. M., Roy, S., Das, S. C., Jha, M. K., Begum, T., Ahsan, Q., & Reza, M. S. (2011). Evaluation of various grades of hydroxypropylmethylcellulose matrix systems as oral sustained release drug delivery systems. *Journal of Pharmaceutical Science and Research, 3*(1), 930.

Rivera-Hernández, G., Antunes-Ricardo, M., Martínez-Morales, P., & Sanchez, M. L. (2021). Polyvinyl alcohol-based drug delivery systems for cancer treatment. *International Journal of Pharmaceutics, 600*, 120478. DOI: 10.1016/j.ijpharm.2021.120478 PMID: 33722756

Roy, S., & Choudhury, S. (2021). Polymeric pH-responsive drug delivery systems for targeted cancer therapy. *Materials Science and Engineering C, 121*, 111799. DOI: 10.1016/j.msec.2021.111799

Sahoo, S., Gupta, A., & Mehta, P. (2023). Recent advancements in multifunctional stimuli-responsive polymers for targeted drug delivery. *Journal of Controlled Release, 354*(2), 120–134.

Shah, N., Hussain, M., Rehan, T., Khan, A., & Khan, Z. U. (2022). Overview of polyethylene glycol-based materials with a special focus on core-shell particles for drug delivery application. *Current Pharmaceutical Design, 28*(5), 352–367. DOI: 10.2174/1381612827666210910104333 PMID: 34514984

Shaibie, N. A., Ramli, N. A., Mohammad Faizal, N. D. F., Srichana, T., & Mohd Amin, M. C. I. (2023). Poly (N-isopropylacrylamide)-Based Polymers: Recent Overview for the Development of Temperature-Responsive Drug Delivery and Biomedical Applications. *Macromolecular Chemistry and Physics, 224*(20), 2300157. DOI: 10.1002/macp.202300157

Shalini, S., & Chaudhury, A. (2021). Biodegradable polymers in drug delivery: A review of clinical applications. *Drug Development and Industrial Pharmacy, 47*(9), 1387–1399. DOI: 10.1080/03639045.2021.1908191

Sharma, V., Agarwal, R., & Sahoo, S. (2022). Advanced drug delivery systems for autoimmune diseases: A focus on stimuli-responsive polymers. *Journal of Pharmaceutical Sciences, 111*(12), 3425–3441.

Shi, D., Beasock, D., Fessler, A., Szebeni, J., Ljubimova, J. Y., Afonin, K. A., & Dobrovolskaia, M. A. (2022). To PEGylate or not to PEGylate: Immunological properties of nanomedicine's most popular component, polyethylene glycol and its alternatives. *Advanced Drug Delivery Reviews, 180*, 114079. DOI: 10.1016/j.addr.2021.114079 PMID: 34902516

Singh, A., Sharma, R., & Kumar, R. (2023). Stimuli-responsive polymers for neurological drug delivery: Innovations and clinical prospects. *Expert Opinion on Drug Delivery, 20*(7), 679–694.

Singh, J., & Nayak, P. (2023). pH-responsive polymers for drug delivery: Trends and opportunities. *Journal of Polymer Science, 61*(22), 2828–2850. DOI: 10.1002/pol.20230403

Smith, A., Johnson, D., & Roberts, F. (2023). Biodegradable polymers for tissue engineering: The role of PPF. *Advanced Healthcare Materials, 12*(7), 2230–2240.

Steichen, S. D. (2016). Design, development and optimization of methacrylic acid-based pH-responsive hydrogels for the oral delivery of model protein therapeutics (Doctoral dissertation).

Suhail, M., Rosenholm, J. M., Minhas, M. U., Badshah, S. F., Naeem, A., Khan, K. U., & Fahad, M. (2019). Nanogels as drug-delivery systems: A comprehensive overview. *Therapeutic Delivery, 10*(11), 697–717. DOI: 10.4155/tde-2019-0010 PMID: 31789106

Tran, T. T., & Tran, P. H. (2019). Controlled release film forming systems in drug delivery: The potential for efficient drug delivery. *Pharmaceutics, 11*(6), 290. DOI: 10.3390/pharmaceutics11060290 PMID: 31226748

Trucillo, P. (2024). Biomaterials for drug delivery and human applications. *Materials (Basel), 17*(2), 456. DOI: 10.3390/ma17020456 PMID: 38255624

Tundisi, L. L., Mostaço, G. B., Carricondo, P. C., & Petri, D. F. S. (2021). Hydroxypropyl methylcellulose: Physicochemical properties and ocular drug delivery formulations. *European Journal of Pharmaceutical Sciences, 159*, 105736. DOI: 10.1016/j.ejps.2021.105736 PMID: 33516807

Tyler, B., Gullotti, D., Mangraviti, A., Utsuki, T., & Brem, H. (2016). Polylactic acid (PLA) controlled delivery carriers for biomedical applications. *Advanced Drug Delivery Reviews, 107*, 163–175. DOI: 10.1016/j.addr.2016.06.018 PMID: 27426411

Ullah, F., Othman, M. B. H., Javed, F., Ahmad, Z., & Akil, H. M. (2015). Classification, processing and application of hydrogels: A review. *Materials Science and Engineering C, 57*, 414–433. DOI: 10.1016/j.msec.2015.07.053 PMID: 26354282

Wang, J., Liu, K., & Chen, X. (2022). Dextran derivatives in sustained drug release: Mechanisms and applications. *Biomaterials Science, 10*(8), 1254–1268.

Wei, P., Cornel, E. J., & Du, J. (2021). Ultrasound-responsive polymer-based drug delivery systems. *Drug Delivery and Translational Research, 11*(4), 1323–1339. DOI: 10.1007/s13346-021-00963-0 PMID: 33761101

Yallapu, M. M., Reddy, M. K., & Labhasetwar, V. (2007). *Nanogels: Chemistry to drug delivery*. Biomedical Applications of Nanotechnology.

Yoshida, T., Lai, T. C., Kwon, G. S., & Sako, K. (2013). pH-and ion-sensitive polymers for drug delivery. *Expert Opinion on Drug Delivery, 10*(11), 1497–1513. DOI: 10.1517/17425247.2013.821978 PMID: 23930949

Zamboulis, A., Nakiou, E. A., Christodoulou, E., Bikiaris, D. N., Kontonasaki, E., Liverani, L., & Boccaccini, A. R. (2019). Polyglycerol hyperbranched polyesters: Synthesis, properties and pharmaceutical and biomedical applications. *International Journal of Molecular Sciences, 20*(24), 6210. DOI: 10.3390/ijms20246210 PMID: 31835372

Zare, M., Bigham, A., Zare, M., Luo, H., Rezvani Ghomi, E., & Ramakrishna, S. (2021). pHEMA: An overview for biomedical applications. *International Journal of Molecular Sciences, 22*(12), 6376. DOI: 10.3390/ijms22126376 PMID: 34203608

Zeb, A., Gul, M., Nguyen, T. T. L., & Maeng, H. J. (2022). Controlled release and targeted drug delivery with poly (lactic-co-glycolic acid) nanoparticles: Reviewing two decades of research. *Journal of Pharmaceutical Investigation, 52*(6), 683–724. DOI: 10.1007/s40005-022-00584-w

Zhang, W., Zhao, Q., & Yuan, J. (2018). Porous polyelectrolytes: The interplay of charge and pores for new functionalities. *Angewandte Chemie International Edition, 57*(23), 6754–6773. DOI: 10.1002/anie.201710272 PMID: 29124842

Zhang, X., & Luo, X. (2020). Thermoresponsive polymers for drug delivery. *Materials Science and Engineering C, 111*, 110758. DOI: 10.1016/j.msec.2020.110758

Zhang, Y., & Lee, J. (2022). Amphiphilic polymers in drug delivery systems. *European Journal of Pharmaceutics and Biopharmaceutics, 174*, 32–45. DOI: 10.1016/j.ejpb.2022.02.006

Zhang, Y., & Liu, Z. (2023). Protein and gene delivery using pH-sensitive polymers. *Journal of Controlled Release, 344*, 185–200.

Zhang, Y., Wang, L., & Liu, X. (2024). Advances in pH-responsive polymers for gene delivery: The case of PDMAEMA. *Macromolecular Bioscience, 24*(3), 1203–1218.

Zhang, Y., Yu, J., Bomba, H. N., Zhu, Y., & Gu, Z. (2016). Mechanical force-triggered drug delivery. *Chemical Reviews*, *116*(19), 12536–12563. DOI: 10.1021/acs.chemrev.6b00369 PMID: 27680291

Zhao, L., Feng, Y., & Huang, J. (2023). Controlled drug release using poly(sebacic acid) for implantable devices. *Materials Today*, *51*, 125–138.

Zheng, M., & Yuan, J. (2022). Polymeric nanostructures based on azobenzene and their biomedical applications: Synthesis, self-assembly and stimuli-responsiveness. *Organic & Biomolecular Chemistry*, *20*(4), 749–767. DOI: 10.1039/D1OB01823J PMID: 34908082

Chapter 8
Designing Smart Polymers for CNS Drug Delivery Stimuli:
Responsive and Biodegradable Systems

Ankur Bhardwaj
https://orcid.org/0000-0003-0687-8810
Shri Vaishnav Vidyapeeth Vishwavidyalaya, Indore, India

ABSTRACT

Smart polymers represent a transformative approach in CNS drug delivery, offering controlled, targeted, and responsive release of therapeutic agents. This chapter explores the design and application of stimuli-responsive and biodegradable polymers tailored to overcome the physiological barriers of the CNS, particularly the BBB. Stimuli-responsive polymers, engineered to respond to internal (pH, redox potential, enzymes) or external (temperature, magnetic field, light) triggers, enable on-demand drug release at the desired site. Concurrently, biodegradable polymers reduce long-term toxicity and eliminate the need for surgical removal, making them ideal for chronic CNS therapies. Finally, the chapter addresses recent advances, current limitations, including stability, scalability, and regulatory challenges, and outlines future directions for research. Integrating material science with neuropharmacology, smart polymers offer a promising platform to revolutionize CNS therapeutics through precision drug delivery.

DOI: 10.4018/979-8-3373-1047-3.ch008

INTRODUCTION

The effective treatment of central nervous system (CNS) disorders remains a major clinical challenge due to the restrictive nature of the blood-brain barrier (BBB), which limits drug penetration into the brain. Smart polymers, engineered to respond to specific physiological or external stimuli, offer promising strategies for targeted and controlled drug delivery across the BBB (Wu et al., 2023). When combined with biodegradability, these systems enhance safety and reduce long-term toxicity, making them ideal candidates for CNS therapeutics. This chapter explores the design, mechanisms, and applications of stimuli-responsive and biodegradable polymers, emphasizing their potential to revolutionize the delivery of drugs for neurological and psychiatric diseases.

Overview of CNS Disorders and Drug Delivery Challenges

Neurodegenerative diseases (Alzheimer's, Parkinson's), neuroinflammatory conditions (multiple sclerosis), brain tumours, and psychiatric illnesses (depression, schizophrenia), pose a significant burden on global health. Despite the availability of potent pharmacological agents, effective treatment remains elusive for many CNS disorders due to the inability of drugs to reach their target sites in the brain at therapeutic concentrations (Wells et al., 2019).

The primary obstacle in CNS drug delivery is the BBB, a selective and tightly regulated interface composed of endothelial cells with tight junctions, supported by astrocytes and pericytes. This barrier efficiently protects the brain from toxins and pathogens but also severely limits the entry of most small-molecule drugs and nearly all large-molecule therapeutics, including peptides, proteins, and nucleic acids (Singh and Vellapandian, 2023).

Traditional drug delivery approaches often suffer from poor bioavailability, systemic side effects, and non-specific distribution. As a result, there is a critical need for innovative delivery platforms that can safely and effectively traverse the BBB, deliver therapeutic agents in a controlled manner, and degrade harmlessly after fulfilling their function (Fagehi et al., 2024). Smart polymers, particularly those that are stimuli-responsive and biodegradable, offer a promising solution to these long-standing challenges.

Importance of Overcoming the Blood-Brain Barrier (BBB)

The blood-brain barrier is a highly specialized and selective physiological barrier that serves to maintain the homeostasis of the CNS. Composed primarily of tightly connected endothelial cells, the BBB regulates the passage of substances from the bloodstream into the brain. While essential for protecting neural tissue from pathogens and toxins, this same barrier significantly limits the entry of therapeutic agents, particularly those that are large, hydrophilic, or not actively transported (Ueno, 2007).

Overcoming the BBB is critical for the effective treatment of a wide range of neurological and psychiatric disorders. Many conventional drugs, even when systemically administered in high doses, fail to reach therapeutic concentrations in the brain due to the restrictive properties of the BBB. This leads to suboptimal treatment outcomes, systemic toxicity, and increased healthcare burdens (Dhiman et al., 2025).

Therefore, innovative drug delivery strategies that can traverse the BBB safely and efficiently are essential. The development of smart polymer-based systems, especially those that are both stimuli-responsive and biodegradable, holds immense potential in addressing this challenge. These systems can decrease systemic adverse effects, increase therapeutic efficacy, and open the door to more individualized, targeted neuropharmacological treatments by facilitating site-specific, controlled drug release within the central nervous system.

Definition and Classification of Smart Polymers

Smart polymers, or stimuli-responsive polymers, are advanced materials engineered to alter their structure, solubility, or conformation in response to specific stimuli (pH, temperature, redox gradients, enzymatic activity) or external triggers like light and magnetic fields. In the treatment of CNS disorders, smart polymers offer substantial promise by enabling controlled, targeted, and site-specific drug delivery that overcomes the limitations posed by the BBB (Fattah-alhosseini et al., 2024).

One of the critical advantages of smart polymers is their ability to respond to the pathological microenvironment of the brain (Table 1). For instance, pH-sensitive polymers such as poly(L-histidine) and chitosan derivatives have been used to release drugs in acidic tumour environments, enhancing treatment specificity. Redox-responsive polymers, such as disulfide-linked PEG-polyesters, have shown effectiveness in responding to the elevated glutathione levels found in certain neuroinflammatory and cancerous tissues (Begum et al., 2025).

A practical example includes the use of poly(N-isopropylacrylamide) (PNIPAAm)-based thermosensitive hydrogels that remain liquid at room temperature but solidify at body temperature, enabling sustained local delivery after intracerebral injection (Wang et al., 2024). Another example is PLGA-PEG nanoparticles functionalized with transferrin or lactoferrin, which target overexpressed receptors on the BBB for receptor-mediated transcytosis, successfully used in models of Alzheimer's disease and glioblastoma (Vashist et al., 2024).

Magnetic-responsive nanoparticles, coated with smart polymers and guided by an external magnetic field, are also being explored for enhanced localization in deep brain tumours. Furthermore, enzyme-responsive systems, such as gelatin-based carriers sensitive to matrix metalloproteinases, are showing potential in targeting sites of neuroinflammation or injury (Rani et al., 2025).

Incorporating biodegradability into these smart polymers adds another layer of clinical relevance by ensuring complete breakdown into non-toxic byproducts, avoiding the risks associated with long-term accumulation in brain tissue. Collectively, these advances exemplify how smart polymer systems are paving the way for more effective, precise, and patient-friendly CNS drug delivery strategies.

Table 1. Application-oriented smart polymer systems for CNS drug delivery

Stimulus Type	Polymer/System	Mechanism	Application Example	Benefit
pH-sensitive	Chitosan derivatives, Poly(L-histidine)	Swelling/ release in acidic environments	Brain tumours, glioblastoma	Selective release in tumour microenvironments
Redox-responsive	PEG-disulfide-PLGA	Disulfide bond cleavage by glutathione	Neuroinflammation, Parkinson's disease	Intracellular-triggered release
Temperature-sensitive	Poly(N-isopropylacrylamide) (PNIPAAm)	Sol-gel transition near body temperature	Injectable hydrogels for local delivery	Minimally invasive, sustained release
Receptor-mediated	PLGA-PEG + Transferrin/Lactoferrin	Ligand-directed BBB transcytosis	Alzheimer's disease, brain metastases	BBB targeting via receptor pathways
Magnetic field	Magnetite nanoparticles + smart coating	Magnetic guidance and localized heating	Brain tumour targeting (glioma)	Remote control of targeting and release
Enzyme-responsive	Gelatin-based systems	Degradation by MMPs in inflamed tissue	Stroke, multiple sclerosis	Activation at sites of neuroinflammation

(Seth et al., 2024; Song et al., 2025; Lu et al., 2024; Dachani et al., 2025)

THE BBB AND ITS IMPLICATIONS

The BBB is a specialized membrane that regulates the passage of substances between the bloodstream and the brain, crucial for maintaining neural homeostasis. Recent studies highlight its role in neuroprotection and mental health, challenging traditional views that oversimplify its functions (Tanaka and Vécsei, 2024; Kruczkowska et al., 2025). Disruption of the BBB, often exacerbated by pharmaceuticals, can lead to severe neurological conditions, including implications in widespread vaccine-related injuries (Grahl et al., 2025). This emerging knowledge emphasizes the need for holistic approaches that prioritize BBB integrity in treatment strategies. In order to create safe, efficient therapies that respect the complexities of the human neurological system, it is essential to comprehend BBB dynamics.

Anatomy and Physiology of the BBB

The BBB is a highly specialized and dynamic interface that regulates the exchange of substances between the systemic circulation and the CNS. Comprised primarily of brain capillary endothelial cells (BCECs) interconnected by tight junctions, and the basal lamina, collectively forming the neurovascular unit (Alahmari, 2021). These components work in concert to preserve CNS homeostasis, protect neural tissue from toxins and pathogens, and maintain a controlled microenvironment essential for synaptic transmission and neural function.

The endothelial cells of the BBB differ significantly from peripheral endothelium by exhibiting minimal pinocytic activity, a high electrical resistance, and specific transport systems for essential nutrients and ions (McConnell and Mishra, 2022). The tight junctions restrict paracellular diffusion, making the BBB both a physical and metabolic barrier (Singh and Vellapandian, 2023). Developing intelligent medication delivery devices that can target CNS illnesses without upsetting the delicate neurovascular balance requires an understanding of this particular physiology.

Transport Mechanisms and Barriers to Drug Penetration

Despite its protective role, the BBB poses a major obstacle to therapeutic intervention in neurological diseases. It restricts over 98% of small-molecule drugs and nearly all biologics from entering the brain, making CNS drug delivery particularly challenging (Van et al., 2022). There are several mechanisms by which substances cross the BBB: passive diffusion, carrier-mediated transport (CMT), receptor-mediated transcytosis (RMT), adsorptive-mediated transcytosis (AMT), and efflux transport systems (P-glycoprotein) (Van et al., 2022; Singh and Vellapandian, 2023).

Passive diffusion favours small, lipophilic, and non-ionized molecules; however, therapeutic agents often do not meet these physicochemical criteria (Ruiz and Scioli-Montoto, 2024). Carrier proteins enable selective uptake of essential molecules such as glucose and amino acids but can be saturated or compete with drugs. RMT and AMT offer pathways for macromolecular transport but are limited by receptor availability and non-specificity. Efflux transporters further hinder drug accumulation by pumping substrates back into circulation (Susa et al., 2024). Thus, a major challenge in CNS pharmacotherapy is not only achieving BBB penetration but also ensuring selective, sustained, and non-toxic drug delivery, a need that modern polymeric systems are poised to address (Rawal et al., 2022).

Strategies for Crossing the BBB Using Polymeric Systems

Smart polymeric systems, especially those that are biodegradable and stimuli-responsive, offer promising strategies for traversing the BBB and delivering therapeutics to the CNS in a controlled manner. These polymers can be engineered to respond to specific stimuli (pH, temperature, redox potential, or enzymatic activity) enabling site-specific and on-demand drug release (Shi et al., 2024). Moreover, their molecular architecture can be tailored to optimize size, surface charge, and hydrophilicity, all of which influence BBB permeability.

Several polymeric delivery platforms have been developed to exploit endogenous transport pathways. Nanoparticles and micelles functionalized with ligands for transferrin or insulin receptors can engage RMT pathways, while cationic polymers enhance AMT via electrostatic interactions with the negatively charged endothelial surface (Dai et al., 2025). Additionally, PEGylation and surface modification with targeting moieties improve circulation time and BBB selectivity (Song et al., 2023).

Emerging designs include dendrimers, hydrogels, and polymer-drug conjugates that combine biodegradability with multifunctionality. These systems not only enhance CNS targeting but also reduce systemic toxicity and immune responses. As polymer science advances, integrating artificial intelligence, high-throughput screening, and biomimetic approaches will further refine these strategies, making smart polymers pivotal tools in the future of CNS drug delivery.

SMART POLYMERS: DEFINITION AND CLASSIFICATION

What are Smart Polymers?

Smart polymers, also known as stimuli-responsive or intelligent polymers, are a class of advanced materials capable of undergoing significant and reversible physicochemical changes in response to external or internal stimuli. These transformations (solubility shifts, conformational changes, phase transitions, or degradation) can be finely tuned to specific environmental cues like pH, temperature, redox conditions, light, enzymes, or magnetic fields (Song et al., 2023).

In the context of CNS drug delivery, smart polymers offer a powerful platform to overcome the physiological barriers of the brain, including the BBB. Drug release can be tailored, regulated, and on-demand by designing these polymers to react to pathological or physiological stimuli in the CNS microenvironment (Nance et al., 2022). Their ability to adapt to the complex and dynamic nature of the central nervous system makes smart polymers a fundamental innovation in neuropharmacology and nanomedicine. Smart polymers are categorized based on the type of stimuli they respond to (Hersh et al., 2022). The table 2, summarizes the major stimuli-responsive polymer systems and their mechanism of action, with relevance to CNS drug delivery. These stimuli can be used individually or in combination (multi-responsive systems) to improve precision, sensitivity, and efficiency in CNS-targeted therapeutics.

Table 2. Types of smart responses: pH, redox, enzyme, temperature, magnetic, light

Stimulus	Mechanism of Action	Examples of Smart Polymers	Applications in CNS Drug Delivery
pH	Protonation/deprotonation affects solubility, swelling, or bond cleavage	Poly(L-histidine), Eudragit, chitosan derivatives	Drug release in acidic tumour regions, inflamed brain tissues
Redox	Cleavage of disulfide bonds in high GSH or ROS environments	Poly(disulfide ester), PEG-SS-PLA	Targeted release inside neurons or tumour cells with high oxidative stress
Enzyme	Enzyme-mediated cleavage of polymer backbone or linkers	Peptide-modified polyesters, gelatin-based hydrogels	Targeting tumours or neuroinflammatory regions with specific proteases
Temperature	LCST/UCST-driven phase transition (gelation or solubilization)	Poly(N-isopropylacrylamide) (PNIPAM), pluronic F127	Injectable thermos-gels for localized brain drug delivery
Magnetic	Movement or heating under an external magnetic field	Superparamagnetic iron oxide nanoparticles (SPIONs) embedded in PLGA	Magnetic targeting and hyperthermia-assisted therapy
Light	Photolysis or photoisomerization releases drug molecules	Azobenzene-modified polymers, coumarin-linked systems	Spatiotemporal drug activation using NIR for deep brain delivery

(Aflori, 2021; Anwar et al., 2023; Parvaresh, et al., 2023; Rezaei et al., 2021)

Advantages Over Conventional Polymers

Smart polymers offer several significant advantages compared to traditional, non-responsive polymers in drug delivery applications, especially for the CNS: (Patkar and Sherje, 2025; Alshamrani et al., 2024; Maddiboyina et al., 2022)

- Targeted Drug Release: Smart polymers enable drug release in response to specific biological triggers, minimizing systemic side effects and enhancing therapeutic efficacy.
- Improved BBB Penetration: Functionalization and responsive behaviour can facilitate crossing or bypassing the BBB, a major challenge in CNS therapy.
- On-Demand and Controlled Delivery: The ability to modulate drug release spatially and temporally ensures sustained therapeutic levels and reduces dosing frequency.
- Biodegradability and Biocompatibility: Many smart polymers are engineered to degrade into non-toxic byproducts, reducing accumulation risks in sensitive brain tissues.
- Multifunctionality: Smart polymers can combine therapeutic delivery with diagnostic capabilities (theranostics), targeted imaging, or stimuli-triggered activation.
- Reduced Drug Resistance: Responsive delivery may bypass efflux transporters and minimize premature degradation or elimination of the drug.

STIMULI-RESPONSIVE POLYMERS FOR CNS DRUG DELIVERY

The development of stimuli-responsive polymers has opened transformative pathways in CNS drug delivery by enabling targeted, controlled, and on-demand therapeutic release. These "intelligent" systems are particularly advantageous in the brain, where physiological and pathological stimuli can be exploited for precise drug deployment (Panchpuri et al., 2025). Here primary categories of stimuli-responsive systems were examined, highlighting their clinical relevance to neurological disorders.

pH-Sensitive Polymers

The extracellular pH in the brain is tightly regulated under normal conditions (~7.4). However, glioblastomas or ischemic sites often present a more acidic microenvironment (pH 6.5–6.8). The pH-sensitive polymers are engineered to respond to this acidic shift, often through protonation-induced solubility changes or hydrolysis of acid-labile linkers (Singh and Nayak, 2023). Some significant polymers are:

- Poly(L-histidine): It becomes hydrophilic in acidic pH, enabling micelle destabilization and drug release (Singh and Nayak, 2023).
- Chitosan derivatives: They exhibit swelling or solubility changes in acidic conditions (Wang et al., 2020).

- Poly(β-amino esters): They undergo rapid degradation in mildly acidic environments (Karlsson et al., 2020).

These systems are highly suitable for glioma-targeted delivery, enabling local release without affecting healthy brain tissue.

Redox-Responsive Systems

Oxidative stress is a hallmark of many CNS disorders including Alzheimer's disease, Parkinson's disease, stroke, and gliomas. These regions typically exhibit elevated levels of glutathione (GSH) or reactive oxygen species (ROS), particularly in the cytosol or mitochondria (Houldsworth, 2024).

Redox-responsive polymers were designed with disulfide linkages, thioketal bonds, or ROS-cleavable segments that break down in these environments, triggering the release of encapsulated drugs. Some significant systems are:

- PEG-SS-PLA nanoparticles: Degrade in high-GSH environments for intracellular drug release (Ren et al., 2025).
- Thioketal-based polyesters: ROS-cleavable backbones for oxidative-stress-induced drug delivery (Xu et al., 2024).
- Di-selenide-containing polymers: Dual-responsive to redox and pH changes (Parvaresh et al., 2023).

These systems are especially relevant for targeting diseased neurons or brain tumours with high intracellular redox potential.

Enzyme-Responsive Polymers

Disease progression in the brain often leads to overexpression of specific enzymes, like, matrix metalloproteinases (MMPs) in gliomas, cathepsins in Alzheimer's disease, and phospholipases in neuroinflammatory conditions. Enzyme-responsive polymers contain cleavable linkers or backbones that are specifically degraded by these enzymes (Ioannou et al., 2025). Some representative materials include:

- Peptide-linked hydrogels: Cleavable by MMP-2 or MMP-9 enzymes (Geranmayeh et al., 2019).
- Gelatin-based nanoparticles: Degraded by proteases in tumour or inflamed tissues (Van et al., 2022).
- Polyesteramides: Incorporating enzyme-cleavable peptide sequences (Anwar et al., 2023).

These systems enable spatially precise drug delivery, ensuring activation only in enzyme-rich diseased areas of the brain.

External Stimuli-Responsive Systems

External stimuli provide a non-invasive means of controlling drug release across the BBB. These methods (Table 3) are particularly useful when paired with nanocarriers capable of responding to magnetic fields, light irradiation (NIR), or ultrasound (Geranmayeh et al., 2019).

Magnetic systems use SPIONs (superparamagnetic iron oxide nanoparticles) embedded in poly(lactic-co-glycolic) acid (PLGA) polymers. They allow magnetic field-guided targeting and magnetothermal drug release (Bhattacharya et al., 2023).

Light-responsive systems include azobenzene- or coumarin-based polymers that undergo structural changes upon light exposure. They enable high spatiotemporal control with NIR light for deep-tissue penetration. These platforms are ideal for non-invasive therapies, particularly in tumour ablation, epilepsy management, or controlled neurostimulation.

Table 3. Selected case studies where stimuli-responsive polymers have been successfully applied in CNS-targeted drug delivery

Study / Application	Stimulus Type	Polymer Used	Target Disease	Delivery Outcome
pH-sensitive micelles (Poly(L-histidine)-PEG)	pH	Poly(L-histidine)-PEG copolymer	Glioblastoma	Triggered doxorubicin release in acidic tumor sites
Redox-responsive nanoparticles (PEG-SS-PLA)	Redox	Disulfide-bridged PEG-PLA	Parkinson's disease	Enhanced drug release in GSH-rich dopaminergic neurons
MMP-2 responsive hydrogel for intracranial injection	Enzyme (MMP-2)	Peptide-crosslinked PEG hydrogel	Brain tumor	Enzyme-triggered breakdown for sustained local delivery
SPION-loaded PLGA nanoparticles for magnetothermal therapy	Magnetic	PLGA + SPION	Glioma	Magnetic targeting + heat-assisted drug release across BBB
NIR light-triggered nanocarriers (azobenzene-linked)	Light	Azobenzene-grafted amphiphilic polymer	Alzheimer's disease	NIR-induced structural shift leading to local drug activation
Dual pH/ROS-sensitive micelles for stroke treatment	pH + Redox	Thioketal- and acetal-containing polymer	Ischemic stroke	Controlled drug release in acidic and oxidative regions

(Wang et al., 2020; Anwar et al., 2023; Houldsworth, 2024; Ren et al., 2025)

BIODEGRADABLE POLYMERS IN CNS DELIVERY

Biodegradable polymers play a pivotal role in advancing the safety, precision, and efficacy of CNS drug delivery. These materials are engineered to degrade into biocompatible byproducts, making them highly suitable for chronic therapies, minimally invasive implants, and injectable systems. In the context of CNS applications, where the risk of long-term foreign material retention is especially critical, the integration of biodegradability with stimuli-responsiveness is a significant innovation (Nance et al., 2022).

Common Biodegradable Polymers

Several biodegradable polymers have been widely employed in CNS drug delivery due to their tenable degradation rates, safety profiles, and functional versatility. The most commonly used polymers include:

- Poly(lactic acid) (PLA): PLA is derived from lactic acid; degrades slowly via hydrolysis and is suitable for long-term controlled release (Bhattacharya et al., 2023).
- Poly(lactic-co-glycolic acid) (PLGA): It is a co-polymer of PLA and PGA. The FDA has approved PLGA with adjustable degradation rates based on monomeric ratio. PLGA is extensively used in CNS-targeted nanoparticles and implants (Bhattacharya et al., 2023).
- Polycaprolactone (PCL): The PCL is a semi-crystalline polymer with slow degradation rate (over months to years). It is ideal for sustained delivery implants and scaffolds (Archer et al., 2023).
- Chitosan: It is naturally derived from chitin and possess biodegradable and mucoadhesive property. It enables intranasal or direct CNS delivery due to its ability to open tight junctions (Begum et al., 2025).

Other materials like gelatine, alginate, poly(ortho esters), and poly(anhydrides) were also explored for their CNS compatibility and functional modifiability.

Degradation Mechanisms and Byproducts

The degradation of biodegradable polymers in the CNS occurs (Table 4) primarily through hydrolysis and/or enzymatic action, depending on polymer composition and implantation site. Predicting drug release kinetics and safety requires an understanding of these mechanisms.

Table 4. Biodegradation of selected polymers

Polymer	Degradation Mechanism	Primary Byproducts	CNS Relevance
PLA	Hydrolytic cleavage of ester bonds	Lactic acid	Metabolized via Krebs cycle; low toxicity
PLGA	Hydrolysis (bulk erosion)	Lactic acid, glycolic acid	Biocompatible; acidification may affect sensitive drugs
PCL	Slow hydrolysis	6-hydroxyhexanoic acid	Very slow degradation beneficial for implants
Chitosan	Enzymatic degradation by lysozyme	Glucosamine, oligosaccharides	Biocompatible, non-toxic; promotes neural repair

(Mukherjee et al., 2023; Archer et al., 2023; Feig et al., 2018; Visan et al., 2021)

Benefits for Long-Term Safety and Chronic Therapies

Biodegradable polymers offer critical benefits for chronic neurological conditions like, Alzheimer's disease, Parkinson's disease, multiple sclerosis, and brain tumours. Implants or nanoparticles degrade over time, eliminating the risks and costs associated with secondary surgeries. Biocompatible degradation products minimize the risk of chronic gliosis or neuroinflammation. Polymers can be engineered to degrade in synchrony with disease states or therapeutic needs. Prolonged degradation allows for weeks to months of drug delivery from a single dose, ideal for conditions requiring steady-state therapy. This makes them particularly attractive for elderly or vulnerable populations where repeat dosing or surgical intervention poses higher risk.

Application in CNS-Targeted Implants and Injectable Systems

Biodegradable polymers are at the forefront of localized and sustained CNS drug delivery platforms. Implantable Systems like Gliadel® wafers (polyanhydride-based) was the first FDA-approved biodegradable implants delivering carmustine to resected glioblastoma cavities. Similarly, PLGA microparticle depots was implanted at lesion sites to provide extended neuroprotective or anti-inflammatory drug delivery (Caraway et al., 2022).

In situ-forming hydrogels based injectable polymers like PLGA-PEG-PLGA form gels upon body temperature contact, enabling minimally invasive delivery. Chitosan- or PLA-based nanoparticles can cross the blood-brain barrier and release payloads over time (Fagehi et al., 2024). Thermo- or pH-sensitive biodegradable hydrogels allow on-demand drug activation and localized delivery. These systems are paving the way for precision neuromodulation, regenerative therapies, and sustained neuropharmacological interventions (Parvaresh et al., 2023).

COMBINED STIMULI-RESPONSIVE AND BIODEGRADABLE PLATFORMS

In the evolving landscape of CNS drug delivery, the convergence of stimuli-responsive and biodegradable properties into a single polymeric system represents a powerful advancement. These dual-function platforms offer the unique capacity to respond to pathological cues while ensuring safe elimination from the body. This synergistic approach addresses the critical need for targeted, temporally controlled, and non-toxic delivery systems within the delicate environment of the brain.

Design of Dual-Function or Multi-Responsive Systems

Dual-function or multi-responsive polymeric systems are engineered to respond to two or more internal or external stimuli while simultaneously undergoing biodegradation. These designs integrate stimuli-sensitivity (pH, redox, enzyme, temperature), degradable linkages (ester, disulfide, thioketal), therapeutic payload (small molecules, peptides, RNA) (Li et al., 2023). These systems allow spatial control through selective responsiveness to disease-specific triggers (acidic tumour microenvironments or oxidative stress), temporal control by synchronizing degradation with drug release over days to weeks and minimal toxicity by breaking down into safe byproducts that are naturally metabolized or excreted (Tao et al., 2020). Some examples include, pH/ROS-responsive micelles made from thioketal-linked polyesters, enzyme-degradable hydrogels with redox-sensitive drug linkers and light-responsive, biodegradable nanoparticle systems for on-demand release. These formulations are often custom-designed using modular polymer architectures, like block copolymers, dendrimers, or star-shaped polymers (Karayianni and Pispas, 2020).

Hybrid Polymeric Nanocarriers

Hybrid systems combine natural and synthetic polymers or merge organic and inorganic components, achieving multifunctionality through structural and compositional diversity. Common hybrid structures include, chitosan-PLGA nanoparticles (merging chitosan's muco-adhesiveness), PEGylated PLGA nanoparticles (provides stealth and solubility, enabling sustained drug release), inorganic-organic hybrids (magnetic nanoparticles, gold nanorods, or silica cores into biodegradable polymeric shells) (Massironi, 2024; Yetisgin et al., 2020; Vasanth et al., 2025).

Functional advantages of hybrid polymeric nanocarriers include improved BBB penetration through surface functionalization, enhanced drug loading and stability through core-shell structuring and capability to co-deliver multiple therapeutic agents (neuroprotectants + anti-inflammatories) (Mohammed et al., 2025; Vasanth et al., 2025). These carriers are especially promising for diseases like glioblastoma, Parkinson's disease, and neuroinflammatory disorders, where multifaceted therapeutic approaches are essential.

Recent Innovations and Synergistic Effects

Recent research showcases the promise of combining stimuli-responsiveness with biodegradability to produce smart systems with synergistic therapeutic benefits. Some significant innovations include injectable thermo-sensitive hydrogels that solidify at body temperature, degrade over time, and release drugs in response to redox conditions (Xu et al., 2024). Dual-responsive micelles (pH and redox) for intracellular delivery of chemotherapeutics and siRNA to glioma cells. Self-assembling polymeric prodrugs that respond to MMP enzymes and degrade into active drugs in situ (Badparvar et al., 2024). Multilayered nanoparticle systems that sequentially respond to external light and internal pH to release neurotrophic agents (Rani et al., 2025).

All these smart systems have synergistic effects like enhanced therapeutic efficacy (due to precise spatiotemporal delivery), reduced systemic toxicity (through localized, responsive release), improved patient compliance (by minimizing dosage frequency and invasiveness), integrated diagnostic and therapeutic functionality (theranostics) in hybrid designs (Liu et al., 2024; Badparvar et al., 2024). These systems represent a shift toward personalized neurotherapy, capable of adapting to complex disease environments and improving the long-term success of CNS drug interventions.

CHALLENGES AND LIMITATIONS

Despite the transformative potential of stimuli-responsive and biodegradable smart polymers in CNS drug delivery, several challenges must be addressed before these systems can be widely translated into clinical applications. These limitations (Table 4) span from fundamental material properties to regulatory and commercial hurdles, underscoring the need for interdisciplinary solutions.

Stability and Premature Drug Release

One of the foremost challenges in smart polymer design is ensuring physiological stability during circulation while allowing timely responsiveness to disease-specific triggers (Table 5). Many stimuli-sensitive carriers suffer from:

- Premature degradation or activation in systemic circulation due to non-specific triggers (e.g., slight pH fluctuations, oxidative stress in non-target tissues).
- Uncontrolled drug leakage during storage or early after administration, reducing therapeutic efficacy and increasing toxicity risk.
- Aggregation or loss of colloidal stability, particularly for nanoparticles exposed to serum proteins and ionic changes in biological fluids.

To combat these issues, PEGylation, core-shell architectures, and multi-layered encapsulation strategies should be used to enhance carrier stability without compromising responsiveness. However, optimizing this balance remains a major design bottleneck.

Polymer Toxicity and Immune Response

Even biodegradable polymers can elicit unintended biological reactions, especially within the immune-privileged environment of the CNS. While many polymers break down into naturally metabolized components (lactic acid), others may yield acidic or reactive intermediates that can disrupt the local microenvironment. Microglia and astrocytes are highly sensitive to foreign materials; improper surface chemistry or particle size can provoke chronic inflammatory responses. Hypersensitivity or allergic reactions could also be elicited, particularly with natural polymers (chitosan or gelatin), if impurities are not properly removed.

Rigorous biocompatibility testing, including long-term *in-vivo* neurotoxicity assessments, is essential to ensure safety for clinical use. New polymer designs also aim to incorporate immunomodulatory properties to mitigate these effects.

Scalability and Reproducibility in Manufacturing

Translating lab-scale smart polymer systems into clinical-grade products involves overcoming significant scale-up challenges:

- Batch-to-batch variability in polymer synthesis can lead to inconsistent drug loading, release kinetics, and particle size.
- Complex fabrication processes (multi-layer encapsulation, stimuli-sensitive linker attachment) are often resource-intensive and difficult to automate.
- Sterility and storage stability: Maintaining carrier integrity and function under pharmaceutical-grade conditions remains a limiting factor.

To move forward, the field must focus on simplifying polymer architectures without sacrificing functionality and adopt GMP-compliant synthesis protocols that ensure reproducibility at scale.

Regulatory and Translational Hurdles

Smart polymer systems present a paradigm shift in drug delivery, but they also face regulatory uncertainty due to their hybrid, multifunctional nature. Regulatory bodies may categorize these systems as drugs, devices, or combination products, complicating the approval process. Current frameworks do not fully address the dynamic behaviour of stimuli-responsive systems, requiring the development of new *in-vitro* and *in-vivo* testing standards. Extended timelines and the need for extensive preclinical data (including biodegradation, immunogenicity, and pharmacokinetics) create high barriers for academic labs and startups.

Translational success will depend on early engagement with regulatory agencies, interdisciplinary collaboration, and robust clinical trial designs tailored to the unique behaviour of these systems. Coordination between the fields of materials science, pharmacology, toxicology, regulatory science, and manufacturing engineering is necessary to meet these issues. Smart biodegradable polymers for CNS drug delivery are becoming more feasible, safe, and scalable, despite the challenging path to clinical deployment.

Table 5. Challenges in smart biodegradable CNS drug delivery systems and proposed solutions

Challenge	Description	Proposed Solutions	Recent Examples	References
Stability and Premature Release	Unintended drug leakage or degradation before reaching CNS targets	PEGylation for stealth and stability- Core-shell or multilayer nanocarriers. Cross-linking for enhanced matrix stability	Dual pH/redox-responsive micelles with disulfide/ester linkers showing minimal leakage	Li et al., 2022
Polymer Toxicity and Immune Response	Degradation byproducts or materials trigger neuroinflammation or toxicity	Use of FDA-approved or metabolite-cleared polymers- Surface modification (zwitterionic coatings) *In vivo* long-term toxicity screening	PLGA/chitosan nanoparticles showed reduced microglial activation when coated with hyaluronic acid	Wang et al., 2024
Scalability and Manufacturing	Complex fabrication limits reproducibility and cost-effectiveness	Emphasis on modular, one-pot synthesis Process optimization for particle uniformity GMP-compliant protocols	Flash nanoprecipitation used for scalable PLGA-based CNS nanoparticles with narrow size distribution	Rawal et al., 2022
Regulatory and Translational Barriers	Lack of clear guidelines for hybrid smart systems	Early regulatory engagement (FDA, EMA)- Standardized biocompatibility and degradation tests Bridging preclinical–clinical data gaps	First-in-human trials of polymeric wafers for glioma using FDA-cleared biodegradable matrix (Gliadel® example)	Caraway et al., 2022
Targeting Efficiency	Inefficient BBB crossing or off-target effects	Ligand-conjugation (transferrin, angiopeptides)- Dual-targeting approaches (receptor + stimuli)	RVG29-modified PLGA nanoparticles showed >3× increase in brain accumulation in Alzheimer's models	Ren et al., 2025

FUTURE PERSPECTIVES

The convergence of materials science, neuroscience, and biomedical engineering is rapidly expanding the frontier of CNS drug delivery using smart, biodegradable polymers. Looking ahead, the future of this field is marked by unprecedented opportunities for precision targeting, real-time responsiveness, and personalized care driven by technological advancements and clinical demands.

Emerging Trends: AI-Assisted Design, Biomimetic Polymers, Theranostics

AI and machine learning are transforming polymer research by enabling predictive modelling of polymer-drug interactions, enhancement of degradation kinetics, and simulation of molecular-level stimuli-response mechanisms. These advancements streamline discovery processes and diminish the need for traditional trial-and-error methods. For instance, ML-driven development has successfully designed ROS-sensitive polymers aimed at stroke therapy, showcasing the potential to create virtual libraries of smart polymers customized for various central nervous system pathologies.

Biomimetic polymers, inspired by natural systems, are being engineered to replicate the extracellular matrix, mimic neuronal membranes, and emulate immune-evasive surfaces, offering enhanced biocompatibility, reduced immune responses, and improved integration with CNS tissue. Recent innovations include self-healing hydrogels, neuron-targeting peptide grafts, and stealth coatings modelled after red blood cells, all aimed at advancing therapeutic performance and tissue compatibility.

Theranostic platforms integrate therapy and diagnostics within a single polymeric system, enabling real-time tracking of drug release and disease progression. Examples include MRI-visible nanoparticles carrying neuroprotective agents and fluorescence-labelled responsive hydrogels, which allow clinicians to simultaneously monitor and treat conditions like tumours, stroke, and neurodegenerative diseases, thereby enhancing treatment precision and therapeutic outcomes.

Potential for Personalized CNS Therapy

Smart polymers are highly suited for personalized medicine, particularly in treating CNS disorders where variations in disease progression, genetic profiles, and drug responses demand individualized approaches. Advances in this field include tuning polymer sensitivity to patient-specific biomarkers such as pH, oxidative stress, or enzyme levels; customizing degradation and drug release profiles to align with therapy duration or circadian rhythms; and employing technologies like 3D printing or microfluidics to create patient-specific implants or drug reservoirs. This level of personalization holds the potential to transform treatment for conditions like glioblastoma, Alzheimer's, Parkinson's, and epilepsy by maximizing therapeutic efficacy while minimizing side effects.

Future Clinical Directions

Despite significant advancements, translating smart biodegradable systems from the laboratory to clinical practice still faces key challenges; however, several promising trends are paving the way forward. Regulatory agencies are increasingly receptive to novel drug-device combination platforms, with efforts underway to establish standardized guidelines for stimuli-responsive and biodegradable systems. Early-phase clinical trials are beginning to evaluate the safety, pharmacokinetics, and targeting capabilities of polymeric CNS therapies, with adaptive designs offering flexibility for personalized treatments. Implantable and injectable CNS devices, like long-acting biodegradable implants for brain tumours and smart depots responsive to events like seizures or neuroinflammation, are showing clinical potential. As these technologies move closer to mainstream care, it will be essential to address neuroethical concerns, long-term safety, affordability, and equitable access.

The future of smart polymer-based CNS drug delivery is bright (Table 6), with AI-guided design, biomimicry, and theranostic integration setting the stage for highly precise, adaptive, and patient-centric therapies. Bridging the remaining scientific, technical, and regulatory barriers will be critical to unlocking their full clinical potential and transforming neurological care in the decades to come.

Table 6. Projected milestones: Smart polymers for CNS drug delivery (2025–2040)

Year	Milestone	Details and Expectations
2025–2027	**AI-driven polymer screening platforms emerge**	Widespread adoption of machine learning for predicting polymer behaviour, drug compatibility, and degradation kinetics. Establishment of open-access databases for CNS-relevant smart polymers.
2026–2028	**First-in-human trials of dual-responsive biodegradable CNS systems**	Phase I/II trials for polymers that respond to pH/ROS or enzyme triggers in glioblastoma or neuroinflammation. Focus on safety, targeting accuracy, and pharmacokinetics.
2028–2030	**Approval of first CNS-targeted smart implant or injectable depot**	Regulatory approval of a long-acting, biodegradable, stimuli-responsive implant for brain tumours or chronic pain. Potential models based on PLGA or chitosan with FDA precedents.
2030–2032	**Commercial availability of biomimetic polymer systems**	Market introduction of polymers mimicking ECM, glial interfaces, or immune-evasive coatings. Enhanced biocompatibility and reduced neuroinflammatory response.
2032–2034	**Theranostic CNS polymers reach clinical trials**	Combined diagnostic and therapeutic polymer systems (MRI-visible, fluorescence-guided release) tested in Alzheimer's and glioma patients. Emphasis on tracking disease and drug response in real time.
2035–2037	**Personalized CNS polymer therapies enter clinical use**	3D-printed or microfluidic-fabricated drug delivery systems tailored to patient-specific biomarkers or therapy durations. Integration into personalized neuro-oncology and epilepsy management.
2038–2040	**Multimodal smart delivery platforms become standard in advanced CNS care**	Widespread clinical use of systems combining biodegradable, multi-stimuli-responsive, and biomimetic functionalities. Approved platforms for Parkinson's, Alzheimer's, glioblastoma, and post-stroke therapy. - Regulatory harmonization across regions.

Footnotes: Significant Enablers for These Milestones

- Development of standardized testing protocols for dynamic stimuli-responsiveness and biodegradability.
- Advances in neuroimaging and biosensing for theranostic integration.
- Growing regulatory flexibility for combination drug-device platforms.
- Public-private partnerships fostering translational research and commercialization.

CONCLUSION

Summary of Key Insights

This chapter has explored the rapidly evolving field of stimuli-responsive and biodegradable polymers for CNS drug delivery, highlighting their potential to overcome the long-standing challenges posed by the BBB and the complex pathophysiology of neurological disorders. Beginning with the anatomy and transport limitations of the BBB, we examined how smart polymers are engineered to respond to specific physiological or pathological stimuli (pH shifts, oxidative stress, enzyme activity, and external triggers like light or magnetic fields) to release therapeutic agents precisely where and when they are needed.

Various types of smart responses were classified and compared, including pH-sensitive systems for tumour microenvironments, redox-sensitive systems targeting oxidative stress, and enzyme-sensitive polymers leveraging disease-specific biomarkers. A wide range of biodegradable polymers (PLA, PLGA, PCL, and chitosan) have demonstrated favourable safety profiles and compatibility with CNS tissues. The integration of stimuli-responsiveness with biodegradability has led to the development of hybrid systems that offer both controlled release and safe clearance from the body.

Case studies illustrated how these advanced systems are being applied to treat diseases such as glioblastoma, Alzheimer's, Parkinson's, and stroke. Despite their promise, challenges such as premature drug release, immune response, scalability, and regulatory hurdles continue to constrain clinical translation.

Outlook on the Role of Smart Biodegradable Polymers in CNS Therapeutics

Smart polymers, also known as stimuli-responsive polymers, are advanced materials that undergo reversible changes in their physical or chemical properties in response to external stimuli such as temperature, pH, light, magnetic fields, or biological signals. These changes can include alterations in shape, solubility, conductivity, or mechanical strength.

In drug delivery and biomedical applications, smart polymers can be engineered to respond to specific physiological conditions (acidic environment of a tumour or the elevated temperature of inflamed tissue) enabling targeted, controlled, and on-demand release of therapeutic agents. Looking ahead, smart biodegradable polymers are poised to play a transformative role in the future of CNS therapy. The integration of AI-assisted design, biomimetic materials, and theranostic capabilities will enable more precise, adaptive, and responsive drug delivery platforms. These systems can be tailored to individual patients, paving the way for personalized neuro-medicine that adapts to disease progression and therapeutic response in real time.

As clinical trials evolve and regulatory pathways mature, we anticipate the emergence of next-generation CNS-targeted implants, injectable depots, and responsive nanocarriers that not only overcome the BBB but also dynamically interact with the neurological microenvironment. Smart biodegradable polymers bridge the gap between molecular precision and neurological treatment, offering new hope for previously incurable diseases. This balance of scientific innovation and clinical necessity marks a pivotal advancement.

REFERENCES

Aflori, M. (2021). Smart nanomaterials for biomedical applications—A review. *Nanomaterials (Basel, Switzerland), 11*(2), 396. DOI: 10.3390/nano11020396 PMID: 33557177

Alahmari, A. (2021). Blood-brain barrier overview: Structural and functional correlation. *Neural Plasticity, 2021*(1), 6564585. DOI: 10.1155/2021/6564585 PMID: 34912450

Alshamrani, A. I., Al Hadi, A. Y., Almutairi, A. H., ALkhweildy, W. S., Aldubiban, G. M., Moafa, A. M., Alshehri, F. A. M., Alnami, M., Alqarni, A. A., Ayoub, Z. H., Banajia, M. A., Almutairi, , Alnami, F. M. A., Safhi, A. M. A., & Aldossary, F. K. M. (2024). Advancing Targeted Drug Delivery: The Role of Smart Polymers in Responsive Therapeutic Systems. *Egyptian Journal of Chemistry, 67*(13), 1549–1556. DOI: 10.21608/ejchem.2024.337880.10836

Anwar, A., Imran, M., & Iqbal, H. M. (2023). Smart chemistry and applied perceptions of enzyme-coupled nano-engineered assemblies to meet future biocatalytic challenges. *Coordination Chemistry Reviews, 493*, 215329. DOI: 10.1016/j.ccr.2023.215329

Archer, E., Torretti, M., & Madbouly, S. (2023). Biodegradable polycaprolactone (PCL) based polymer and composites. *Physical Sciences Reviews, 8*(11), 4391–4414. DOI: 10.1515/psr-2020-0074

Badparvar, F., Marjani, A. P., Salehi, R., & Ramezani, F. (2024). Dual pH/redox-responsive hyperbranched polymeric nanocarriers with TME-trigger size shrinkage and charge reversible ability for amplified chemotherapy of breast cancer. *Scientific Reports, 14*(1), 8567. DOI: 10.1038/s41598-024-57296-4 PMID: 38609391

Begum, R. F., Singh, S., Prajapati, B., Sumithra, M., & Patel, R. J. (2025). Advanced targeted drug delivery of bioactive agents fortified with graft chitosan in management of cancer: A review. *Current Medicinal Chemistry, 32*(19), 3759–3789. Advance online publication. DOI: 10.2174/0109298673285334240112104709 PMID: 38415441

Bhattacharya, S., Prajapati, B. G., Ali, N., Mohany, M., Aboul-Soud, M. A., & Khan, R. (2023). Therapeutic potential of methotrexate-loaded superparamagnetic iron oxide nanoparticles coated with poly (lactic-co-glycolic acid) and polyethylene glycol against breast cancer: Development, characterization, and comprehensive in vitro investigation. *ACS Omega, 8*(30), 27634–27649. DOI: 10.1021/acsomega.3c03430 PMID: 37546601

Caraway, C. A., Gaitsch, H., Wicks, E. E., Kalluri, A., Kunadi, N., & Tyler, B. M. (2022). Polymeric nanoparticles in brain cancer therapy: A review of current approaches. *Polymers, 14*(14), 2963. DOI: 10.3390/polym14142963 PMID: 35890738

Dachani, S. R., Vashi, A., Mundada, A. B., Mundada, P. A., Rudrangi, S. R. S., Rudrangi, S., & Tiwari, R. (2025). Innovative polymers in pharmaceutical chemistry: Revolutionizing drug delivery systems. *Polymer-Plastics Technology and Materials, 64*(7), 911–933. DOI: 10.1080/25740881.2024.2440531

Dai, X. J., Li, W. J., Xie, D. D., Liu, B. X., Gong, L., & Han, H. H. (2025). Stimuli-Responsive Nano Drug Delivery Systems for the Treatment of Neurological Diseases. *Small, 2410030*(9). Advance online publication. DOI: 10.1002/smll.202410030 PMID: 39840482

Dhiman, A., Rana, D., Benival, D., & Garkhal, K. (2025). Comprehensive insights into glioblastoma multiforme: Drug delivery challenges and multimodal treatment strategies. *Therapeutic Delivery, 16*(1), 87–115. DOI: 10.1080/20415990.2024.2415281 PMID: 39445563

Fagehi, T. H. M., Alanazi, H. S., Alshehri, A. A. M., Alanazi, M. M., Matari, H. A. Y., Ayoub, T. M., Alrashedi, E. A., Alotaibi, A. U. S., Alrasheedi, S. M. M., & Zawli, H. M. N. (2024). Innovative Drug Delivery Systems: Nanotechnology, Liposomes, or Hydrogels for Targeted Drug Delivery. *Journal of International Crisis and Risk Communication Research, 7*(S11), 368. DOI: 10.63278/jicrcr.vi.353

Fattah-alhosseini, A., Chaharmahali, R., Alizad, S., Kaseem, M., & Dikici, B. (2024). A review of smart polymeric materials: Recent developments and prospects for medicine applications. *Hybrid Advances, 100178*, 100178. Advance online publication. DOI: 10.1016/j.hybadv.2024.100178

Feig, V. R., Tran, H., & Bao, Z. (2018). Biodegradable polymeric materials in degradable electronic devices. *ACS Central Science, 4*(3), 337–348. DOI: 10.1021/acscentsci.7b00595 PMID: 29632879

Geranmayeh, M. H., Rahbarghazi, R., & Farhoudi, M. (2019). Targeting pericytes for neurovascular regeneration. *Cell Communication and Signaling : CCS, 17*(1), 26. DOI: 10.1186/s12964-019-0340-8 PMID: 30894190

Grahl, M. V., Hohl, K. S., Smaniotto, T., & Carlini, C. R. (2025). Microbial Trojan Horses: Virulence Factors as Key Players in Neurodegenerative Diseases. *Molecules (Basel, Switzerland), 30*(3), 687. DOI: 10.3390/molecules30030687 PMID: 39942791

Hersh, A. M., Alomari, S., & Tyler, B. M. (2022). Crossing the blood-brain barrier: Advances in nanoparticle technology for drug delivery in neuro-oncology. *International Journal of Molecular Sciences*, *23*(8), 4153. DOI: 10.3390/ijms23084153 PMID: 35456971

Houldsworth, A. (2024). Role of oxidative stress in neurodegenerative disorders: A review of reactive oxygen species and prevention by antioxidants. *Brain Communications*, *6*(1), fcad356. Advance online publication. DOI: 10.1093/braincomms/fcad356 PMID: 38214013

Ioannou, P., Katsoulieris, E., & Afratis, N. A. (2025). Matrix Dynamics and Microbiome Crosstalk: Matrix Metalloproteinases as Key Players in Disease and Therapy. *International Journal of Molecular Sciences*, *26*(8), 3621. DOI: 10.3390/ijms26083621 PMID: 40332093

Karayianni, M., & Pispas, S. (2021). Block copolymer solution self-assembly: Recent advances, emerging trends, and applications. *Journal of Polymer Science*, *59*(17), 1874–1898. DOI: 10.1002/pol.20210430

Karlsson, J., Rhodes, K. R., Green, J. J., & Tzeng, S. Y. (2020). Poly (beta-amino ester) s as gene delivery vehicles: Challenges and opportunities. *Expert Opinion on Drug Delivery*, *17*(10), 1395–1410. DOI: 10.1080/17425247.2020.1796628 PMID: 32700581

Kruczkowska, W., Gałęziewska, J., Grabowska, K. H., Gromek, P., Czajkowska, K., Rybicki, M., Kciuk, M., & Kłosiński, K. K. (2025). From Molecules to Mind: The Critical Role of Chitosan, Collagen, Alginate, and Other Biopolymers in Neuroprotection and Neurodegeneration. *Molecules (Basel, Switzerland)*, *30*(5), 1017. DOI: 10.3390/molecules30051017 PMID: 40076240

Li, C., Deng, Z., & Gillies, E. R. (2023). Designing polymers with stimuli-responsive degradation for biomedical applications. *Current Opinion in Biomedical Engineering*, *25*, 100437. DOI: 10.1016/j.cobme.2022.100437

Liu, J., Liu, J., Mu, W., Ma, Q., Zhai, X., Jin, B., Liu, Y., & Zhang, N. (2024). Delivery strategy to enhance the therapeutic efficacy of Liver Fibrosis via Nanoparticle Drug Delivery systems. *ACS Nano*, *18*(32), 20861–20885. DOI: 10.1021/acsnano.4c02380 PMID: 39082637

Lu, H., Cai, Z., & Hu, P. (2024). Recent Advances in Polymeric Delivery Vehicles for Controlled and Sustained Drug Release. *Pharmaceutics*, *16*(9), 1184. DOI: 10.3390/pharmaceutics16091184 PMID: 39339220

Maddiboyina, B., Desu, P. K., Vasam, M., Challa, V. T., Surendra, A. V., Rao, R. S., Alagarsamy, S., & Jhawat, V. (2022). An insight of nanogels as novel drug delivery system with potential hybrid nanogel applications. *Journal of Biomaterials Science. Polymer Edition*, *33*(2), 262–278. DOI: 10.1080/09205063.2021.1982643 PMID: 34547214

Massironi, A. (2024). Hybrid organic polymer/inorganic nano-materials for biomedical applications: Where we are and where to go? *Current Nanoscience*, *20*(2), 188–205. DOI: 10.2174/1573413719666230410113733

McConnell, H. L., & Mishra, A. (2022). Cells of the blood–brain barrier: An overview of the neurovascular unit in health and disease. *The Blood-Brain Barrier. Methods and Protocols*, *2492*, 3–24. DOI: 10.1007/978-1-0716-2289-6_1 PMID: 35733036

Mohammed, P. N., Hussen, N. H., Hasan, A. H., Salh, H. J. H., Jamalis, J., Ahmed, S., Bhat, A. R., & Kamal, M. A. (2025). A review on the role of nanoparticles for targeted brain drug delivery: Synthesis, characterization, and applications. *EXCLI Journal*, *24*, 34. DOI: 10.17179/excli2024-7163 PMID: 39967907

Mukherjee, C., Varghese, D., Krishna, J. S., Boominathan, T., Rakeshkumar, R., Dineshkumar, S., Brahmananda Rao, C. V. S., & Sivaramakrishna, A. (2023). Recent advances in biodegradable polymers–properties, applications and future prospects. *European Polymer Journal*, *192*, 112068. DOI: 10.1016/j.eurpolymj.2023.112068

Nance, E., Pun, S. H., Saigal, R., & Sellers, D. L. (2022). Drug delivery to the central nervous system. *Nature Reviews. Materials*, *7*(4), 314–331. DOI: 10.1038/s41578-021-00394-w PMID: 38464996

Panchpuri, M., Painuli, R., & Kumar, C. (2025). Artificial Intelligence in Smart Drug Delivery Systems: A Step Toward Personalized Medicine. *RSC Pharmaceutics*. Advance online publication. DOI: 10.1039/D5PM00089K

Parvaresh, A., Izadi, Z., Nemati, H., Derakhshankhah, H., & Jaymand, M. (2023). Redox-and pH-responsive alginate-based magnetic hydrogel:"Smart" drug delivery and protein corona studies. *Journal of Molecular Liquids*, *382*, 121990. DOI: 10.1016/j.molliq.2023.121990

Patkar, S. V., & Sherje, A. (2025). Recent Advances in Smart Hydrogel for Drug Delivery in Cancer Therapeutics. *Indian Journal of Pharmaceutical Education and Research*, *59*(3), 854–864. DOI: 10.5530/ijper.20250454

Rani, E. E., Sanjana, D. S., Karthikeyan, E., & Nandhini, J. (2025). Smart Nanomaterials: Current State and Future Prospects in Drug Delivery and Tissue Engineering. *Biomedical Materials & Devices*, 1–28. DOI: 10.1007/s44174-025-00350-0

Rawal, S. U., Patel, B. M., & Patel, M. M. (2022). New drug delivery systems developed for brain targeting. *Drugs, 82*(7), 749–792. DOI: 10.1007/s40265-022-01717-z PMID: 35596879

Ren, Y., Li, P., Xie, Y., Xu, J., Luo, Q., Chen, M., Liu, R., Feng, H., Chen, Y., Liu, Y., Bao, C., Duan, J., Li, J., & Lu, W. (2025). Dual-responsive nanoparticles for enhanced drug delivery in breast Cancer chemotherapy. *Journal of Controlled Release, 377*, 146–161. DOI: 10.1016/j.jconrel.2024.11.026 PMID: 39549730

Rezaei, N., Akbarzadeh, I., Kazemi, S., Montazeri, L., Zarkesh, I., Hossein-Khannazer, N., Hassan, M., & Vosough, M. (2021). Smart materials in regenerative medicine. *Modern Medical Laboratory Journal, 4*(1), 39–51. DOI: 10.30699/mmlj17.4.1.39

Ruiz, M. E., & Scioli-Montoto, S. (2024). Routes of drug administration. In *ADME processes in pharmaceutical sciences: dosage, design, and pharmacotherapy* (pp. 129–172). Springer Nature Switzerland. DOI: 10.1007/978-3-031-50419-8_7

Seth, A. K., Parmar, G., Aundhia, C., Shah, N., & Gohil, D. (2024). Smart polymer systems: A futuristic approach to enhance therapeutic efficacy. *Current Organic Chemistry, 28*(15), 1164–1178. DOI: 10.2174/0113852728305580240429100851

Shi, J., Tan, C., Ge, X., Qin, Z., & Xiong, H. (2024). Recent advances in stimuli-responsive controlled release systems for neuromodulation. *Journal of Materials Chemistry. B, 12*(24), 5769–5786. Advance online publication. DOI: 10.1039/D4TB00720D PMID: 38804184

Singh, J., & Nayak, P. (2023). pH-responsive polymers for drug delivery: Trends and opportunities. *Journal of Polymer Science, 61*(22), 2828–2850. DOI: 10.1002/pol.20230403

Singh, S. A., & Vellapandian, C. (2023). Structure of the blood brain barrier and its role in the transporters for the movement of substrates across the barriers. *Current Drug Metabolism, 24*(4), 250–269. DOI: 10.2174/1389200224666230608110349 PMID: 37291784

Song, B., Wu, M., Qin, L., Liang, W., & Wang, X. (2025). Smart Design of Targeted Drug Delivery System for Precise Drug Delivery and Visual Treatment of Brain Gliomas. *Advanced Healthcare Materials, 14*(4), 2402967. DOI: 10.1002/adhm.202402967 PMID: 39707642

Song, Y. H., De, R., & Lee, K. T. (2023). Emerging strategies to fabricate polymeric nanocarriers for enhanced drug delivery across blood-brain barrier: An overview. *Advances in Colloid and Interface Science, 320*, 103008. DOI: 10.1016/j.cis.2023.103008 PMID: 37776736

Susa, F., Arpicco, S., Pirri, C. F., & Limongi, T. (2024). An Overview on the Physiopathology of the Blood–Brain Barrier and the Lipid-Based Nanocarriers for Central Nervous System Delivery. *Pharmaceutics*, *16*(7), 849. DOI: 10.3390/pharmaceutics16070849 PMID: 39065547

Tanaka, M., & Vécsei, L. (2024). A decade of dedication: Pioneering perspectives on neurological diseases and mental illnesses. *Biomedicines*, *12*(5), 1083. DOI: 10.3390/biomedicines12051083 PMID: 38791045

Tao, Y., Chan, H. F., Shi, B., Li, M., & Leong, K. W. (2020). Light: A magical tool for controlled drug delivery. *Advanced Functional Materials*, *30*(49), 2005029. DOI: 10.1002/adfm.202005029 PMID: 34483808

Ueno, M. (2007). Molecular anatomy of the brain endothelial barrier: An overview of the distributional features. *Current Medicinal Chemistry*, *14*(11), 1199–1206. DOI: 10.2174/092986707780597943 PMID: 17504140

Van, D. B. S. L., Shalgunov, V., & Herth, M. M. (2022). Transport of nanomedicines across the blood-brain barrier: Challenges and opportunities for imaging and therapy. *Biomaterials Advances*, *141*, 213125. DOI: 10.1016/j.bioadv.2022.213125 PMID: 36182833

Vasanth, S., Misiriya, A., Thahsin, F., & Priya, S. (2025). Nonmagnetic Inorganic/Organic Core-Shell Nanoconstructs for Cancer Theranostics. In *Core-Shell Nano Constructs for Cancer Theragnostic: Current Scenario, Challenges and Regulatory Aspects* (pp. 365-388). Singapore: Springer Nature Singapore. DOI: 10.1007/978-981-96-3025-7_13

Vashist, R., Sangave, S. D., & Alexander, A. (2024). Surface active ligands for enhanced brain targeting of nanoparticles. In *Application of Nanocarriers in Brain Delivery of Therapeutics* (pp. 241–262). Springer Nature Singapore. DOI: 10.1007/978-981-97-2859-6_9

Visan, A. I., Popescu-Pelin, G., & Socol, G. (2021). Degradation behavior of polymers used as coating materials for drug delivery—A basic review. *Polymers*, *13*(8), 1272. DOI: 10.3390/polym13081272 PMID: 33919820

Wang, W., Meng, Q., Li, Q., Liu, J., Zhou, M., Jin, Z., & Zhao, K. (2020). Chitosan derivatives and their application in biomedicine. *International Journal of Molecular Sciences*, *21*(2), 487. DOI: 10.3390/ijms21020487 PMID: 31940963

Wang, Z., Zhai, B., Sun, J., Zhang, X., Zou, J., Shi, Y., & Guo, D. (2024). Recent advances of injectable in situ-forming hydrogels for preventing postoperative tumor recurrence. *Drug Delivery*, *31*(1), 2400476. DOI: 10.1080/10717544.2024.2400476 PMID: 39252545

Wells, C. M., Harris, M., Choi, L., Murali, V. P., Guerra, F. D., & Jennings, J. A. (2019). Stimuli-responsive drug release from smart polymers. *Journal of Functional Biomaterials*, *10*(3), 34. DOI: 10.3390/jfb10030034 PMID: 31370252

Wu, D., Chen, Q., Chen, X., Han, F., Chen, Z., & Wang, Y. (2023). The blood–brain barrier: Structure, regulation and drug delivery. *Signal Transduction and Targeted Therapy*, *8*(1), 217. DOI: 10.1038/s41392-023-01481-w PMID: 37231000

Xu, L., Cao, Y., Xu, Y., Li, R., & Xu, X. (2024). Redox-Responsive Polymeric Nanoparticle for Nucleic Acid Delivery and Cancer Therapy: Progress, Opportunities, and Challenges. *Macromolecular Bioscience*, *24*(3), 2300238. DOI: 10.1002/mabi.202300238 PMID: 37573033

Yetisgin, A. A., Cetinel, S., Zuvin, M., Kosar, A., & Kutlu, O. (2020). Therapeutic nanoparticles and their targeted delivery applications. *Molecules (Basel, Switzerland)*, *25*(9), 2193. DOI: 10.3390/molecules25092193 PMID: 32397080

Chapter 9
Polymer-Based Microneedles in Transdermal Drug Delivery Systems

Sajid Raza
https://orcid.org/0009-0003-6675-5553
Gomal University, Pakistan

Palwasha Bibi
University of Science and Technology Bannu Kpk, Pakistan

Jawad Riaz
University of Agriculture, Peshawar, Pakistan

Ali Shan
University of Science and Technology Bannu Kpk, Pakistan

Abul Kalam Azad
https://orcid.org/0000-0003-4874-1169
University MAIWP International, Malaysia

ABSTRACT

Polymeric microneedles (MNs) constitute an advanced technology in transdermal drug delivery that overcomes the limitations of traditional patches bypassing the stratum corneum to facilitate precise and minimum penetrative administration. By employing biodegradable polymers, such as PLGA, PCL, and hyaluronic acid, MNs offer enhanced biocompatibility, sustainability, and controlled drug release, addressing various medical applications, including vaccines, cosmeceuticals, and

DOI: 10.4018/979-8-3373-1047-3.ch009

gene therapy. Significant advancements include biosensor integration for real-time monitoring, nanotechnology for improved drug solubility and stability, and hydrogel systems for sustained drug release. Despite these advantages, challenges persist in addressing skin irritation, brittleness, and immunogenicity while ensuring product consistency through 3D printing. Advancements in regulatory compliance, machine learning-optimized trial designs, and safety protocols are crucial for the evolution of MNs in personalized medicine and for enhancing patient care and therapeutic outcomes.

INTRODUCTION

Microneedle Technology: An Overview

Microneedle technology is the principal system for drug delivery among all transdermal products because of its high efficiency. Microneedles as microscopic structures use invasive processes to penetrate only stratum corneum skin layers while creating temporary channels for medication delivery into the epidermis layer. An expanding body of research documents the advancements in microneedle technology that have occurred, especially in cosmetic and therapeutic development, during the previous five years. Artificial collagen stimulation techniques launched initially to address facial scarring and skin rejuvenation needs have evolved to become one of the treatment modalities with extensive dermatology and dermatologic surgery, despite the lack of verified clinical data. The miniaturization of hypodermic needles down to the micron scale has attracted great interest among researchers because the diminution of the physical characteristics of particulates and devices has a significant impact on their physicochemical and biological characteristics as well as their performance (McCrudden et al., 2015). Research efforts and publications investigating the design, development, and applications of microneedles have increased exponentially in recent years, particularly in cosmetic and therapeutic applications. The collagen-induction therapy of microneedling began treating facial scars and improving skin appearance, but modern scientific research has extended these applications throughout the dermatology field (Houshmand, 2021). Research interest in micron-scale hypodermic needles has grown extensively because decreasing the physical dimensions influences the properties and functions of both biological agents and medical devices simultaneously. Researchers have used microneedles across the cosmeceutical market to alter skin cells through tissue alteration, which triggers elastin and collagen production while transporting pharmaceutical products across dermal tissue barriers.

Figure 1. Polymer microneedles for transdermal drug delivery

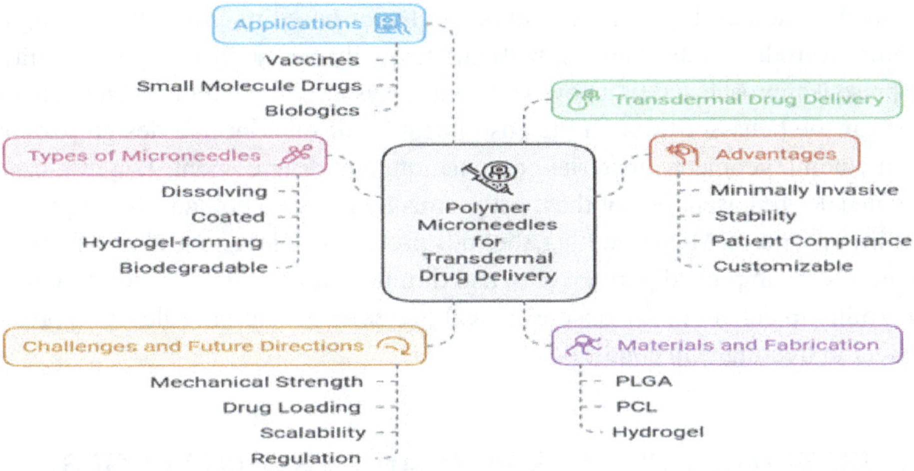

An Overview on The Potential of Polymer Based Microneedle System

Current transdermal therapy requires polymer-based microneedles, which show promise for replacing traditional sharp needle delivery systems. The microneedle platform has continued to transform over recent years through needle size reductions that lead to better patient compliance, reduced pain experience, and better drug delivery outcomes. The reduction in needle dimensions has important effects on both physiological characteristics and biological traits, which help produce better therapeutic results and decrease patient discomfort while improving drug delivery precision. The use of 3D printing technologies for microneedle fabrication has led to progress in manufacturing capabilities and product versatility according to studies by (Bhatnagar et al., 2017). The cosmetic and therapeutic applications of microneedles have been extensively researched for prolonged periods, and multiple products using this technology have now received clinician approval. The newly generated microneedles effectively penetrated the protective stratum corneum to open skin microchannels that allowed the application of dermal and epidermal active ingredients. This unique mechanism allows for the delivery of a wide range of molecules, including therapeutic drugs and pharmaceuticals, together with active cosmetic substances. Researchers have extensively explored the use of polymers in microneedles because they offer properties to those of bulk materials. The properties required for mechanical flexibility can be engineered in biocompatible polymers that offer potential use in minimally invasive transdermal delivery. 3D printing technology has enabled the development of complex polymer-based microneedles through a system that produces multiple designed and accurately measured microneedle arrays. The present approach acts as a powerful control system for regulating drug delivery and cosmetic behavior through the customization of microneedle design parameters, including geometry, material composition, and drug content. Polymer-based microneedles represent one of the most promising research domains for improving the skin appearance. Microneedling, the body produces collagen and elastin for better skin elasticity, improved pore texture, and diminished lines or wrinkles. Owing to their ability to deliver active cosmeceutical ingredients, microneedles have proven efficient in treating skin conditions.

TRANSDERMAL DRUG DELIVERY AND ITS CHALLENGES

Transdermal drug delivery has been identified as a novel approach to conventional oral and injectable drug delivery systems and has the following advantages: improved patient compliance, fewer side effects, and sustained drug concentration in the body. Transdermal drug delivery systems are difficult to manage after installation, even though their implementation faces specific problems. The protective properties of the skin layers will challenge any transdermal drug delivery system because the skin acts as an important problem for these methods. The stratum corneum forms the outermost layer of the skin and is primarily composed of flat dead cells intertwined with lipids and protein fibers packed together. Three alternative methods have been investigated for stratum corneum adjustment, including chemical permeation enhancers and microneedle patches, in conjunction with iontophoresis devices, to facilitate drug delivery through the skin. Transdermal drug delivery systems extremely limited range of drugs. A successful transdermal drug must be medium-sized in molecular weight with the minimum required concentration for optimal delivery through the skin. However, large natural drug molecules with high water affinities remain particularly resistant to penetration (Prausnitz & Langer, 2008). Transdermal drug delivery methods are difficult because both the intricate structure of the skin and the wide range of drug properties present complications to innovation. In addition, skin moisture permeability and blood flow density of hair follicles and sweat glands affect skin barrier characteristics. Researchers now employ essential oils that affect the skin barrier efficiency alongside permeation enhancers along with microneedle devices and transmucosal applications for medication administration as potential delivery methods for complicated skin barriers. Despite the technical hurdles, the outlook for enhancing drug delivery through transdermal administration appears promising.

CHALLENGES ASSOCIATED WITH EFFECTIVE TRANSDERMAL DRUG ABSORPTION

The transdermal drug delivery system has become the ideal choice because it benefits from improved patient compliance along with the avoidance of reduced drug efficacy from first-pass metabolism and can maintain the drug at steady-state concentration levels inside the body. Effective transdermal drug delivery encounters technical obstacles because of the skin barrier, which starts at the stratum corneum, covering the outermost three to five layers of the skin. According to the present data, the stratum corneum presents a significant challenge for transdermal drug absorption. The layer has dead compressed cells combined with intercellular lipids, which provides selective and efficient diffusion barriers for both small and large molecular structures. Drug properties, including molecular size, solubility, polarity, and charge, limit their passage through the stratum corneum (Khan et al., 2023). Scientists have experienced permeation enhancers to overcome this obstacle. Chemical agents temporarily modify the SC structure or function to improve drug diffusion rates through the skin. Aromatherapy elements have enhanced the transdermal delivery of metoprolol succinate through intercellular lipids in the SC. Key formulation delivery methods, including lipid-based nanocarriers, are effective for increasing skin permeability and accelerating transdermal drug delivery. These nanocarriers protect pharmaceutical compounds and aid transdermal delivery, as affirmed by Tran (2013).

Role of the Stratum Corneum and Epidermal Barrier in Molecular Absorption

Skin the body largest organ protects against further pathological and environmental attacks, while managing thermoregulation. The outermost part of the epidermis, which contains interposed flattened cornified keratinocytes within the extracellular lipid lamellae, defines the stratum corneum. Because of its unique structure and contents, the stratum corneum demonstrates strong resistance to penetration, resulting in skin impermeability to transdermal drug delivery. The molecular structure and chemical characteristics of penetrants establish the barrier qualities of the skin SC. Two basic physicochemical characteristics can be distinguished among candidates for skin permeation: lipophilia and molecular size. Compounds with lipophilic characteristics dissolve much more effectively within the intercellular lipids of the stratum corneum than larger molecular structures, while smaller molecules move easily through spaces between corneocytes (Scheuplein, 2015).

Limitations of Transdermal Patches

Transdermal patches are widespread delivery methods for broadly used drugs because they administer drugs through skin application outside oral methods without causing pain. The primary limitations of these patches are the multiple disadvantages that compromise their utilization and performance. Transdermal systems to deliver more than a limited selection of drugs because of developmental constraints. The barrier function of the skin originates from the stratum corneum, which blocks many molecules but permits passage through only lipophilic molecules of small size. Only a limited number of medications have suitable physicochemical properties that enable them to be contained within transdermal patch due to their low molecular weight and high lipophilicity. (Larrañeta et al., 2016). Transdermal patches make it challenging to maintain exact levels of drug delivery rates and durations. Skin conditions, such as temperature and moisture, combined with drug physicochemical properties, determine how drugs pass through the skin. These specific characteristics lead to different drug concentration levels inside the body but make the maintenance of necessary therapeutic levels problematic (Sorathia et al., 2020).

POLYMER-BASED MICRONEEDLES: TYPES OF MATERIALS AND MODES OF MANUFACTURING

The field of drug delivery has recently introduced polymer-based microneedles, which function as novel minimally invasive devices and provide better bioavailability and reduced discomfort during patient use. Micron-scale devices with linear dimensions of 50–900 μm operate at the skin's stratum corneum level to avoid the dermis containing sensitive nerves, because reaching this layer could cause patient discomfort and create open pathways to infection. The design of microneedle systems requires careful selection of the appropriate polymeric materials. Poly (lactic-co-glycolic acid), poly (vinyl alcohol), and chitosan are biocompatible bioresorbable polymers, and research has demonstrated desirable mechanical properties, drug compatibility, and effective drug delivery rates from these materials. The device performance and practical usability of polymer-based microneedles are significantly dependent on their production and processing technologies. Fabrication of multiple hydrogel polymers takes place through three widely adopted methods: micromolding and micropatterning alongside ink jetting with respective pros and cons, as explained later (Zheng et al., 2020). Researchers utilize three-dimensional printing to establish optimal microneedle shapes alongside multiple material applications. Polymer-based microneedles have demonstrated greater suitability for industrial applications after recent research revealed their effectiveness in therapeutic treatments and aesthetic uses involving drug delivery through skin penetrating, vaccine administration, and cosmetic distribution (Badnikar et al., 2020).

Polymers Used in the Fabrication of Polymer Microneedles

Recent advancements in cosmetic treatments demonstrate that polymer microneedle usage has gained traction because it can break through the skin cell structure and boost collagen and elastic fiber production, while delivering cosmetic pharmaceutical molecules enhanced by combining microneedles with additional light or external treatments. The alternative polymer microneedle system benefits from current advances in nanofabrication technologies. According to Bennet and Kim (2014), targeted drug delivery systems and skin treatment applications benefit from modern technologies that create particles and structures with consistent size and shape characteristics.

Biodegradable Polymers: PLGA and PCL

Scientists have shown great research interest in biodegradable polymers because they display appropriate features for multiple applications, making them ideal for medical applications. Poly (lactic-co-glycolic acid) coupled with polycaprolactone is an effective material for medical applications because of its promising capabilities across medical devices and drug delivery fields. Poly lactic-co-glycolic acid, which scientists often refer to as poly forms when lactic acid is chemically bonded with glycolic acid. Due to its array of beneficial properties, which include in situ degradation potential, inherent biocompatibility, and attributes, this synthetic biodegradable polymer occupies a essential position as a commonly used biomaterial in the biomedical field. Human physiological systems and urination processes safely remove biocompatible, non-toxic hydrolytic degradation products from PLGA materials, which enables their use in biomedical applications. Polycaprolactone is an aliphatic polyester, both in medical research domains and industry, due to intensive contemporary interest in its use. Because PCL is both biodegradable and biocompatible, it remains an industrially processable material that services numerous medical devices and pharmaceutical delivery systems. These biodegradable polymers serve various purposes in the biomedical industry because of their ability to fulfill numerous unmet needs, as described previously.

Table 1. Polymers commonly used in polymer-based microneedles for transdermal drug delivery systems: Properties, applications, and references

Polymer	Type	Properties	Applications	Reference
Hyaluronic Acid	Biodegradable	Biocompatible, hydrophilic, non-toxic, and skin-friendly	Enhances transdermal delivery of macromolecules and small drugs; dissolving microneedles	Kim et al. (2012)
Polyvinylpyrrolidone (PVP)	Synthetic	High water solubility, flexible, and thermoplastic	Dissolvable microneedles for rapid drug release	Donnelly et al. (2010)
Polyvinyl Alcohol (PVA)	Synthetic	Biocompatible, flexible, and forms robust microneedle structures	Delivery of vaccines and biologics	Larrañeta et al. (2016)
Poly(lactic-co-glycolic acid) (PLGA)	Biodegradable	Biocompatible, tunable degradation, and mechanical strength	Sustained drug release systems using microneedles	Prausnitz et al. (2004)
Chitosan	Natural	Biocompatible, bioadhesive, and biodegradable	Controlled drug delivery; antimicrobial microneedle coatings	Sullivan et al. (2008)
Gelatin	Natural	Biocompatible, flexible, and forms hydrogels	Transdermal delivery of peptides and proteins	Yu et al. (2015)
Carboxymethylcellulose (CMC)	Semi-synthetic	Hydrophilic, non-toxic, and biodegradable	Dissolving microneedles for delivery of hydrophilic drugs	Larrañeta et al. (2015)
Polyurethane (PU)	Synthetic	High mechanical strength, elastic, and non-biodegradable	Microneedles for diagnostic sensing and minimally invasive drug delivery	Lee et al. (2017)

MECHANISMS OF DRUG DELIVERY

Targeted delivery of therapeutic drugs to specific body sites in modern pharmacotherapy, and this capability influences both treatment efficacy and safety. Drug transportation challenges within various bodily barriers have led previous studies on pharmaceutical development to focus on their research efforts. Modern drug delivery systems primarily improve the water solubility and chemical stability of APIs in advanced pharmaceutical compounds to enhance drug effectiveness and minimize toxic reactions. Drugs stored in nanoparticle carriers, such as micellar systems, liposomes, and dendritic forms, along with nanocapsules, demonstrate great potential through their ability to protect drugs against breakdown and assist precise delivery to specific cells or tissues. (Li et al., 2019). The features of nanoparticles that include increased surface area to volume aspect ratio, and the ability to be coated or functionalized makes them ideal for drug delivery. Nanoparticles can bind with biomolecules in a way that allows for internalization in cells; with a large surface area that can allow for the conjugation of target ligands or antibodies, contributes to directing drug delivery to certain locations within the body (Nakagawa et al., 2023). To overcome various physiological barriers that can impede the effective transport and distribution of drugs. The key objective of drug delivery systems is to optimize the aqueous solubility and chemical stability of active pharmaceutical compounds, leading to improved pharmacological activity and diminished adverse side effects. Drug protection from degradation, together with targeted delivery modifications, has been enabled through nanoparticle-based carriers, such as micelle liposome dendrimers and nanocapsules. (Li et al., 2019). Nanoparticles possess unique physical and chemical properties, along with distinct geometric configurations, including a high surface-area-to-volume ratio and modular design options that enable an array of benefits when used in drug delivery applications. The extensive surface area of nanoparticles enables biomolecular interactions, which promote cellular uptake while allowing the attachment of targeting ligands or antibodies to increase drug delivery efficiency to targeted body sites, as shown in Nakagawa et al. 's 2023 research, which may affect the transport and distribution of drugs. The principal aims of current drug delivery systems are to increase the low water solubility and poor chemical stability of the API of value-added pharmaceuticals to boost their pharmacological activity and minimize adverse effects. Nanoparticle-based carriers, such as micellar, liposomal, dendritic, and nanocapsular systems, appear to have enormous potential because they can protect drugs from degradation and deliver drugs selectively to target cells or tissues (Li et al., 2019). The features of nanoparticles that include increased surface area to volume aspect ratio, and the ability to be coated or functionalized makes them ideal for drug delivery. Nanoparticles can bind with biomolecules in a way that allows for internalization in cells; with a large surface area that can allow for the conjugation of target ligands or antibodies, contributes to directing drug delivery to certain locations within the body (Nakagawa et al., 2023). used to overcome various physiological barriers that can impede the effective transport and distribution of drugs. One of the primary goals of modern drug delivery systems is to improve the aqueous solubility and chemical stability of active pharmaceutical ingredients, thereby enhancing their pharmacological activity while reducing undesirable side effects. Nanoparticle-based carriers, such as micelles, liposomes, dendrimers, and nanocapsules, have shown great potential in this regard, as they can protect drugs from degradation and facilitate their targeted delivery to specific cells or tissues. (Li et al., 2019). The unique physicochemical and geometric characteristics of nanoparticles, including their high surface-area-to-volume ratio and ability to be chemically and structurally tailored, confer several advantages for drug delivery applications. Nanoparticles can interact with biomolecules to facilitate cellular uptake, and their large surface area enables the attachment of targeting ligands or antibodies, further enhancing their ability to deliver drugs to desired sites within the body (Nakagawa et al., 2023). Drug molecules must be directly bonded to both drug carriers and targeting agents to improve drug delivery. The advantages of the drug delivery method include altered pharmacokinetics and biodistribution profiles, which extend plasma half-life while preventing drug degradation and metabolism, together with precise organ and subcellular drug targeting.

EVALUATION OF POLYMERIC MICRONEEDLES

Polymeric microneedles (MNs) are suitable for transdermal drug delivery because they penetrate the stratum corneum to administer drugs directly into the dermal layer. Evaluation of critical parameters, including morphology, mechanical strength, and drug release kinetics, is essential for determining the effectiveness of polymeric MNs. Tests of mechanical toughness ensured that rigid Mn needles could penetrate the skin without cracking during the evaluation of the drug release profile using Franz diffusion cells. The ability of drug delivery systems to traverse skin layers can be demonstrated through basic permeation assays utilizing microscopy ex vivo human or animal skin. Biodegradable polymers have shown recent progress in micro-needles by lowering the toxic effects on the environment while enhancing the sustainability of medication delivery. According to Ita (2021), in vivo experiments confirm the safety and pharmacokinetic performance of MNs, while demonstrating their capacity for patient-friendly and pain-free drug administration through therapeutic applications.

Figure 2. Polymeric microneedle evaluation

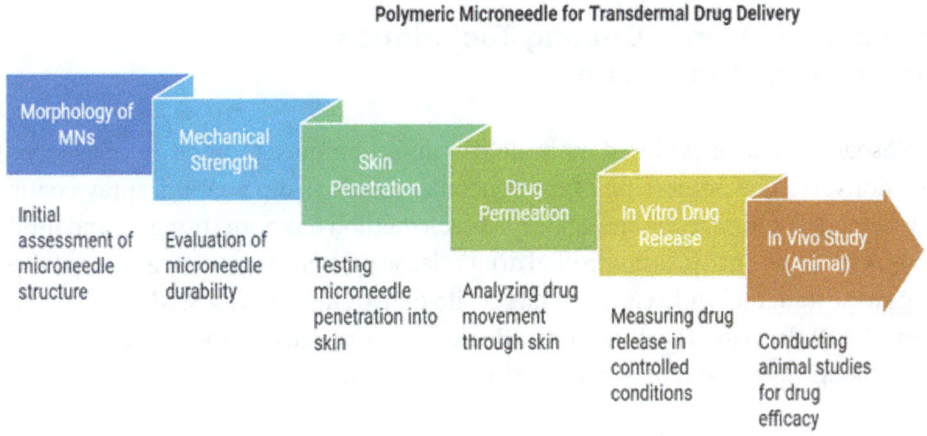

Benefits and Applications

Several advantages make biodegradable microneedles superior to the current transdermal drug delivery systems. Microneedles result in increased patient compliance and comfort because the repeated use of traditional hypodermic needles leads to patient discomfort. The temporary disruption of the skin's stratum corneum layer enables numerous therapeutic substances, ranging from small molecules to biomolecules, to penetrate the skin for delivery when these therapeutic molecules normally cannot pass through. In addition to drug delivery applications, dissolving microneedles can be used as delivery systems for both vaccine administration and cosmeceuticals aimed at skin recovery, together with treatments for different skin disorders. These microneedles needles approach medicine delivery in a focused manner because they combine drug formula flexibility with precise delivery regulation capabilities.

Coated Microneedles: Delivery Mechanisms and Challenges

The microneedle drug delivery method is an effective solution for distributing pharmaceutical agents past the skin surface barrier, known as the stratum corneum, so that they can enter directly through skin tissues. A fundamental characteristic of microneedle-based drug delivery systems is their ability to load and release diverse therapeutic agents.

Application of Drug Coating Techniques and Delivery Mechanism

Researchers have explored multiple coating techniques to improve microneedle drug delivery. The processing of microneedles employs dip coating, spray coating, and vapor deposition techniques because each method has both strengths and limitations. As coatings possess controlled drug release properties, researchers can design drugs to sustain their release at the target site for extended periods. Mishra and their team (2019) described that microneedle delivery depends on their insertion where their coatings break down to release drugs into skin tissues.

Figure 3. Development process of polymer based microneedle

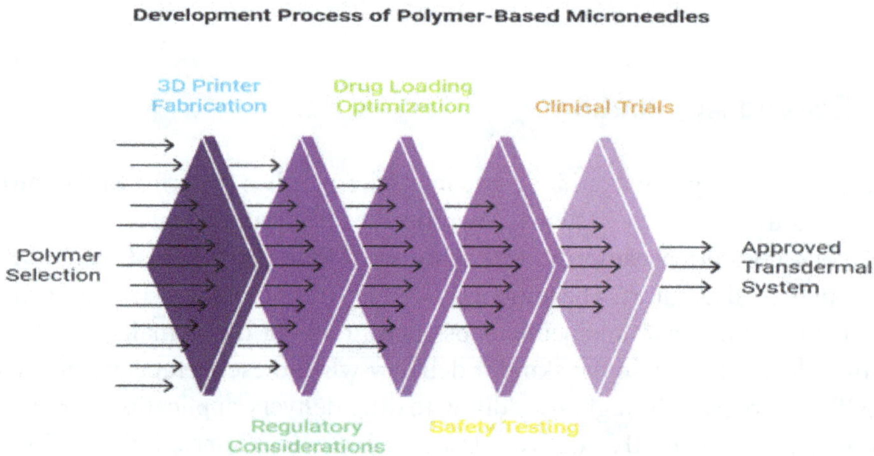

Hydrogel-Forming Microneedles: Controlled Release and Long-Term Drug Delivery Study

Both their protective and encapsulation functions, along with their ability to sustain and remotely adapt spatial-temporal drug release, allow hydrogels to serve as a flexible drug delivery medium. The water-swollen three-dimensional structures of these networks originate from polymers alongside proteins and colloids or small molecular components and possess customizable responsiveness to particular stimuli that establish specific release kinetics and durations matching the prescribed treatment needs. The swelling behavior of the hydrogel was the dominant factor in determining the release of drugs from the hydrogel material. By modifying both the composition and structure, researchers can retain drugs within porous hydrogel matrices for controlled release throughout a specified time period. Hydrogels show sustained drug release, which extends therapeutic outcomes and reduces dosage requirements, while promoting better patient adherence. (Nguyen et al., 2023). The field of long-term drug delivery has shown strong interest in minimally invasive delivery systems, such as the hydrogel-forming microneedles discussed in this thesis. Researchers can now structure these devices to detect specific triggers that enable remote control of drug delivery systems that release medication only in particular locations and moments, as needed (Sosnik & Seremeta, 2017). The process of embedding drugs into hydrogel matrices contained in microneedles enables both sustained payload delivery and precise drug control, which has the potential to increase treatment effectiveness and ease patient adherence to medication regimens. A careful balance of the swelling behavior of hydrogel-forming microneedles with the drug release kinetics and mechanical properties is necessary when designing systems for extended drug delivery.

Table 2. Summary of categories of microneedles (MNs)

Category	Material	Fabrication	Advantage	Limitation	Reference
Solid MNs	Metals, silicon, ceramics	Laser cutting, etching, photolithography	High mechanical strength	Infection, inflammation	Johnson et al. (2020); Kim et al. (2019)
Coated MNs	Metals, silicon, ceramics	Spray coating, dip coating, piezoelectric inkjet printing	High mechanical strength, single-step application	Lower drug capacity	Park et al. (2020); Yan et al. (2021)
Hollow MNs	Metals, silicon, polymers	Microelectromechanical systems, 3D printing	High scalability, high stability	Complex and expensive fabrication method	Prausnitz et al. (2019); Lee et al. (2021)
Hydrogel MNs	Crosslinking polymers	Micromolding, 3D printing	Unbreakable, multifunctional	Insufficient mechanical strength and toxicity	Singh et al. (2022); Zhao et al. (2021)
Dissolvable MNs	Biodegradable and biocompatible polymers	Micromolding, drawing lithography, 3D printing	Scale-up fabrication, high biocompatibility	Limit dosing and inconsistent pharmacokinetics	Jamaledin et al. (2020); Wang et al. (2022)
Ceramic MNs	Silica-based ceramics, hydroxyapatite	Micromolding, 3D printing	Excellent biocompatibility, stability	Brittle nature	Zhao et al. (2022); Yang et al. (2021)

continued on following page

Table 2. Continued

Category	Material	Fabrication	Advantage	Limitation	Reference
Metal MNs	Stainless steel, titanium	Photolithography, etching, laser cutting	High strength, precise skin penetration	Risk of corrosion, cost	Larrañeta et al. (2020); Gittard et al. (2019)
Polymeric MNs	Biodegradable polymers	Micromolding, injection molding, 3D printing	Biocompatibility, controlled drug release	Limited drug loading	Donnelly et al. (2021); Park et al. (2021)
Hybrid MNs	Combination of ceramics, polymers, and metals	Multistep fabrication methods	Improved mechanical and drug release properties	High production cost	Xu et al. (2020); Zhang et al. (2021)

Innovative Vaccine Delivery Mechanisms

Various studies have focused on improving the convenience and effectiveness of vaccine delivery methods. An effective new technique involves the utilization of natural immune system components, such as polymorphonuclear leukocytes, which carry influenza vaccines. Microneedle arrays serve as an innovative strategy by which DNA vaccines show improved safety, while maintaining high levels of efficacy. According to Williams (2013), neutrophils perform key roles in the innate immune system as polymorphonuclear leukocytes. The migratory properties of these cells to sites of infection and inflammation make them appealing options for influenza vaccine delivery systems. In this study, we developed a delivery system for vaccine antigens that achieved precise targeting, while producing stronger immune responses that lasted longer. The method of using polymorphonuclear leukocytes for vaccine transport eliminates several difficulties associated with traditional intramuscular or subcutaneous vaccine delivery, such as diminished dendritic cell density in muscled tissue. (Liang et al., 2012) Two powerful areas of investigation for vaccine delivery include DNA vaccine development with microneedle array platforms. DNA vaccines exhibit advantages, including potential safety enhancements, as well as their ability to stimulate both immune system responses and their fast development and production timelines. Microscopic microneedle arrays are modern delivery platforms that demonstrate promising capabilities for DNA vaccines by penetrating the epidermal layer of the skin. New minimally invasive vaccine delivery systems offer improved antigen transport abilities while enhancing the effectiveness of vaccines and overcoming the acceptance barriers associated with conventional needles.

MICRONEEDLE DELIVERY OF SMALL MOLECULE DRUGS: OVERCOMING CHALLENGES

Modern medicine, small-molecule drugs have emerged as foundational elements that deliver target-specific solutions to all disease categories. Drug delivery systems struggle when delivering small-molecule drugs because these treatments suffer from low bioavailability combined with narrow therapeutic windows and unintended side effects at non-target sites. Microneedle-based delivery systems are a potential solution for overcoming these pharmaceutical delivery challenges. Microneedles are micron-scale inserts that pass through the outer skin layer of the stratum corneum, enabling them to transport small-molecule medications alongside other therapeutic elements. Drug bioavailability improves when this approach avoids skin barrier mechanisms and facilitates drug entry into systemic circulation. Microneedle drug delivery of small-molecule treatments shows promising potential for patients with hypertension, as well as those suffering from diabetes or cancer. Microneedle-based delivery enables antihypertensive drugs to improve their pharmacokinetic profiles, resulting in more reliable blood pressure management across stable dosages. Microneedle-based delivery systems boost antidiabetic drug absorption and bioavailability to improve glycemic control in patients, and chemotherapy using microneedle delivery methods can overcome both poor tumor penetration and restricted therapeutic windows that occur during standard intravenous or oral drug delivery. Nanocarrier-based microneedle systems allow healthcare professionals to direct potent cytotoxic drugs specifically to tumors without affecting the surrounding healthy tissue levels.

Delivery of Biologics: Challenges and Innovative Approaches to Overcoming these Challenges

Delivery systems for biologics containing proteins together with peptides and antibodies must overcome greater obstacles compared to small-molecule drug delivery. The complex architecture of biologics manifests as several delivery barriers, including degradation, aggregation, and immunogenic behavior. Administration of biologics is a fundamental problem in drug delivery. Traditionally, the delivery of most biologics has required invasive parenteral administration methods, such as intravenous infusion or subcutaneous injection, which results in patient dissatisfaction and low compliance rates for people receiving chronic disease treatments. In an effort to improve user experience, scientists have developed alternative delivery systems that use hydrogel microneedles to administer insulin through a much less invasive process. The delivery of biologics is an important obstacle in maintaining their structure and stability throughout the manufacturing process as well as throughout the storage period. Biologics face degradation through unfolding and aggregation, and laboratory results show that they lose biological activity when exposed to temperature changes, pH extremes, and increased shear forces (Bizeau & Mertz, 2020). The stability of biologics has been enhanced through new manufacturing methods that utilize stabilizing excipients and processing optimizations after researchers respond to manufacturing challenges that also affect monoclonal antibody and growth hormone deliveries. Biologics that require high dosing concentrations embody problems from manufacturing high viscosity levels that affect both production and delivery paths. The biological delivery field remains innovatively active because continuous technological advancements help researchers overcome existing operational challenges.

Innovations in Pain Management and Gene Therapy through Microneedle Based Devices

Researchers in microneedle technology have seen substantial field advancements with new hot topic applications focusing on pain management and gene therapy.

Microneedle Pain Management

Pain management problems appear to be effectively addressed by scientists exploring microneedle technology. The transdermal delivery functionality of microneedles requires researchers to create new methodologies for the targeted delivery of analgesics and similar agents while eliminating reliance on oral or parental routes. This approach may deliver better pain management results because microneedles deliver controlled sustained drug release throughout the treatment duration and target drugs directly at their desired sites. The current study demonstrates the application of microneedle-based nanoparticle delivery systems for the treatment of central nervous system disorders, as specified by Li et al. (2023). Nanoparticulate delivery systems achieve direct therapeutic agent delivery to brain tissue when merged with microneedle systems, bypassing the blood-brain barrier limitations. The treatment approach presented promising potential applications for multiple neurological conditions, including chronic pain disorder treatment.

mRNA Delivery with Gene Therapy Using Microneedles

The fields of gene therapy and mRNA delivery demonstrate exciting potential applications of microneedle technology. DNA modifications and additional structural changes represent powerful techniques to explore proteins and serve advanced applications in characterizing protein behavior and creating protein capsules. The use of nanoparticles to distribute mRNA vaccines has demonstrated vital importance during the COVID-19 pandemic, which represents an emerging infectious disease because these vaccines support worldwide virus containment efforts. Microneedle systems have been merged with nanoparticles, such as PLGA, to create new opportunities in gene delivery techniques for effective outcomes. Microneedle systems deliver therapeutic genetic material through nanoparticles, which both protect and ensure precise cellular delivery. Researchers must continue to optimize nanoparticle-based gene delivery methods while developing materials and techniques that enhance the safety and manufacturing capabilities of these systems for clinical applications. (Piperno et al., 2021).

Polymeric Microneedles Advantages

There are multiple benefits to choosing polymers during microneedle manufacturing. Polymer microneedles have adjustable mechanical properties suitable for specific applications owing to their flexibility and rigidity. Health professionals can produce microneedle arrays that can target multiple skin compositions and specific application areas. Scientists can engineer polymers with active pharmaceutical ingredients that enable dual-function drug delivery and skin rejuvenation therapy.

Microneedle Arrays for Mechanical Strength and Penetration

Skin penetration effectiveness and drug delivery success depend on the unbreakable microneedle device construction. To achieve successful microneedle insertion, two crucial considerations stand among which are the stiffness of the microneedle for skin penetration, while minimizing the risk of structural failure during insertion. Microneedles designed for successful stratum corneum penetration must prevent buckling, which requires a minimum level of stiffness. Precise control over the microneedle geometry and material composition achieved through photopolymerization and extrusion-based 3D printing enables the optimization of their mechanical strength through these methods. Stiffness enhancement of microneedles depends on meticulous adjustments of the polymer material, whereas strategies involving reinforcing agents that resist skin entry forces and prevent needle breakage remain critical issues. The minimal dimensions, together with the breakable nature of microneedles, make them especially vulnerable to damage during penetration through skin areas with different mechanical strengths. Scientists have investigated how microneedle devices composed of composite materials (composed of polymers infused with inorganic powders) function to extend their mechanical robustness.

Drug-Loading Capacity and Optimization

Drug delivery systems remain a critical problem in the pharmaceutical industry. Microneedle devices in the form of tiny needles function as an advanced method for transdermal drug delivery because they can painlessly enter the skin to directly transport drugs to their intended location. Despite progress in the field, the optimization of microneedle drug-loading capacity represents one of the largest remaining obstacles to creating fully efficient delivery systems. Recent research has focused on multiple approaches to improve drug encapsulation within microneedle structures. Magneto-electric nanocarriers deliver promising improvements in biocompatibility, together with adjustable surface modifications that balance drug loading and preservation as well as enabling controlled drug release through electrical-magnetic adjustments. An extra treatment approach utilizes non-lamellar liquid crystalline nanoparticles, known as cubosomes and hexosomes, which are recognized as superior carriers for poorly water-soluble drugs with inadequate bioavailability and pharmacokinetic properties (Yaghmur & Mu, 2021). Research efforts to improve microneedle drug loading have incorporated formulation strategies beyond those of nanocarrier systems. Previous research has shown that nanostructured lipid carriers enhance drug encapsulation and release drugs differently, making them suitable structural platforms for transdermal drug delivery, according to Jaiswal et al. (2014). Current developments towards microneedles with high drug loads show possibility but call for deeper refinement because of existing challenges and available prospects for enhancement.

CHALLENGES OF REGULATORY AND SAFETY WITH POLYMER MICRONEEDLE TECHNOLOGY

The development of polymer-based microneedle delivery systems has received extensive research attention owing to their potential utility in transdermal drug delivery and multiple biomedical applications. The successful implementation of these new devices requires them to clear significant regulatory challenges, along with methodological safety evaluations.

Polymer Microneedle Safety Protocols

Combine skin penetration with electronic monitoring functionality, render safety a critical concern. The selection of biocompatible materials and risk assessment of skin irritation potential while verifying microneedle stability must form essential design elements for these medical devices. Polymer microneedle medical devices require several procedural steps, thus demanding safety and efficacy assessments through in vitro and in vivo experimentation. Safety evaluations of devices need to measure skin penetration depth as well as quantify skin irritation reactions, systemic absorption of substances, and track the distribution of active substances employed. Firms need effective quality control measures, along with solid post-market monitoring systems, to detect emergent safety threats during the post-market life cycle. Through strict regulatory oversight combined with comprehensive safety procedures, researchers and manufacturers can identify safe pathways for transferring polymer microneedle technology through clinical translation and deployment, benefiting both patients and healthcare providers. Within the evolving health technology space, smart microneedles now stand at the forefront, as they present new possibilities for real-time monitoring systems along with direct drug delivery applications. Miniaturized devices which integrate drug release controls with biosensing capabilities can respond to undetectable skin penetration. The latest research acknowledges smart microneedles as a promising tool for therapeutic drug monitoring because they serve as a critical element of personalized medicine performance. These biosensors may stop all repeated blood draws throughout the development of specific drugs owing to technology and the lowest analyte measurement capabilities noted by Ranamukhaarachchi et al. (2017). Microneedle fabrication methods are developed using multiple materials and manufacturing approaches that permit the creation of more advanced and bespoke structures.

CONCLUSION

The drug delivery system, along with patient monitoring practices, currently operates on a new framework driven by technological improvements in polymer microneedles. Microneedles represent a pain-free substitute for conventional injection techniques through transdermal delivery with a controlled release function that enables personalized medical therapies for drug monitoring purposes. The widespread adoption of microneedle technology in commercial markets requires overcoming multiple regulatory barriers owing to device safety and functionality standards and further mechanical integrity concerns during biocompatible evaluations. By adding biosensing features to smart microneedles, these medical devices could achieve new functionality for real-time monitoring, in addition to creating tailored treatment plans. Current research, together with clinical trials on microneedles, shows promising opportunities to transform healthcare. Applications combining practical manufacturing solutions, such as 3D printing and AI-driven clinical studies, enhance device tailoring and improve medication scalability and cost benefits. The development and deployment of polymer-based microneedles will succeed through enhanced partnerships among scientists, industrial producers, and regulatory organizations to both evaluate safety and compliance requirements and expand the potential benefits for medical providers and patients.

REFERENCES

Badnikar K., Jayadevi S. N., Pahal S., Sripada S., Nayak M. M., Vemula P. K., & Subrahmanyam D. N. (2020). Generic molding platform for simple, low-cost fabrication of polymeric microneedles. *Macromolecular Materials and Engineering*, 305(5). DOI: 10.1002/mame.202000072

Bennet, D., & Kim, S. (2014). Polymer Nanoparticles for Smart Drug Delivery. In *InTech eBooks*. DOI: 10.5772/58422

Bhatnagar S., Dave K., & Venuganti V. V. K. (2017). Microneedles in the clinic. *Journal of Controlled Release*, 260, 164. DOI: 10.1016/j.jconrel.2017.05.029

Bizeau, J., & Mertz, D. (2020). Design and applications of protein delivery systems in nanomedicine and tissue engineering. *Advances in Colloid and Interface Science*, 287, 102334. DOI: 10.1016/j.cis.2020.102334

Casalini, T., Bassas-Galià, M., Girard, H., Castrovinci, A., Carolis, A. D., Brianza, S., Zinn, M., & Perale, G. (2019). A Systematic Experimental and Computational Analysis of Commercially Available Aliphatic Polyesters. *Applied Sciences*, 9(16), 3397. DOI: 10.3390/app9163397

Choudhary, U., Bhinchhar, B. K., Paswan, V. K., Kharkwal, S., Yadav, S. P., & Singh, P. (2021). Utilization of Agro-Industrial Wastes as Edible Coating and Films for Food Packaging Materials. In *IntechOpen eBooks*. IntechOpen. DOI: 10.5772/intechopen.99786

Donnelly, R., Alkilani, A., & McCrudden, M. (2021). Microneedle-based systems for drug delivery: A review. *European Journal of Pharmaceutics and Biopharmaceutics, 161*, 100–118. DOI: 10.1016/j.ejpb.2020.10.014

Donnelly, R. F., Moffatt, K., & Alkilani, A. Z. (2010). Hydrogel-forming microneedles prepared from "super swelling" polymers for transdermal drug delivery. *Pharmaceutical Research, 27*(3), 399–408.

Gittard, S., Bohnsack, J., & Hsia, Y. (2019). Development of microneedle arrays for controlled release applications. *Journal of Drug Targeting*, *27*(8), 732–740. DOI: 10.1080/1061186X.2019.1624443

Houshmand, E. B. (2021). *Introduction to Microneedling*. DOI: 10.1002/9781119431817

Ita, K. (2021). Current status of microneedles for transdermal drug delivery: Recent advances and future perspectives. *Acta Pharmaceutica Sinica. B*, *11*(3), 738–755. DOI: 10.1016/j.apsb.2020.10.014 PMID: 33777679

Jaiswal, P., Gidwani, B., & Vyas, A. (2014). Nanostructured lipid carriers and their current application in targeted drug delivery. *Artificial Cells Nanomedicine and Biotechnology*, 44(1), 27. DOI: 10.3109/21691401.2014.90982

Jamaledin, R., Ahmed, H., & Liu, T. (2020). Development of microneedles for drug delivery. *Journal of Pharmaceutical Development*, *18*(2), 117–130. DOI: 10.1016/j.jpd.2020.07.005

Johnson, R., Smith, L., & Williams, K. (2020). Advances in transdermal drug delivery systems. *Journal of Pharmaceutical Sciences*, *109*(5), 1489–1498. DOI: 10.1016/j.xphs.2020.01.010

Khan M. S., Mohapatra S., Gupta V., Ali A., Naseef P. P., Kuruniyan M. S., Alshadidi A. A. F., Alam M. S., Mirza Mohd. A., & Iqbal Z. (2023). Potential of lipid-based nanocarriers against two major barriers to drug delivery—skin and blood–brain barrier. *Membranes*, 13(3), 343. DOI: 10.3390/membranes13030343

Kim, H., Lee, J., & Park, M. (2019). Nanotechnology for drug delivery: Trends and applications. *Journal of Nanomedicine*, *15*(2), 72–84. DOI: 10.2217/nnm.19.3

Kim, Y.-C., Park, J.-H., & Prausnitz, M. R. (2012). Microneedles for drug and vaccine delivery. *Advanced Drug Delivery Reviews*, *64*(14), 1575–1586. DOI: 10.1016/j.addr.2012.04.005 PMID: 22575858

Larrañeta, E., Lutton, R. E. M., & Woolfson, A. D. (2016). Biodegradable polymer microneedles: In vitro experiments and finite element analysis modeling. *Journal of Controlled Release*, *225*, 193–204.

Larrañeta E., Lutton R. E. M., Woolfson A. D., & Donnelly R. F. (2016). Microneedle arrays as transdermal and intradermal drug delivery systems: Materials science, manufacture and commercial development. *Materials Science and Engineering R Reports*, 104, 1. DOI: 10.1016/j.mser.2016.03.001

Larrañeta, E., Moore, J., & Vicente-Perez, E. M. (2015). A novel biopolymer formulation for drug delivery via dissolvable microneedles. *Materials Science and Engineering C*, *45*, 214–221. DOI: 10.1016/j.msec.2014.09.070

Larrañeta, E., Stewart, S., & Donnelly, R. (2020). Microneedles for transdermal drug delivery: Current state of the art. *Journal of Controlled Release*, *321*, 143–157. DOI: 10.1016/j.jconrel.2020.03.013

Lasprilla, A. J. R., Martínez, G. A., Lunelli, B. H., Jardini, A. L., & Filho, R. M. (2011). Poly-lactic acid synthesis for application in biomedical devices — A review. *Biotechnology Advances*, 30(1), 321. DOI: 10.1016/j.biotechadv.2011.06.019

Lee, J. W., Park, J.-H., & Prausnitz, M. R. (2017). Microneedle-mediated delivery of nanoparticles and DNA vaccines: Progress and challenges. *Advanced Functional Materials, 27*(10), 1605281. DOI: 10.1002/adfm.201605281

Lee, Y., Zhang, X., & Wang, J. (2021). Biodegradable polymers in drug delivery systems. *Biomaterials Science, 9*(5), 1895–1908. DOI: 10.1039/d0bm01667c PMID: 33432933

Li B., Geng L., Liu W., Liao L., Ban J., & Lu Z. (2023). Formulation and evaluation of PLGA nanoparticulate-based microneedle system for potential treatment of neurological diseases. *International Journal of Nanomedicine*, 3745. DOI: 10.2147/IJN.S415728

Li C., Wang J., Wang Y., Gao H., Wei G., Huang Y., Yu H., Gan Y., Wang Y., Mei L., Chen H., Hu H., Zhang Z., & Jin Y. (2019). Recent progress in drug delivery. *Acta Pharmaceutica Sinica B*, 9(6), 1145. DOI: 10.1016/j.apsb.2019.08.003

Liang F., Sandgren K., Fausther-Bovendo H., O'Hagan D. T., Seubert A., Gregorio E. D., Rao S., Sullivan N., Seder R. A., Koup R., & Loré K. (2012). Infiltration of dendritic cells and antigen uptake in the muscle after injection of HIV-1 Env gp120 in adjuvant. *Retrovirology*, 9. DOI: 10.1186/1742-4690-9-S2-O16

Lu, Q., Cui, Z., Wu, Y., Wang, H., Zhang, S., Guan, J., & Mao, S. (2022). Challenges and Strategies to Enhance the Systemic Absorption of Inhaled Peptides and Proteins. *Pharmaceutical Research*, 40(5), 1037. DOI: 10.1007/s11095-022-03435-3

McCrudden M., McAlister E., Courtenay A. J., González-Vázquez P., Singh T. R. R., & Donnelly R. F. (2015). Microneedle applications in improving skin appearance. *Experimental Dermatology*, 24(8), 561. DOI: 10.1111/exd.12723

Mishra, D. K., Pandey, V., Maheshwari, R., Ghode, P., & Tekade, R. K. (2019). *Cutaneous and Transdermal Drug Delivery. Elsevier eBooks, 595.* Elsevier BV. DOI: 10.1016/B978-0-12-817909-3.00015-7

Modi, K., Modi, M., Mishra, D., Mittal, P., Sorathiya, U., & Shelat, P. (2013). Oral Controlled Release Drug Delivery System: An Overview. *International Research Journal of Pharmacy, 4*(3), 70–76. DOI: 10.7897/2230-8407.04312

Nakagawa Y., Ushidome K., Masuda K., Igarashi K., Matsumoto Y., Yamasoba T., Anraku Y., Takai M., & Cabral H. (2023). Multi-armed star-shaped block copolymers of poly(ethylene glycol)-poly(furfuryl glycidol) as long circulating nanocarriers. *Polymers*, 15(12), 2626. DOI: 10.3390/polym15122626

Nayak, A. K., Maji, R., & Das, B. (2010). Gastroretentive drug delivery systems: A review. *Asian Journal of Pharmaceutical and Clinical Research*, *3*(1). https://ajpcr.com/Vol3Issue1/250.pdf

Nguyen, H. T., Chien, T. B., & Cư ng, Đ. X. (2023). Polymer-Based Hydrogels Applied in Drug Delivery: An Overview. *Gels*, 9(7), 523. DOI: 10.3390/gels9070523

Park, C., Jeong, S., & Kim, Y. (2021). Advanced delivery systems for transdermal applications. *Drug Development and Industrial Pharmacy*, *47*(10), 1620–1632. DOI: 10.1080/03639045.2021.1895573

Park, S., Lee, J., & Kim, Y. (2020). Enhanced skin penetration with lipid nanoparticles for transdermal delivery. *International Journal of Pharmaceutics*, *585*, 119440. DOI: 10.1016/j.ijpharm.2020.119440

Piperno, A., Sciortino, M. T., Giusto, E., Montesi, M., Panseri, S., & Scala, A. (2021). Recent Advances and Challenges in Gene Delivery Mediated by Polyester-Based Nanoparticles. *International Journal of Nanomedicine*, 5981. DOI: 10.2147/IJN.S321329

Prausnitz, M., Langer, R., & Chen, X. (2019). Transdermal drug delivery: Past, present, and future. *Advanced Drug Delivery Reviews*, *64*(1), 40–57. DOI: 10.1016/j.addr.2012.07.008

Prausnitz M. R., & Langer R. (2008). Transdermal drug delivery. *Nature Biotechnology*, 26(11), 1261. DOI: 10.1038/nbt.1504

Prausnitz, M. R., Mikszta, J. A., & Andrianov, A. K. (2004). Microneedle-based vaccines. *Journal of Controlled Release*, *97*(3), 503–511. PMID: 15212882

Ranamukhaarachchi, S., Padeste, C., Häfeli, U. O., Stoeber, B., & Cadarso, V. J. (2017). Design considerations of a hollow microneedle-optofluidic biosensing platform incorporating enzyme-linked assays. *Journal of Micromechanics and Microengineering*, 28(2), 24002. DOI: 10.1088/1361-6439/aa9c9c

Rodríguez-Fragoso L., Gutiérrez-Sancha I., Rodríguez-Fragoso P., Rodríguez-López A., & Reyes-Esparza J. (2014). Pharmacokinetic properties and safety of cadmium-containing quantum dots as drug delivery systems. *InTech eBooks*. DOI: 10.5772/58553

Scheuplein, R. J. (2015). Permeability of the Skin: A Review of Major Concepts. *Current Problems in Dermatology*, 172. DOI: 10.1159/000401285

Sharma S., Hatware K., Bhadane P., Sindhikar S., & Mishra D. K. (2019). Recent advances in microneedle composites for biomedical applications: Advanced drug delivery technologies. *Materials Science and Engineering C*, 103, 109717. DOI: 10.1016/j.msec.2019.05.002

Singh, S., Kumar, A., & Sharma, R. (2022). Nanoparticles in transdermal drug delivery: Current research and future perspectives. *Pharmaceutical Research*, *39*(10), 1562–1577. DOI: 10.1007/s11095-022-03322-z

Sorathia, K., Patel, M., Soni, T., Vaghasiya, V., Patel, J., & Suhagıa, B. N. (2020). Effect of essential oils on transdermal permeation of metoprolol succinate. *Bulletin of Pharmaceutical Sciences*, *0*(0), 0. Advance online publication. DOI: 10.21608/bfsa.2020.41240.1028

Sosnik, A., & Seremeta, K. P. (2017). Polymeric Hydrogels as Technology Platform for Drug Delivery Applications. *Gels*, 3(3), 25. DOI: 10.3390/gels3030025

Sullivan, S. P., Koutsonanos, D. G., & Del Pilar Martin, M. (2008). Dissolving polymer microneedle patches for influenza vaccination. *Biomaterials*, *29*(13), 1928–1941.

Sun Y., & Davis E. (2021). Nanoplatforms for targeted stimuli-responsive drug delivery: A review of platform materials and stimuli-responsive release and targeting mechanisms. *Nanomaterials*, 11(3), 746. DOI: 10.3390/nano11030746

Tran, T.-N. T. (2013). Cutaneous drug delivery: An update. *Journal of Investigative Dermatology Symposium Proceedings*, 16(1). DOI: 10.1038/jidsymp.2013.28

Wang, Y., Zhang, Y., & Liu, X. (2022). Microneedle-based drug delivery systems: A review. *International Journal of Pharmaceutics*, *617*, 114694. DOI: 10.1016/j.ijpharm.2022.114694

Wilczewska A. Z., Niemirowicz K., Markiewicz K. H., & Car H. (2012). Nanoparticles as drug delivery systems. *Pharmacological Reports*, 64(5), 1020. DOI: 10.1016/S1734-1140(12)70901-5

Williams J. (2013). Vector design for improved DNA vaccine efficacy, safety and production. *Vaccines*, 1(3), 225. DOI: 10.3390/vaccines1030225

Yaghmur, A., & Mu, H. (2021). Recent advances in drug delivery applications of cubosomes, hexosomes, and solid lipid nanoparticles. *Acta Pharmaceutica Sinica B*, 11(4), 871. DOI: 10.1016/j.apsb.2021.02.013

Yan, X., Liu, Z., & Zhang, J. (2021). Mechanisms and strategies for improving transdermal drug delivery. *Journal of Controlled Release*, *332*, 201–211. DOI: 10.1016/j.jconrel.2021.03.001

Yang, H., Shan, C., & Kolen, A. F. (2020). Medical Instrument Detection in Ultrasound-Guided Interventions: A Review. *arXiv (Cornell University).* Cornell University. /arxiv.2007.04807DOI: 10.48550

Yang, L., Wang, L., & Zhang, Q. (2021). Nanocarriers for transdermal drug delivery: Opportunities and challenges. *Pharmaceutical Nanocarriers*, *17*(3), 245–256. DOI: 10.1016/j.phnan.2021.03.002

Yu, W., Jiang, G., & Li, P. (2015). Gelatin methacryloyl-based microneedles for controlled release. *Acta Biomaterialia*, *12*, 220–230. DOI: 10.1016/j.actbio.2015.08.013

Zhao, J., Li, Y., & Zhang, Z. (2022). Skin penetration enhancement by electroporation for transdermal drug delivery. *Journal of Controlled Release*, *338*, 183–190. DOI: 10.1016/j.jconrel.2022.07.027

Zhao, W., Xu, X., & Chen, L. (2021). Controlled release formulations for transdermal applications. *International Journal of Nanomedicine*, *16*, 1373–1389. DOI: 10.2147/IJN.S308135

Zheng X., Xie J., Zhang X., Sun W., Zhao H., Li Y., & Wang C. (2020). An overview of polymeric nanomicelles in clinical trials and on the market. *Chinese Chemical Letters*, 32(1), 243. DOI: 10.1016/j.cclet.2020.11.029

Compilation of References

Abaee, A., Mohammadian, M., & Jafari, S. M. (2017). Whey and soy protein-based hydrogels and nano-hydrogels as bioactive delivery systems. *Trends in Food Science & Technology*, *70*, 69–81. DOI: 10.1016/j.tifs.2017.10.011

Abdl Aali, R. A. K., & Al-Sahlany, S. T. G. (2024). Gellan gum as a unique microbial polysaccharide: Its characteristics, synthesis, and current application trends. *Gels (Basel, Switzerland)*, *10*(3), 183. DOI: 10.3390/gels10030183 PMID: 38534601

Abdollahi, S., & Lotfipour, F. (2012). PLGA-and PLA-based polymeric nanoparticles for antimicrobial drug delivery. *Biomed Int*, *3*(1), 1–11.

Aflori, M. (2021). Smart nanomaterials for biomedical applications—A review. *Nanomaterials (Basel, Switzerland)*, *11*(2), 396. DOI: 10.3390/nano11020396 PMID: 33557177

Afshar, M., Dini, G., Vaezifar, S., Mehdikhani, M., & Movahedi, B. (2020). Preparation and characterization of sodium alginate/polyvinyl alcohol hydrogel containing drug-loaded chitosan nanoparticles as a drug delivery system. *Journal of Drug Delivery Science and Technology*, *56*, 101530. DOI: 10.1016/j.jddst.2020.101530

Aftab, M., Ikram, S., Ullah, M., Khan, N., Naeem, M., Khan, M. A., O'g'li, R. B. B., Qizi, K. S. S., Ugli, O. O. E., Abdurasulovna, B. M., & Qizi, O. K. A. (2025). Recent trends and future directions in 3D printing of biocompatible polymers. *Journal of Manufacturing and Materials Processing*, *9*(4), 129. DOI: 10.3390/jmmp9040129

Agarwal, R., Gupta, A., & Mehta, P. (2023). Personalized medicine and stimuli-responsive polymers: Innovations for on-demand drug delivery. *Trends in Biotechnology*, *41*(3), 290–302.

Agnihotri, S. A., Mallikarjuna, N. N., & Aminabhavi, T. M. (2004). Recent advances on chitosan-based micro- and nanoparticles in drug delivery. *Journal of Controlled Release*, *100*(1), 5–28. DOI: 10.1016/j.jconrel.2004.08.010 PMID: 15491807

Agrawal, A., Rellegadla, S., & Jain, S. (2019). Biomedical applications of PLGA particles. In *Elsevier eBooks* (p. 87). Elsevier BV., DOI: 10.1016/B978-0-12-816913-1.00004-0

Aguirre, M., Ballard, N., Gonzalez, E., Hamzehlou, S., Sardon, H., Calderon, M., Paulis, M., Tomovska, R., Dupin, D., Bean, R. H., Long, T. E., Leiza, J. R., & Asua, J. M. (2023). Polymer Colloids: Current Challenges, Emerging Applications, and New. *Macromolecules*, *56*(7), 2579–2607. DOI: 10.1021/acs.macromol.3c00108 PMID: 37066026

Ahmed, T. A., & Aljaeid, B. M. (2016). Preparation, characterization, and potential application of chitosan, chitosan derivatives, and chitosan metal nanoparticles in pharmaceutical drug delivery. *Drug Design, Development and Therapy*, *483*, 483. Advance online publication. DOI: 10.2147/DDDT.S99651 PMID: 26869768

Ahuja, V., Bhatt, A. K., Banu, J. R., Kumar, V., Kumar, G., Yang, Y. H., & Bhatia, S. K. (2023). Microbial exopolysaccharide composites in biomedicine and healthcare: Trends and advances. *Polymers*, *15*(7), 1801. DOI: 10.3390/polym15071801 PMID: 37050415

Akolade, J. O., Oloyede, H. O. B., Salawu, M. O., Amuzat, A. O., Ganiyu, A. I., & Onyenekwe, P. C. (2018). Influence of formulation parameters on encapsulation and release characteristics of curcumin loaded in chitosan-based drug delivery carriers. *Journal of Drug Delivery Science and Technology*, *45*, 11–19. DOI: 10.1016/j.jddst.2018.02.001

Al Khateb, K., Ozhmukhametova, E. K., Mussin, M. N., Seilkhanov, T. M., Rakhypbekov, T. K., & Khutoryanskiy, V. V. (2016). In situ gelling systems based on Pluronic F127/alginate: Development and ocular delivery of timolol maleate. *International Journal of Pharmaceutics*, *502*(1-2), 70–79. DOI: 10.1016/j.ijpharm.2016.02.027 PMID: 26899977

Alahmari, A. (2021). Blood-brain barrier overview: Structural and functional correlation. *Neural Plasticity*, *2021*(1), 6564585. DOI: 10.1155/2021/6564585 PMID: 34912450

Alaniz, L., Cabrera, P. V., Blanco, G., Ernst, G., Rimoldi, G., Alvarez, E., & Hajos, S. E. (2016). Interaction of CD44 with different forms of hyaluronic acid. Its role in adhesion and migration of tumor cells. *Cell Communication & Adhesion*, *9*(3), 117–130. DOI: 10.1080/15419060214522 PMID: 12521133

Al-Hussainawy, M. K., & Jasim Al-Hayder, L. S. (2024). Drug carriers in the delivery and release of hydroxychloroquine by biopolymer. *Chemicke Zvesti*, *78*(9), 5255–5266. DOI: 10.1007/s11696-024-03463-w

Ali, A., & Ahmed, S. (2018). A review on chitosan and its nanocomposites in drug delivery. *International Journal of Biological Macromolecules, 109*, 273–286. DOI: 10.1016/j.ijbiomac.2017.12.078 PMID: 29248555

Ali, S., Ranjha, N. M., Ahmad, B., Khan, A. A., Hassan, F. U., Aziz, T., Alharbi, M., Alshammari, A., Alasmari, A. F., & Alharbi, M. E. (2023). Sustained Release of Drug Facilitated Through Chemically Crosslinked Polyvinyl Alcohol-Gelatin (PVA-GE) Hydrogels. A sustainable biomedical approach. *Polish Journal of Chemical Technology, 25*(2), 56–65. DOI: 10.2478/pjct-2023-0017

Alkholief, M., Kalam, M. A., Raish, M., Ansari, M. A., Alsaleh, N. B., Almomen, A., Ali, R., & Alshamsan, A. (2023). Topical Sustained-Release Dexamethasone-Loaded Chitosan Nanoparticles: Assessment of Drug Delivery Efficiency in a Rabbit Model of Endotoxin-Induced Uveitis. *Pharmaceutics, 15*(9), 2273. DOI: 10.3390/pharmaceutics15092273 PMID: 37765242

Alshamrani, A. I., Al Hadi, A. Y., Almutairi, A. H., ALkhweildy, W. S., Aldubiban, G. M., Moafa, A. M., Alshehri, F. A. M., Alnami, M., Alqarni, A. A., Ayoub, Z. H., Banajia, M. A., Almutairi, , Alnami, F. M. A., Safhi, A. M. A., & Aldossary, F. K. M. (2024). Advancing Targeted Drug Delivery: The Role of Smart Polymers in Responsive Therapeutic Systems. *Egyptian Journal of Chemistry, 67*(13), 1549–1556. DOI: 10.21608/ejchem.2024.337880.10836

Alvarez-Lorenzo, C., & Concheiro, A. (2013). *Handbook of biodegradable polymeric materials and their applications*. American Scientific Publishers.

Alzoubi, L., Aljabali, A. A., & Tambuwala, M. M. (2023). Empowering Precision Medicine: The impact of 3D printing on personalized Therapeutic. *AAPS PharmSciTech, 24*(8), 228. Advance online publication. DOI: 10.1208/s12249-023-02682-w PMID: 37964180

Andersen, T., Auk-Emblem, P., & Dornish, M. (2015). 3D cell culture in alginate hydrogels. *Microarrays (Basel, Switzerland), 4*(2), 133–161. DOI: 10.3390/microarrays4020133 PMID: 27600217

Anderson, J. M., & Shive, M. S. (2012). Biodegradation and biocompatibility of PLA and PLGA microspheres. *Advanced Drug Delivery Reviews, 64*, 72–82. DOI: 10.1016/j.addr.2012.09.004

Anirudhan, T. S., Nima, J., & Divya, P. L. (2015). Synthesis, characterization and in vitro 5-fluorouracil, an anticancer drug. *Applied Surface Science, 355*, 64–73. DOI: 10.1016/j.apsusc.2015.07.077

Anwar, A., Imran, M., & Iqbal, H. M. (2023). Smart chemistry and applied perceptions of enzyme-coupled nano-engineered assemblies to meet future biocatalytic challenges. *Coordination Chemistry Reviews*, *493*, 215329. DOI: 10.1016/j.ccr.2023.215329

Arabiyat, S. A. (2024). Prospectus and Concerns of Immunomodulatory Nanotechnologies and Nanoparticles Biocompatibility and Toxicity. In *Nanotechnology Based Microbicides and Immune Stimulators* (pp. 165–189). https://doi.org/DOI: 10.1007/978-981-96-0397-8_10

Aranaz, I., Alcántara, A. R., Civera, M. C., Arias, C., Elorza, B., Heras Caballero, A., & Acosta, N. (2021). Chitosan: An Overview of Its Properties and Applications. *Polymers*, *13*(19), 3256. DOI: 10.3390/polym13193256 PMID: 34641071

Archer, E., Torretti, M., & Madbouly, S. (2023). Biodegradable polycaprolactone (PCL) based polymer and composites. *Physical Sciences Reviews*, *8*(11), 4391–4414. DOI: 10.1515/psr-2020-0074

Atanase, L. I. (2021). Micellar drug delivery systems based on natural biopolymers. *Polymers*, *13*(3), 477. DOI: 10.3390/polym13030477 PMID: 33540922

Augst, A. D., Kong, H. J., & Mooney, D. J. (2006). Alginate hydrogels as biomaterials. *Macromolecular Bioscience*, *6*(8), 623–633. DOI: 10.1002/mabi.200600069 PMID: 16881042

Badnikar K., Jayadevi S. N., Pahal S., Sripada S., Nayak M. M., Vemula P. K., & Subrahmanyam D. N. (2020). Generic molding platform for simple, low-cost fabrication of polymeric microneedles. *Macromolecular Materials and Engineering*, 305(5). DOI: 10.1002/mame.202000072

Badparvar, F., Marjani, A. P., Salehi, R., & Ramezani, F. (2024). Dual pH/redox-responsive hyperbranched polymeric nanocarriers with TME-trigger size shrinkage and charge reversible ability for amplified chemotherapy of breast cancer. *Scientific Reports*, *14*(1), 8567. DOI: 10.1038/s41598-024-57296-4 PMID: 38609391

Bala, R., Rana, R. K., & Madaan, R. (2021). Formulation Design and Optimization of Sustained Released Matrix Tablets of Propranolol HCl Using Natural and Synthetic Polymers. *ACTA Pharmaceutica Sciencia*, *59*(2), 321–341. DOI: 10.23893/1307-2080.APS.05920

Balcerak-Woźniak, A., Dzwonkowska-Zarzycka, M., & Kabatc-Borcz, J. (2024). A Comprehensive Review of Stimuli-Responsive Smart Polymer Materials—Recent Advances and Future Perspectives. *Materials (Basel)*, *17*(17), 4255. DOI: 10.3390/ma17174255 PMID: 39274645

Bami, M. S., Estabragh, M. A. R., Khazaeli, P., Ohadi, M., & Dehghannoudeh, G. (2022). pH-responsive drug delivery systems as intelligent carriers for targeted drug therapy: Brief history, properties, synthesis, mechanism and application. *Journal of Drug Delivery Science and Technology*, *70*, 102987. DOI: 10.1016/j.jddst.2021.102987

Baryakova, T. H., Pogostin, B. H., Langer, R., & McHugh, K. J. (2023). Overcoming barriers to patient adherence: The case for developing innovative drug delivery systems. *Nature Reviews. Drug Discovery*, *22*(5), 387–409. DOI: 10.1038/s41573-023-00670-0 PMID: 36973491

Beach, M. A., Nayanathara, U., Gao, Y., Zhang, C., Xiong, Y., Wang, Y., & Such, G. K. (2024). Polymeric nanoparticles for drug delivery. *Chemical Reviews*, *124*(9), 5505–5616. DOI: 10.1021/acs.chemrev.3c00705 PMID: 38626459

Beck, A., Goetsch, L., Dumontet, C., & Corvaïa, N. (2017). Strategies and challenges for the next generation of antibody–drug conjugates. *Nature Reviews. Drug Discovery*, *16*(5), 315–337. DOI: 10.1038/nrd.2016.268 PMID: 28303026

Begum, R. F., Singh, S., Prajapati, B., Sumithra, M., & Patel, R. J. (2025). Advanced targeted drug delivery of bioactive agents fortified with graft chitosan in management of cancer: A review. *Current Medicinal Chemistry*, *32*(19), 3759–3789. Advance online publication. DOI: 10.2174/0109298673285334240112104709 PMID: 38415441

Bennet, D., & Kim, S. (2014). Polymer Nanoparticles for Smart Drug Delivery. In *InTech eBooks*. DOI: 10.5772/58422

Berillo, D., Zharkinbekov, Z., Kim, Y., Raziyeva, K., Temirkhanova, K., & Saparov, A. (2021). Stimuli-responsive polymers for transdermal, transmucosal and ocular drug delivery. *Pharmaceutics*, *13*(12), 2050. DOI: 10.3390/pharmaceutics13122050 PMID: 34959332

Bernkop-Schnürch, A., & Dünnhaupt, S. (2012). Chitosan-based drug delivery systems. *European Journal of Pharmaceutics and Biopharmaceutics*, *81*(3), 463–469. DOI: 10.1016/j.ejpb.2012.04.007 PMID: 22561955

Bhatnagar S., Dave K., & Venuganti V. V. K. (2017). Microneedles in the clinic. *Journal of Controlled Release*, 260, 164. DOI: 10.1016/j.jconrel.2017.05.029

Bhattacharya, S., Prajapati, B. G., Ali, N., Mohany, M., Aboul-Soud, M. A., & Khan, R. (2023). Therapeutic potential of methotrexate-loaded superparamagnetic iron oxide nanoparticles coated with poly (lactic-co-glycolic acid) and polyethylene glycol against breast cancer: Development, characterization, and comprehensive in vitro investigation. *ACS Omega*, *8*(30), 27634–27649. DOI: 10.1021/acsomega.3c03430 PMID: 37546601

Binauld, S., & Stenzel, M. H. (2012). Acid-degradable polymers for drug delivery: A decade of innovation. *Chemical Communications*, *49*(21), 2082–2102. DOI: 10.1039/c2cc36589h PMID: 23320254

Biswas, R., Mondal, S., Ansari, M. A., Sarkar, T., Condiuc, I. P., Trifas, G., & Atanase, L. I. (2025). Chitosan and Its Derivatives as Nanocarriers for Drug Delivery. *Molecules (Basel, Switzerland)*, *30*(6), 1297. DOI: 10.3390/molecules30061297 PMID: 40142072

Bizeau, J., & Mertz, D. (2020). Design and applications of protein delivery systems in nanomedicine and tissue engineering. *Advances in Colloid and Interface Science*, 287, 102334. DOI: 10.1016/j.cis.2020.102334

Boateng, J. S., Matthews, K. H., Stevens, H. N. E., & Eccleston, G. M. (2008). Wound healing dressings and drug delivery systems: A review. *Journal of Pharmaceutical Sciences*, *97*(8), 2892–2923. DOI: 10.1002/jps.21210 PMID: 17963217

Bobo, D., Robinson, K. J., Islam, J., Thurecht, K. J., & Corrie, S. R. (2016). Nanoparticle-based medicines: A review of FDA-approved materials and clinical trials to date. *Pharmaceutical Research*, *33*(10), 2373–2387. DOI: 10.1007/s11095-016-1958-5 PMID: 27299311

Bot, P. T., Hoefer, I. E., Piek, J. J., & Pasterkamp, G. (2008). Hyaluronic acid: Targeting immune modulatory components of the extracellular matrix in atherosclerosis. *Current Medicinal Chemistry*, *15*(8), 786–791. DOI: 10.2174/092986708783955554 PMID: 18393848

Brown, M. B., & Jones, S. A. (2005). Hyaluronic acid: A unique topical vehicle for the localized delivery of drugs to the skin. *Journal of the European Academy of Dermatology and Venereology*, *19*(3), 308–318. DOI: 10.1111/j.1468-3083.2004.01180.x PMID: 15857456

Brugnoli, M., Carvalho, J. P. F., Arena, M. P., Oliveira, H., Vilela, C., Freire, C. S. R., & Gullo, M. (2025). Co-cultivation of Komagataeibacter sp. and Lacticaseibacillus sp. strains to produce bacterial nanocellulose-hyaluronic acid nanocomposite membranes for skin wound healing applications. *International Journal of Biological Macromolecules, 140208*, 140208. Advance online publication. DOI: 10.1016/j.ijbiomac.2025.140208 PMID: 39848389

Bupphathong, , Quiroz, C., Huang, W., Chung, P.-F., Tao, H.-Y., & Lin, C.-H. (2022). Gelatin Methacrylate Hydrogel for Tissue Engineering Applications—A Review on Material Modifications. *Pharmaceuticals (Basel, Switzerland)*, *15*(2), 171. DOI: 10.3390/ph15020171 PMID: 35215284

Caccavo, D., Cascone, S., Lamberti, G., Barba, A. A., & Larsson, A. (2016). Swellable Hydrogel-based Systems for Controlled Drug Delivery. IntechOpen eBooks. https://doi.org/DOI: 10.5772/61792

Caraway, C. A., Gaitsch, H., Wicks, E. E., Kalluri, A., Kunadi, N., & Tyler, B. M. (2022). Polymeric nanoparticles in brain cancer therapy: A review of current approaches. *Polymers*, *14*(14), 2963. DOI: 10.3390/polym14142963 PMID: 35890738

Casalini, T., Bassas-Galià, M., Girard, H., Castrovinci, A., Carolis, A. D., Brianza, S., Zinn, M., & Perale, G. (2019). A Systematic Experimental and Computational Analysis of Commercially Available Aliphatic Polyesters. *Applied Sciences*, 9(16), 3397. DOI: 10.3390/app9163397

Chai, Q., Jiao, Y., & Yu, X. (2017). Hydrogels for Biomedical Applications: Their Characteristics and the Mechanisms behind Them. *Gels (Basel, Switzerland)*, *3*(1), 6. DOI: 10.3390/gels3010006 PMID: 30920503

Chang, M.-C., Chiang, P.-F., Kuo, Y.-J., Peng, C.-L., Chen, K.-Y., & Chiang, Y.-C. (2021). Hyaluronan-Loaded Liposomal Dexamethasone–Diclofenac Nanoparticles for Local Osteoarthritis Treatment. *International Journal of Molecular Sciences*, *22*(2), 665. DOI: 10.3390/ijms22020665 PMID: 33440880

Chaudhari, V., Buttar, H. S., Bagwe-Parab, S., Tuli, H. S., Vora, A., & Kaur, G. (2021). Therapeutic and industrial applications of curdlan with overview on its recent patents. *Frontiers in Nutrition*, *8*, 646988. DOI: 10.3389/fnut.2021.646988 PMID: 34262922

Chaudhury, A., & Das, S. (2011). Recent advancement of chitosan-based nanoparticles for oral controlled delivery of insulin and other therapeutic agents. *AAPS PharmSciTech*, *12*(1), 10–20. DOI: 10.1208/s12249-010-9561-2 PMID: 21153572

Chen, H., Zhao, Q., & Li, T. (2023). Polyethyleneimine for efficient gene therapy: Exploring the proton sponge effect. *Journal of Controlled Release*, *351*(1), 45–59.

Chen, X., Xu, X., Li, W., Sun, B., Yan, J., Chen, C., Liu, J., Qian, J., & Sun, D. (2018). Effective drug carrier based on polyethylenimine-functionalized bacterial cellulose with controllable release properties. *ACS Applied Bio Materials*, *1*(1), 42–50. DOI: 10.1021/acsabm.8b00004

Chen, Y., Gao, Y., da Silva, L. P., Pirraco, R. P., Ma, M., Yang, L., Reis, R. L., & Chen, J. (2018). A thermo-/pH-responsive hydrogel (PNIPAM-PDMA-PAA) with diverse nanostructures and gel behaviors as a general drug carrier for drug release. *Polymer Chemistry*, *9*(29), 4063–4072. DOI: 10.1039/C8PY00838H

Chopin, T.. (2012). Seaweed aquaculture for integrated multi-trophic aquaculture (IMTA): A key towards sustainability. [ander]. *Journal of Phycology*, *48*(5), 983–992.

Choudhary, U., Bhinchhar, B. K., Paswan, V. K., Kharkwal, S., Yadav, S. P., & Singh, P. (2021). Utilization of Agro-Industrial Wastes as Edible Coating and Films for Food Packaging Materials. In *IntechOpen eBooks*. IntechOpen. DOI: 10.5772/intechopen.99786

Chuah, L. H., Loo, H. L., Goh, C. F., Fu, J. Y., & Ng, S. F. (2023). Chitosan-based drug delivery systems for skin atopic dermatitis: Recent advancements and patent trends. *Drug Delivery and Translational Research*, *13*(5), 1436–1455. DOI: 10.1007/s13346-023-01307-w PMID: 36808298

Chyzy, A., Tomczykowa, M., & Płońska-Brzezińska, M. E. (2020). Hydrogels as Potential Nano-, Micro- and Macro-Scale Systems for Controlled Drug Delivery Materials, 13(1): 188. https://doi.org/DOI: 10.3390/ma13010188

Costa, D. F., Mendes, L. P., & Torchilin, V. P. (2018). The effect of low- and high-penetration light on localized cancer therapy. *Advanced Drug Delivery Reviews*, *138*, 105–116. DOI: 10.1016/j.addr.2018.09.004 PMID: 30217518

D'Amora, U., Ronca, A., Raucci, M. G., Dozio, S. M., Lin, H., Fan, Y., Zhang, X., & Ambrosio, L. (2019). In situ sol-gel synthesis of hyaluronan derivatives bio-nanocomposite hydrogels. *Regenerative Biomaterials*, *6*(5), 249–258. DOI: 10.1093/rb/rbz029 PMID: 31620307

D'souza, A. A., & Shegokar, R. (2016). Polyethylene glycol (PEG): A versatile polymer for pharmaceutical applications. *Expert Opinion on Drug Delivery*, *13*(9), 1257–1275. DOI: 10.1080/17425247.2016.1182485 PMID: 27116988

Dachani, S. R., Vashi, A., Mundada, A. B., Mundada, P. A., Rudrangi, S. R. S., Rudrangi, S., & Tiwari, R. (2025). Innovative polymers in pharmaceutical chemistry: Revolutionizing drug delivery systems. *Polymer-Plastics Technology and Materials*, *64*(7), 911–933. DOI: 10.1080/25740881.2024.2440531

Dai, T., Tanaka, M., Huang, Y. Y., & Hamblin, M. R. (2011). Chitosan preparations for wounds and burns: Antimicrobial and wound-healing effects. *Expert Review of Anti-Infective Therapy*, *9*(7), 857–879. DOI: 10.1586/eri.11.59 PMID: 21810057

Dai, X. J., Li, W. J., Xie, D. D., Liu, B. X., Gong, L., & Han, H. H. (2025). Stimuli-Responsive Nano Drug Delivery Systems for the Treatment of Neurological Diseases. *Small*, *2410030*(9). Advance online publication. DOI: 10.1002/smll.202410030 PMID: 39840482

Dalei, G., & Das, S. (2022). Polyacrylic acid-based drug delivery systems: A comprehensive review on the state-of-art. *Journal of Drug Delivery Science and Technology*, *78*, 103988. DOI: 10.1016/j.jddst.2022.103988

Dash, T. K., & Konkimalla, V. B. (2012). Poly-ϵ-caprolactone based formulations for drug delivery and tissue engineering: A review. *Journal of Controlled Release*, *158*(1), 15–33. DOI: 10.1016/j.jconrel.2011.09.064 PMID: 21963774

De Geest, B. G., De Koker, S., Sukhorukov, G. B., Kreft, O., Parak, W. J., & Sanders, N. N. (2007). Polyelectrolyte microcapsules for biomedical applications. *Soft Matter*, *3*(6), 711–719. DOI: 10.1039/B617074H

de Vos, P., Faas, M. M., Spasojevic, M., & Sikkema, J. (2010). Encapsulation for preservation of functionality and targeted delivery of bioactive food components. *International Dairy Journal*, *20*(4), 292–302. DOI: 10.1016/j.idairyj.2009.11.008

Dheer, D., Nicolas, J., & Shankar, R. (2019). Cathepsin-sensitive nanoscale drug delivery systems for cancer therapy and other diseases. *Advanced Drug Delivery Reviews*, *151*, 130–151. DOI: 10.1016/j.addr.2019.01.010 PMID: 30690054

Dhiman, A., Rana, D., Benival, D., & Garkhal, K. (2025). Comprehensive insights into glioblastoma multiforme: Drug delivery challenges and multimodal treatment strategies. *Therapeutic Delivery*, *16*(1), 87–115. DOI: 10.1080/20415990.2024.2415281 PMID: 39445563

Dicker, K. T., Gurski, L. A., Pradhan-Bhatt, S., Witt, R. L., Farach-Carson, M. C., & Jia, X. (2013). Hyaluronan: A simple polysaccharide with diverse biological functions. *Acta Biomaterialia*, *10*(4), 1558–1570. DOI: 10.1016/j.actbio.2013.12.019 PMID: 24361428

Ding, D., & Zhu, Q. (2018). Recent advances of PLGA micro/nanoparticles for the delivery of biomacromolecular therapeutics. *Materials Science and Engineering C*, *92*, 1041–1060. DOI: 10.1016/j.msec.2017.12.036 PMID: 30184728

Ding, L., Agrawal, P., Singh, S. K., Chhonker, Y. S., Sun, J., & Murry, D. J. (2024). Polymer-Based Drug Delivery Systems for Cancer Therapeutics. *Polymers*, *16*(6), 843. DOI: 10.3390/polym16060843 PMID: 38543448

Doberenz, F., Zeng, K., Willems, C., Zhang, K., & Groth, T. (2020). Thermoresponsive polymers and their biomedical application in tissue engineering–a review. *Journal of Materials Chemistry. B*, *8*(4), 607–628. DOI: 10.1039/C9TB02052G PMID: 31939978

Domard, A. (1987). pH and cd measurements on a fully deacetylated chitosan: Application to CuII—polymer interactions. *International Journal of Biological Macromolecules*, *9*(2), 98–104. DOI: 10.1016/0141-8130(87)90033-X

Donnelly, R. F., Moffatt, K., & Alkilani, A. Z. (2010). Hydrogel-forming microneedles prepared from "super swelling" polymers for transdermal drug delivery. *Pharmaceutical Research*, *27*(3), 399–408.

Donnelly, R., Alkilani, A., & McCrudden, M. (2021). Microneedle-based systems for drug delivery: A review. *European Journal of Pharmaceutics and Biopharmaceutics*, *161*, 100–118. DOI: 10.1016/j.ejpb.2020.10.014

Draget, K. I., Skjåk-Bræk, G., & Smidsrød, O. (2006). Alginates from algae. In Phillips, G. O., & Williams, P. A. (Eds.), *Handbook of Hydrocolloids* (pp. 379–395). Woodhead Publishing., DOI: 10.1533/9781845693149.3.379

Dreiss, C. A. (2020). Hydrogel design strategies for drug delivery. *Current Opinion in Colloid & Interface Science*, *48*, 1–17. DOI: 10.1016/j.cocis.2020.02.001

Duan, Y., Freyburger, A., Kunz, W., & Zollfrank, C. (2018). Cellulose and chitin composite materials from an ionic liquid and a green co-solvent. *Carbohydrate Polymers*, *192*, 159–165. DOI: 10.1016/j.carbpol.2018.03.045 PMID: 29691008

Dubashynskaya, N. V., Raik, S. V., Dubrovskii, Y. A., Demyanova, E. V., Shcherbakova, E. S., Poshina, D. N., & Skorik, Y. A. (2021). Hyaluronan/diethylaminoethyl chitosan polyelectrolyte complexes as carriers for improved colistin delivery. *International Journal of Molecular Sciences*, *22*(16), 8381. DOI: 10.3390/ijms22168381 PMID: 34445088

Duceac, I. A., Lobiuc, A., Coseri, S., & Vere tiuc, L. (2019). *Tunable Hydrogels based on Chitosan, Collagen and Poly(Acrylic Acid) for Regenerative Medicine. In 2022 E-Health and Bioengineering Conference*. EHB., DOI: 10.1109/EHB47216.2019.8969886

Eckmann, D. M., Composto, R. J., Tsourkas, A., & Muzykantov, V. R. (2014). Nanogel carrier design for targeted drug delivery. *Journal of Materials Chemistry. B*, *2*(46), 8085–8097. DOI: 10.1039/C4TB01141D PMID: 25485112

Elezaby, R. S., Gad, H. A., Metwally, A. A., Geneidi, A. S., & Awad, G. A. (2017). Self-assembled amphiphilic core-shell nanocarriers in line with the modern strategies for brain delivery. *Journal of Controlled Release*, *261*, 43–61. DOI: 10.1016/j.jconrel.2017.06.019 PMID: 28648865

El-Naggar, M. E., Abdelgawad, A. M., Salas, C., & Rojas, O. J. (2016). Curdlan in fibers as carriers of tetracycline hydrochloride: Controlled release and antibacterial activity. *Carbohydrate Polymers*, *154*, 194–203. DOI: 10.1016/j.carbpol.2016.08.042 PMID: 27577910

Eskandari, S., Guerin, T., Toth, I., & Stephenson, R. J. (2017). Recent advances in self-assembled peptides: Implications for targeted drug delivery and vaccine engineering. *Advanced Drug Delivery Reviews*, *110*, 169–187. DOI: 10.1016/j.addr.2016.06.013 PMID: 27356149

Esmaeili, J., Pirzadeh, K., Pakrooyan, M., Lukolayeh, M. E., & Kırboğa, K. K. (2024). Synthesis of cellulose nanocrystals from spinach waste for insulin delivery: Comparison to chitosan nanoparticles. *New Journal of Chemistry*, *48*(17), 7953–7963. DOI: 10.1039/D4NJ00073K

Fagehi, T. H. M., Alanazi, H. S., Alshehri, A. A. M., Alanazi, M. M., Matari, H. A. Y., Ayoub, T. M., Alrashedi, E. A., Alotaibi, A. U. S., Alrasheedi, S. M. M., & Zawli, H. M. N. (2024). Innovative Drug Delivery Systems: Nanotechnology, Liposomes, or Hydrogels for Targeted Drug Delivery. *Journal of International Crisis and Risk Communication Research*, *7*(S11), 368. DOI: 10.63278/jicrcr.vi.353

Fang, J., Kawano, S., Tajima, K., & Kondo, T. (2015). In vivo curdlan/cellulose bionanocomposite synthesis by genetically modified Gluconacetobacter xylinus. *Biomacromolecules*, *16*(10), 3154–3160. DOI: 10.1021/acs.biomac.5b01075 PMID: 26360299

Fang, Z., Hu, J., Yuan, Q., Zhou, W., & Fan, L. (2022). Biological responses to PMMA-based bone cements and potential improvements. *Materials (Basel)*, *15*(1), 145. DOI: 10.3390/ma15010145

Far, B. F.. (2022). A Review on Biomedical Application of Polysaccharide-Based Hydrogels with a Focus on Drug Delivery System. *Polymers, 14*(24), 5432. DOI: 10.3390/polym14245432 PMID: 36559799

Fattah-alhosseini, A., Chaharmahali, R., Alizad, S., Kaseem, M., & Dikici, B. (2024). A review of smart polymeric materials: Recent developments and prospects for medicine applications. *Hybrid Advances, 100178*, 100178. Advance online publication. DOI: 10.1016/j.hybadv.2024.100178

Fazal, T., Murtaza, B. N., Shah, M., Iqbal, S., Rehman, M. U., Jaber, F., & Ibrahium, H. A. (2023). Recent developments in natural biopolymer based drug delivery systems. *RSC Advances, 13*(33), 23087–23121. DOI: 10.1039/D3RA03369D PMID: 37529365

Feig, V. R., Tran, H., & Bao, Z. (2018). Biodegradable polymeric materials in degradable electronic devices. *ACS Central Science, 4*(3), 337–348. DOI: 10.1021/acscentsci.7b00595 PMID: 29632879

Feng, J., Zhao, L., Huang, Z., & Du, Y. (2013). Alginate-based microcapsules for drug delivery: Effect of preparation method on drug release. *International Journal of Pharmaceutics, 453*(2), 497–503. DOI: 10.1016/j.ijpharm.2013.06.034

Folkman, J., & Long, D. M. (1964). The use of silicone rubber as a carrier for prolonged drug therapy. *The Journal of Surgical Research, 4*(3), 139–142. DOI: 10.1016/S0022-4804(64)80040-8 PMID: 14130164

Foox, M., & Zilberman, M. (2015). Drug delivery from gelatin-based systems. *Expert Opinion on Drug Delivery, 12*(9), 1547–1563. DOI: 10.1517/17425247.2015.1037272 PMID: 25943722

Franco, P., & De Marco, I. (2020). The Use of Poly (N-vinyl pyrrolidone) in the Delivery of Drugs: A Review. *Polymers, 12*(5), 1114. DOI: 10.3390/polym12051114 PMID: 32414187

Fredenberg, S., Wahlgren, M., Reslow, M., & Axelsson, A. (2011). The mechanisms of drug release in poly(lactic-co-glycolic acid)-based drug delivery systems—A review. *International Journal of Pharmaceutics, 415*(1–2), 34–52. DOI: 10.1016/j.ijpharm.2011.05.049 PMID: 21640806

Fu, C. P., Cai, X. Y., Chen, S. L., Yu, H. W., Fang, Y., Feng, X. C., & Li, C. Y. (2023). Hyaluronic acid-based nanocarriers for anticancer drug delivery. *Polymers, 15*(10), 2317. DOI: 10.3390/polym15102317 PMID: 37242892

Fu, L., Zhang, J., & Yang, G. (2013). Present status and applications of bacterial cellulose-based materials for skin tissue repair. *Carbohydrate Polymers*, *92*(2), 1432–1442. DOI: 10.1016/j.carbpol.2012.10.071 PMID: 23399174

Gabizon, A. A., Gabizon-Peretz, S., Modaresahmadi, S., & La-Beck, N. M. (2025). Thirty years from FDA approval of pegylated liposomal doxorubicin (Doxil/Caelyx): An updated analysis and future perspective. *BMJ Oncology*, *4*(1), e000573. DOI: 10.1136/bmjonc-2024-000573 PMID: 39885941

Gandhi, A., Paul, A., Sen, S. O., & Sen, K. K. (2015). Studies on thermoresponsive polymers: Phase behaviour, drug delivery and biomedical applications. *Asian Journal of Pharmaceutical Sciences*, *10*(2), 99–107. DOI: 10.1016/j.ajps.2014.08.010

Ganie, S. A., Rather, L. J., & Li, Q. (2022). Review on anti-cancer and anti-microbial applications of curdlan biomaterials. *Journal of Polymers and the Environment*, *30*(4), 1284–1299. DOI: 10.1007/s10924-021-02299-9

George, M., & Abraham, T. E. (2006). Polyionic hydrocolloids for the intestinal delivery of protein drugs: Alginate and chitosan—a review. *Journal of Controlled Release*, *114*(1), 1–14. DOI: 10.1016/j.jconrel.2006.04.017 PMID: 16828914

Geranmayeh, M. H., Rahbarghazi, R., & Farhoudi, M. (2019). Targeting pericytes for neurovascular regeneration. *Cell Communication and Signaling : CCS*, *17*(1), 26. DOI: 10.1186/s12964-019-0340-8 PMID: 30894190

Ghavimi, S. A., Gehret, P. M., Aronson, M. R., Schipani, R., Smith, K. W., Borek, R. C., Germiller, J. A., Jacobs, I. N., Zur, K. B., & Gottardi, R. (2021). Drug delivery to the pediatric upper airway. *Advanced Drug Delivery Reviews*, *174*, 168–189. DOI: 10.1016/j.addr.2021.04.004 PMID: 33845038

Gholap, A. D., Uddin, M. J., Faiyazuddin, M., Omri, A., Gowri, S., & Khalid, M. (2024). Advances in artificial intelligence for drug delivery and development: A comprehensive review. *Computers in Biology and Medicine*, *178*, 108702. DOI: 10.1016/j.compbiomed.2024.108702 PMID: 38878397

Gil, E. S., & Hudson, S. M. (2004). Stimuli-responsive polymers and their bioconjugates. *Progress in Polymer Science*, *29*(12), 1173–1222. DOI: 10.1016/j.progpolymsci.2004.08.003

Girish, K. S., & Kemparaju, K. (2007). The magic glue hyaluronan and its eraser hyaluronidase: A biological overview. *Life Sciences*, *80*(21), 1921–1943. DOI: 10.1016/j.lfs.2007.02.037 PMID: 17408700

Gittard, S., Bohnsack, J., & Hsia, Y. (2019). Development of microneedle arrays for controlled release applications. *Journal of Drug Targeting*, *27*(8), 732–740. DOI: 10.1080/1061186X.2019.1624443

Gombotz, W. R., & Wee, S. F. (1998). Protein release from alginate matrices. *Advanced Drug Delivery Reviews*, *31*(3), 267–285. DOI: 10.1016/S0169-409X(97)00124-5 PMID: 10837629

Gomte, S. S., Agnihotri, T. G., Khopade, S., & Jain, A. (2024). Exploring the potential of pH-sensitive polymers in targeted drug delivery. *Journal of Biomaterials Science. Polymer Edition*, *35*(2), 228–268. DOI: 10.1080/09205063.2023.2279792 PMID: 37927045

Goyanes, A., Buanz, A. B., Basit, A. W., & Gaisford, S. (2015). Fused-filament 3D printing (3DP) for fabrication of tablets. *International Journal of Pharmaceutics*, *476*(1–2), 88–92. DOI: 10.1016/j.ijpharm.2014.09.044 PMID: 25275937

Grahl, M. V., Hohl, K. S., Smaniotto, T., & Carlini, C. R. (2025). Microbial Trojan Horses: Virulence Factors as Key Players in Neurodegenerative Diseases. *Molecules (Basel, Switzerland)*, *30*(3), 687. DOI: 10.3390/molecules30030687 PMID: 39942791

Grant, G. T., Morris, E. R., Rees, D. A., Smith, P. J. C., & Thom, D. (1973). Biological interactions between polysaccharides and divalent cations: The egg-box model. *FEBS Letters*, *32*(1), 195–198. DOI: 10.1016/0014-5793(73)80770-7

Grant, S. M., & Maleki, H., & GhavamiNejad, A. (2022). Translation of alginate-based delivery systems: From bench to market. *Journal of Drug Delivery Science and Technology*, *69*, 103095.

Gref, R., Minamitake, Y., Peracchia, M. T., Trubetskoy, V., Torchilin, V., & Langer, R. (1994). Biodegradable long-circulating polymeric nanospheres. *Science*, *263*(5153), 1600–1603. DOI: 10.1126/science.8128245 PMID: 8128245

Grigoras, A. G. (2019). Drug delivery systems using pullulan, a biocompatible polysaccharide produced by fungal fermentation of starch. *Environmental Chemistry Letters*, *17*(3), 1209–1223. DOI: 10.1007/s10311-019-00862-4

Grishkewich, N., Mohammed, N., Tang, J., & Tam, K. C. (2017). Recent advances in the application of cellulose nanocrystals. *Current Opinion in Colloid & Interface Science*, *29*, 32–45. DOI: 10.1016/j.cocis.2017.01.005

Gulfam, M., Jo, S., Jo, S., Vu, T. T., Park, S.-H., & Lim, K. T. (2022). Highly porous and injectable hydrogels derived from cartilage acellularized matrix exhibit reduction and NIR light dual-responsive drug release properties for application in antitumor therapy. *NPG Asia Materials*, *14*(1), 8. Advance online publication. DOI: 10.1038/s41427-021-00354-4

Gulfam, M., Sahle, F. F., & Lowe, T. L. (2019). Design strategies for chemical-stimuli-responsive programmable nanotherapeutics. *Drug Discovery Today*, *24*(1), 129–147. DOI: 10.1016/j.drudis.2018.09.019 PMID: 30292916

Gupta, A., Mehta, P., & Sharma, S. (2022). Nanotechnology in stimuli-responsive drug delivery: Promises and challenges. *International Journal of Pharmaceutics*, *619*, 121702.

Gupta, P., Vermani, K., & Garg, S. (2002). Hydrogels: From controlled release to pH-responsive drug delivery. *Drug Discovery Today*, *7*(10), 569–579. DOI: 10.1016/S1359-6446(02)02255-9 PMID: 12047857

Guzmán Rodríguez, A., Sablón Carrazana, M., Rodríguez Tanty, C., Malessy, M. J., Fuentes, G., & Cruz, L. J. (2022). Smart polymeric micelles for anticancer hydrophobic drugs. *Cancers (Basel)*, *15*(1), 4. DOI: 10.3390/cancers15010004 PMID: 36612002

Hardenia, A., Maheshwari, N., Hardenia, S. S., Dwivedi, S., Maheshwari, R., & Tekade, R. K. (2019). *Scientific Rationale for Designing Controlled Drug Delivery Systems. eBooks*. Elsevier BV., DOI: 10.1016/B978-0-12-817909-3.00001-7

Hariyadi, D. M., & Islam, N. (2020). Current strategies to improve the bioavailability of poorly water-soluble drugs. *Journal of Drug Delivery Science and Technology*, *55*, 101458. DOI: 10.1016/j.jddst.2019.101458

Harris, J. M., & Chess, R. B. (2003). Effect of PEGylation on pharmaceuticals. *Nature Reviews. Drug Discovery*, *2*(3), 214–221. DOI: 10.1038/nrd1033 PMID: 12612647

Hersh, A. M., Alomari, S., & Tyler, B. M. (2022). Crossing the blood-brain barrier: Advances in nanoparticle technology for drug delivery in neuro-oncology. *International Journal of Molecular Sciences*, *23*(8), 4153. DOI: 10.3390/ijms23084153 PMID: 35456971

He, Y., Xu, Y., Huang, Y., Quang, H., Xia, X., Zhao, H., & Yu, J. (2022). Redox sensitive nano-capsules self-assembled from hyaluronic acid-hydroxychloroquine conjugates for CD44-targeted delivery of hydroxychloroquine to combat breast cancer metastasis in vitro and in vivo. *Colloids and Surfaces. B, Biointerfaces*, *210*, 112249. DOI: 10.1016/j.colsurfb.2021.112249 PMID: 34864404

Hines, D. J., & Kaplan, D. L. (2013). Poly(lactic-co-glycolic) Acid-Controlled-Release Systems: Experimental and Modeling Insights. *Critical Reviews in Therapeutic Drug Carrier Systems*, *30*(3), 257–276. DOI: 10.1615/CritRevTherDrugCarrierSyst.2013006475 PMID: 23614648

Hoare, T. R., & Kohane, D. S. (2014). Hydrogels in drug delivery: Progress and challenges. *Polymer*, *49*(8), 1993–2007. DOI: 10.1016/j.polymer.2008.01.027

Hogan, K. J., & Mikos, A. G. (2020). Biodegradable thermoresponsive polymers: Applications in drug delivery and tissue engineering. *Polymer*, *211*, 123063. DOI: 10.1016/j.polymer.2020.123063

Ho, T., Chang, C.-C., Chan, H.-P., Chung, T., Shu, C., Chuang, K., Duh, T., Yang, M., & Tyan, Y. (2022). Hydrogels: Properties and Applications in Biomedicine. *Molecules (Basel, Switzerland)*, *27*(9), 2902. DOI: 10.3390/molecules27092902 PMID: 35566251

Houldsworth, A. (2024). Role of oxidative stress in neurodegenerative disorders: A review of reactive oxygen species and prevention by antioxidants. *Brain Communications*, *6*(1), fcad356. Advance online publication. DOI: 10.1093/braincomms/fcad356 PMID: 38214013

Houshmand, E. B. (2021). *Introduction to Microneedling*. DOI: 10.1002/9781119431817

Hou, X., Zaks, T., Langer, R., & Dong, Y. (2021). Lipid nanoparticles for mRNA delivery. *Nature Reviews. Materials*, *6*(12), 1078–1094. DOI: 10.1038/s41578-021-00358-0 PMID: 34394960

Huang, C.-C. (2021). Characteristics and Preparation of Designed Alginate-Based Composite Scaffold Membranes with Decellularized Fibrous Micro-Scaffold Structures from Porcine Skin. *Polymers*, *13*(20), 3464. DOI: 10.3390/polym13203464 PMID: 34685220

Huang, G., & Huang, H. (2018). Application of hyaluronic acid as carriers in drug delivery. *Drug Delivery*, *25*(1), 766–772. DOI: 10.1080/10717544.2018.1450910 PMID: 29536778

Huang, H., Qi, X., Chen, Y., & Wu, Z. (2019). Thermo-sensitive hydrogels for delivering biotherapeutic molecules: A review. *Saudi Pharmaceutical Journal : SPJ : the Official Publication of the Saudi Pharmaceutical Society*, *27*(7), 990–999. DOI: 10.1016/j.jsps.2019.08.001 PMID: 31997906

Huang, Y., Leobandung, W., Foss, A., & Peppas, N. A. (2010). Molecular aspects of muco- and bioadhesion: Tethered structures and site-specific drug delivery. *Journal of Controlled Release*, *65*(1-2), 63–71. DOI: 10.1016/S0168-3659(99)00233-3 PMID: 10699271

Huang, Y., Seitz, D., König, F., Müller, P. E., Jansson, V., & Klar, R. M. (2019). Induction of articular chondrogenesis by chitosan/hyaluronic-acid-based biomimetic matrices using human adipose-derived stem cells. *International Journal of Molecular Sciences*, *20*(18), 4487. DOI: 10.3390/ijms20184487 PMID: 31514329

Illum, L. (2007). Nanoparticulate systems for nasal delivery of drugs: A real improvement over simple systems? *Journal of Pharmaceutical Sciences*, *96*(3), 473–483. DOI: 10.1002/jps.20718 PMID: 17117404

Indulekha, S., Arunkumar, P., Bahadur, D., & Srivastava, R. (2016). Thermoresponsive polymeric gel as an on-demand transdermal drug delivery system for pain management. *Materials Science and Engineering C*, *62*, 113–122. DOI: 10.1016/j.msec.2016.01.021 PMID: 26952404

Ioannou, P., Katsoulieris, E., & Afratis, N. A. (2025). Matrix Dynamics and Microbiome Crosstalk: Matrix Metalloproteinases as Key Players in Disease and Therapy. *International Journal of Molecular Sciences*, *26*(8), 3621. DOI: 10.3390/ijms26083621 PMID: 40332093

Ishihara, T., & Mizushima, T. (2010). Techniques for efficient entrapment of pharmaceuticals in biodegradable solid micro/nanoparticles. *Expert Opinion on Drug Delivery*, *7*(5), 565–575. DOI: 10.1517/17425241003713486 PMID: 20230305

Ita, K. (2021). Current status of microneedles for transdermal drug delivery: Recent advances and future perspectives. *Acta Pharmaceutica Sinica. B*, *11*(3), 738–755. DOI: 10.1016/j.apsb.2020.10.014 PMID: 33777679

Jacob, S., Nair, A. B., Shah, J., Sreeharsha, N., Gupta, S., & Shinu, P. (2021). Emerging Role of Hydrogels in Drug Delivery Systems, Tissue Engineering and Wound Management. *Pharmaceutics*, *13*(3), 357. DOI: 10.3390/pharmaceutics13030357 PMID: 33800402

Jafernik, K., Ładniak, A., Blicharska, E., Czarnek, K., Ekiert, H., Wiącek, A. E., & Szopa, A. (2023). Chitosan-based nanoparticles as effective drug delivery systems—A review. *Molecules (Basel, Switzerland)*, *28*(4), 1963. DOI: 10.3390/molecules28041963 PMID: 36838951

Jahan, A., Lonare, M. K., Khatoon, S., & Devi, K. K. (2025). Polymer-Based Drug Delivery Systems: Design and Characterization. In *Methods in pharmacology and toxicology* (pp. 111–131). https://doi.org/DOI: 10.1007/978-1-0716-4554-3_6

Jain, D., & Jain, V. (2016). Novel microencapsulation techniques for controlled drug delivery—A review. *International Journal of Pharmaceutical Sciences and Research*, *7*(11), 4373–4380.

Jain, S., Patel, V., & Sharma, S. (2021). Poly(ethylene oxide)-based formulations in pulmonary drug delivery: A review. *International Journal of Pharmaceutics*, *606*, 120892.

Jaiswal, P., Gidwani, B., & Vyas, A. (2014). Nanostructured lipid carriers and their current application in targeted drug delivery. *Artificial Cells Nanomedicine and Biotechnology*, 44(1), 27. DOI: 10.3109/21691401.2014.90982

Jamaledin, R., Ahmed, H., & Liu, T. (2020). Development of microneedles for drug delivery. *Journal of Pharmaceutical Development*, *18*(2), 117–130. DOI: 10.1016/j.jpd.2020.07.005

Jana, P., Shyam, M., Singh, S., Jayaprakash, V., & Dev, A. (2021). Biodegradable polymers in drug delivery and oral vaccination. *European Polymer Journal*, *142*, 110155. DOI: 10.1016/j.eurpolymj.2020.110155

Javanbakht, S., & Shaabani, A. (2019). Carboxymethyl cellulose-based oral delivery systems. *International Journal of Biological Macromolecules*, *133*, 21–29. DOI: 10.1016/j.ijbiomac.2019.04.079 PMID: 30986470

Jiang, D., Zou, J., Huang, L., Shi, Q., Zhu, X., Wang, G., & Yang, H. (2010). Efficacy of intra-articular injection of celecoxib in a rabbit model of osteoarthritis. *International Journal of Molecular Sciences*, *11*(10), 4106–4113. DOI: 10.3390/ijms11104106 PMID: 21152323

Johnson, R., Smith, L., & Williams, K. (2020). Advances in transdermal drug delivery systems. *Journal of Pharmaceutical Sciences*, *109*(5), 1489–1498. DOI: 10.1016/j.xphs.2020.01.010

Juncan, A. M., Moisă, D. G., Santini, A., Morgovan, C., Rus, L. L., Vonica- incu, A. L., & Loghin, F. (2021). Advantages of hyaluronic acid and its combination with other bioactive ingredients in cosmeceuticals. *Molecules (Basel, Switzerland)*, *26*(15), 4429. DOI: 10.3390/molecules26154429 PMID: 34361586

Kaith, A., Garg, U., Jain, N., Pandey, M., Kaul, S., Gorain, B., & Amin, M. C. I. M. (2025). Pullulan as a sustained release carrier for ocular drug delivery: A review. *International Journal of Biological Macromolecules, 143146*, 143146. Advance online publication. DOI: 10.1016/j.ijbiomac.2025.143146 PMID: 40233906

Kamaly, N., Yameen, B., Wu, J., & Farokhzad, O. C. (2016). Degradable controlled-release polymers and polymeric nanoparticles: Mechanisms of controlling drug release. *Chemical Reviews, 116*(4), 2602–2663. DOI: 10.1021/acs.chemrev.5b00346 PMID: 26854975

Kapoor, D. U., Sharma, J. B., Gandhi, S. M., Prajapati, B. G., Thanawuth, K., Limmatvapirat, S., & Sriamornsak, P. (2024). AI-driven design and optimization of nanoparticle-based drug delivery systems. *Science. Engineering and Health Studies, 24010003*, 24010003. Advance online publication. DOI: 10.69598/sehs.18.24010003

Karayianni, M., & Pispas, S. (2021). Block copolymer solution self-assembly: Recent advances, emerging trends, and applications. *Journal of Polymer Science, 59*(17), 1874–1898. DOI: 10.1002/pol.20210430

Karimi, M., Sahandi Zangabad, P., Ghasemi, A., Amiri, M., Bahrami, M., Malekzad, H., Ghahramanzadeh Asl, H., Mahdieh, Z., Bozorgomid, M., Ghasemi, A., Rahmani Taji Boyuk, M. R., & Hamblin, M. R. (2016). Temperature-responsive smart nanocarriers for delivery of therapeutic agents: Applications and recent advances. *ACS Applied Materials & Interfaces, 8*(33), 21107–21133. DOI: 10.1021/acsami.6b00371 PMID: 27349465

Karlsson, J., Rhodes, K. R., Green, J. J., & Tzeng, S. Y. (2020). Poly (beta-amino ester) s as gene delivery vehicles: Challenges and opportunities. *Expert Opinion on Drug Delivery, 17*(10), 1395–1410. DOI: 10.1080/17425247.2020.1796628 PMID: 32700581

Kataoka, K., Harada, A., & Nagasaki, Y. (2012). Block copolymer micelles for drug delivery: Design, characterization, and biological significance. *Advanced Drug Delivery Reviews, 64*, 37–48. DOI: 10.1016/j.addr.2012.09.013 PMID: 11251249

Kaveramma, U. P., Brindha, M. H., Nayac, R. A., Pauline, A. R. R., & Kumar, N. A. (2025). Poly (N-isopropyl acrylamide): a thermo-responsive polymer for biotherapeutic applications. In *Synthetic Polymers in Drug and Biotherapeutics Delivery* (pp. 235-267). Woodhead Publishing.

Kean, T., & Thanou, M. (2010). Biodegradation, biodistribution and toxicity of chitosan. *Advanced Drug Delivery Reviews, 62*(1), 3–11. DOI: 10.1016/j.addr.2009.09.004 PMID: 19800377

Khan M. S., Mohapatra S., Gupta V., Ali A., Naseef P. P., Kuruniyan M. S., Alshadidi A. A. F., Alam M. S., Mirza Mohd. A., & Iqbal Z. (2023). Potential of lipid-based nanocarriers against two major barriers to drug delivery—skin and blood–brain barrier. *Membranes*, 13(3), 343. DOI: 10.3390/membranes13030343

Kim, D., Lee, S., & Park, H. (2023). Poly(organophosphazenes) as injectable hydrogels: Temperature and pH responsiveness for localized drug delivery. *Journal of Materials Chemistry. B*, *11*(4), 775–788.

Kim, H., Lee, J., & Park, M. (2019). Nanotechnology for drug delivery: Trends and applications. *Journal of Nanomedicine*, *15*(2), 72–84. DOI: 10.2217/nnm.19.3

Kim, H., Park, H., & Kim, S. H. (2014). Enhanced stability and controlled release of alginate hydrogel beads through carbodiimide-mediated covalent crosslinking. *Carbohydrate Polymers*, *101*, 98–105. DOI: 10.1016/j.carbpol.2013.08.076

Kim, Y.-C., Park, J.-H., & Prausnitz, M. R. (2012). Microneedles for drug and vaccine delivery. *Advanced Drug Delivery Reviews*, *64*(14), 1575–1586. DOI: 10.1016/j.addr.2012.04.005 PMID: 22575858

Klemm, D., Kramer, F., Moritz, S., Lindström, T., Ankerfors, M., Gray, D., & Dorris, A. (2011). Nanocelluloses: A new family of nature-based materials. *Angewandte Chemie International Edition*, *50*(24), 5438–5466. DOI: 10.1002/anie.201001273 PMID: 21598362

Knop, K., Hoogenboom, R., Fischer, D., & Schubert, U. S. (2010). Poly(ethylene glycol) in drug delivery: Pros and cons as well as potential alternatives. *Angewandte Chemie International Edition*, *49*(36), 6288–6308. DOI: 10.1002/anie.200902672 PMID: 20648499

Kolawole, O. M., & Cook, M. T. (2023). In situ gelling drug delivery systems for topical drug delivery. *European Journal of Pharmaceutics and Biopharmaceutics*, *184*, 36–49. DOI: 10.1016/j.ejpb.2023.01.007 PMID: 36642283

Krasaekoopt, W., Bhandari, B., & Deeth, H. (2004). The influence of coating materials on some properties of alginate beads and survivability of microencapsulated probiotic bacteria. *International Dairy Journal*, *14*(8), 737–743. DOI: 10.1016/j.idairyj.2004.01.004

Krishnan, A., Roy, S., & Menon, S. (2022). Amphiphilic block copolymers: From synthesis including living polymerization methods to applications in drug delivery. *European Polymer Journal*, *172*, 111224. DOI: 10.1016/j.eurpolymj.2022.111224

Kruczkowska, W., Gałęziewska, J., Grabowska, K. H., Gromek, P., Czajkowska, K., Rybicki, M., Kciuk, M., & Kłosiński, K. K. (2025). From Molecules to Mind: The Critical Role of Chitosan, Collagen, Alginate, and Other Biopolymers in Neuroprotection and Neurodegeneration. *Molecules (Basel, Switzerland)*, *30*(5), 1017. DOI: 10.3390/molecules30051017 PMID: 40076240

Kumar, L., Kukreti, G., Rana, R., Chaurasia, H., Sharma, A., Sharma, N., & Komal, . (2023). Poly (lactic-co-glycolic) Acid (PLGA) nanoparticles and transdermal drug delivery: An overview. *Current Pharmaceutical Design*, *29*(37), 2940–2953. DOI: 10.2174/0113816128275385231027054743 PMID: 38173050

Kumar, M. R., Muzzarelli, R., Muzzarelli, C., Sashiwa, H., & Domb, A. J. (2004). Chitosan chemistry and pharmaceutical perspectives. *Chemical Reviews*, *104*(12), 6017–6084. DOI: 10.1021/cr030441b PMID: 15584695

Kumar, N., Gupta, P., & Sharma, A. (2024). Poly(β-amino esters) for pH-responsive delivery systems: Applications in gene and protein delivery. *ACS Applied Materials & Interfaces*, *16*(2), 854–868.

Kumar, N., & Kumar, A. (2023). Environmentally sensitive nanocomposite hydrogels for biomedical applications. In *Functional Nanocomposite Hydrogels* (pp. 517–540). Elsevier. DOI: 10.1016/B978-0-323-99638-9.00021-6

Kumar, P., Goel, A., Dash, B., Rishishwar, P. S. J., Suryawanshi, M. V., Rishishwar, S., & Asokan, N. (2025). Innovations in drug delivery systems for Biologics: Enhancing stability and targeted delivery for Next-Generation therapeutics. *Chinese. Journal of Applied Physiology (Bethesda, Md.)*, *20250001*, e20250001. Advance online publication. DOI: 10.62958/j.cjap.2025.001 PMID: 39748757

Kumar, R., Sharma, V., & Patel, P. (2023). Smart drug delivery systems: Innovations in multifunctional polymers. *Advanced Drug Delivery Reviews*, *210*, 114826.

Kuo, S. H., & Lee, W. H. (2023). Responsive polymers in cancer and gene therapy. *International Journal of Pharmaceutics*, *501*(1-2), 178–191.

Kurakula, M., & Rao, G. K. (2020). Pharmaceutical assessment of polyvinylpyrrolidone (PVP): As excipient from conventional to controlled delivery systems with a spotlight on COVID-19 inhibition. *Journal of Drug Delivery Science and Technology*, *60*, 102046. DOI: 10.1016/j.jddst.2020.102046 PMID: 32905026

Kutner, N., Kunduru, K. R., Rizik, L., & Farah, S. (2021). Recent advances for improving functionality, biocompatibility, and longevity of implantable medical devices and deliverable drug delivery systems. *Advanced Functional Materials*, *31*(44), 2010929. DOI: 10.1002/adfm.202010929

Lammerding, L. C., Arora, A., Braun, S., & Breitkreutz, J. (2024). Spatial separation of different drug substances in one microneedle array patch by combining inkjet printing and micromolding technology. *International Journal of Pharmaceutics, 125102*. Advance online publication. DOI: 10.1016/j.ijpharm.2024.125102 PMID: 39708846

Langer, R. (1998). Drug delivery and targeting. *Nature*, *392*(6679, Suppl), 5–10. DOI: 10.1038/32380 PMID: 9579855

Laredj-Bourezg, F., Bolzinger, M. A., Pelletier, J., Valour, J. P., Rovère, M. R., Smatti, B., & Chevalier, Y. (2015). Skin delivery by block copolymer nanoparticles (block copolymer micelles). *International Journal of Pharmaceutics*, *496*(2), 1034–1046. DOI: 10.1016/j.ijpharm.2015.11.031 PMID: 26602293

Larrañeta E., Lutton R. E. M., Woolfson A. D., & Donnelly R. F. (2016). Microneedle arrays as transdermal and intradermal drug delivery systems: Materials science, manufacture and commercial development. *Materials Science and Engineering R Reports*, 104, 1. DOI: 10.1016/j.mser.2016.03.001

Larrañeta, E., Lutton, R. E. M., & Woolfson, A. D. (2016). Biodegradable polymer microneedles: In vitro experiments and finite element analysis modeling. *Journal of Controlled Release*, *225*, 193–204.

Larrañeta, E., Moore, J., & Vicente-Perez, E. M. (2015). A novel biopolymer formulation for drug delivery via dissolvable microneedles. *Materials Science and Engineering C*, *45*, 214–221. DOI: 10.1016/j.msec.2014.09.070

Larrañeta, E., Stewart, S., & Donnelly, R. (2020). Microneedles for transdermal drug delivery: Current state of the art. *Journal of Controlled Release*, *321*, 143–157. DOI: 10.1016/j.jconrel.2020.03.013

Lasprilla, A. J. R., Martínez, G. A., Lunelli, B. H., Jardini, A. L., & Filho, R. M. (2011). Poly-lactic acid synthesis for application in biomedical devices — A review. *Biotechnology Advances*, 30(1), 321. DOI: 10.1016/j.biotechadv.2011.06.019

Layek, B. (2024). A Comprehensive Review of Xanthan Gum-Based Oral Drug Delivery Systems. *International Journal of Molecular Sciences*, *25*(18), 10143. DOI: 10.3390/ijms251810143 PMID: 39337626

Layek, B., & Mandal, S. (2020). Natural polysaccharides for controlled delivery of oral therapeutics: A recent update. *Carbohydrate Polymers*, *230*, 115617. DOI: 10.1016/j.carbpol.2019.115617 PMID: 31887888

Lee, J. W., Park, J.-H., & Prausnitz, M. R. (2017). Microneedle-mediated delivery of nanoparticles and DNA vaccines: Progress and challenges. *Advanced Functional Materials, 27*(10), 1605281. DOI: 10.1002/adfm.201605281

Lee, J., & Kim, H. (2018). Emerging properties of hydrogels in tissue engineering. *Journal of Tissue Engineering, 9*, 2041731418768285. Advance online publication. DOI: 10.1177/2041731418768285 PMID: 29623184

Lee, K. Y., & Mooney, D. J. (2012). Alginate: Properties and biomedical applications. *Progress in Polymer Science, 37*(1), 106–126. DOI: 10.1016/j.progpolymsci.2011.06.003 PMID: 22125349

Lee, M., & Kim, J. (2023). Amphiphilic polymers in oral drug delivery. *Advanced Drug Delivery Reviews, 173*, 12–30.

Lee, Y., Zhang, X., & Wang, J. (2021). Biodegradable polymers in drug delivery systems. *Biomaterials Science, 9*(5), 1895–1908. DOI: 10.1039/d0bm01667c PMID: 33432933

Lévesque, S. G., & Shoichet, M. S. (2006). Synthesis of cell-adhesive dextran hydrogels and macroporous scaffolds. *Biomaterials, 27*(30), 5277–5285. DOI: 10.1016/j.biomaterials.2006.06.004 PMID: 16793132

Li B., Geng L., Liu W., Liao L., Ban J., & Lu Z. (2023). Formulation and evaluation of PLGA nanoparticulate-based microneedle system for potential treatment of neurological diseases. *International Journal of Nanomedicine*, 3745. DOI: 10.2147/IJN.S415728

Li C., Wang J., Wang Y., Gao H., Wei G., Huang Y., Yu H., Gan Y., Wang Y., Mei L., Chen H., Hu H., Zhang Z., & Jin Y. (2019). Recent progress in drug delivery. *Acta Pharmaceutica Sinica B*, 9(6), 1145. DOI: 10.1016/j.apsb.2019.08.003

Liakos, I., Rizzello, L., Scurr, D. J., Pompa, P. P., & Bayer, I. S. (2016). Alginate-based antimicrobial films and fibers: Current advances and future perspectives. *Advances in Colloid and Interface Science, 237*, 55–64.

Liang F., Sandgren K., Fausther-Bovendo H., O'Hagan D. T., Seubert A., Gregorio E. D., Rao S., Sullivan N., Seder R. A., Koup R., & Loré K. (2012). Infiltration of dendritic cells and antigen uptake in the muscle after injection of HIV-1 Env gp120 in adjuvant. *Retrovirology*, 9. DOI: 10.1186/1742-4690-9-S2-O16

Liang, Y., Zhao, X., Hu, T., Chen, B., Yin, Z., Ma, P. X., & Guo, B. (2021). Flexible and Highly Sensitive Wearable Strain Sensors Fabricated by 3D Printing for Personalized Health Monitoring. *Advanced Functional Materials, 31*(7), 2008722. DOI: 10.1002/adfm.202008722

Li, C., Deng, Z., & Gillies, E. R. (2023). Designing polymers with stimuli-responsive degradation for biomedical applications. *Current Opinion in Biomedical Engineering, 25*, 100437. DOI: 10.1016/j.cobme.2022.100437

Liechty, W. B., Kryscio, D. R., Slaughter, B. V., & Peppas, N. A. (2010). Polymers for drug delivery systems. *Annual Review of Chemical and Biomolecular Engineering, 1*(1), 149–173. DOI: 10.1146/annurev-chembioeng-073009-100847 PMID: 22432577

Lien, S.-M., Ko, L.-Y., & Huang, T.-J. (2009). Effect of pore size on ECM secretion and cell growth in gelatin scaffold for articular cartilage tissue engineering. *Acta Biomaterialia, 5*(2), 670–679. DOI: 10.1016/j.actbio.2008.09.020 PMID: 18951858

Li, F., Qin, Y., Lee, J., Liao, H., Wang, N., Davis, T. P., Qiao, R., & Ling, D. (2020). Stimuli-responsive nano-assemblies for remotely controlled drug delivery. *Journal of Controlled Release, 322*, 566–592. DOI: 10.1016/j.jconrel.2020.03.051 PMID: 32276006

Li, J., & Mooney, D. J. (2016). Designing hydrogels for controlled drug delivery. *Nature Reviews. Materials, 1*(12), 16071. DOI: 10.1038/natrevmats.2016.71 PMID: 29657852

Li, J., Zhang, Y., Yang, Y., & Sun, T. (2024). Advancing mRNA therapeutics: The role and future of nanoparticle delivery systems. *Molecular Pharmaceutics, 21*(8), 3743–3763. DOI: 10.1021/acs.molpharmaceut.4c00276 PMID: 38953708

Li, L., Ni, R., Shao, Y., & Mao, S. (2013). Carrageenan and its applications in drug delivery. *Carbohydrate Polymers, 103*(1), 1–11. Advance online publication. DOI: 10.1016/j.carbpol.2013.12.008 PMID: 24528694

Li, L., Wang, N., Jin, X., Deng, R., Nie, S., Sun, L., & Gong, C. (2014). Biodegradable and injectable in situ cross-linking chitosan-hyaluronic acid based hydrogels for postoperative adhesion prevention. *Biomaterials, 35*(12), 3903–3917. DOI: 10.1016/j.biomaterials.2014.01.050 PMID: 24507411

Lin, C., & Metters, A. T. (2006). Hydrogels in controlled release formulations: Network design and mathematical modelling. *Advanced Drug Delivery Reviews, 58*(12), 1379–1408. DOI: 10.1016/j.addr.2006.09.004 PMID: 17081649

Lin, N., & Dufresne, A. (2014). Nanocellulose in biomedicine: Current status and future prospect. *European Polymer Journal, 59*, 302–325. DOI: 10.1016/j.eurpolymj.2014.07.025

Lin, Z., Zhang, H., & Wu, Y. (2023). Elastin-like polypeptides: Phase transitions for injectable drug delivery. *Nature Communications, 14*, 3345.

Liu, J., Liu, J., Mu, W., Ma, Q., Zhai, X., Jin, B., Liu, Y., & Zhang, N. (2024). Delivery strategy to enhance the therapeutic efficacy of Liver Fibrosis via Nanoparticle Drug Delivery systems. *ACS Nano*, *18*(32), 20861–20885. DOI: 10.1021/acsnano.4c02380 PMID: 39082637

Liu, L., Yao, W., Rao, Y., Lu, X., & Gao, J. (2017). pH-Responsive carriers for oral drug delivery: Challenges and opportunities of current platforms. *Drug Delivery*, *24*(1), 569–581. DOI: 10.1080/10717544.2017.1279238 PMID: 28195032

Liu, Q., Liu, H., Yuan, Z., Yang, Y., & Liu, C. (2020). Ligand-modified alginate microspheres for targeted drug delivery. *International Journal of Pharmaceutics*, *579*, 119181. DOI: 10.1016/j.ijpharm.2020.119181

Liu, Y., Sun, M., Wang, T., Chen, X., & Wang, H. (2021). Chitosan-based self-assembled nanomaterials: Their application in drug delivery. *VIEW*, *2*(1), 20200069. DOI: 10.1002/VIW.20200069

Li, X., & Zhang, Y. (2023). Biodegradable polymers in injectable drug delivery. *Biomaterials Science*, *10*(3), 745–758.

Li, Y., Jiang, H., Zheng, W., Gong, N., Chen, L., Jiang, X., & Yang, G. (2015). Bacterial cellulose–hyaluronan nanocomposite biomaterials as wound dressings for severe skin injury repair. *Journal of Materials Chemistry. B*, *3*(17), 3498–3507. DOI: 10.1039/C4TB01819B PMID: 32262233

Li, Y., Liu, J., He, J., Dey, A., Bui, V. D., & Park, J. H. (2024). Recent Advances in Stimuli-Responsive Self-Immolative Polymers for Drug Delivery and Molecular Imaging. *Chemistry of Materials*, *36*(9), 4054–4077. DOI: 10.1021/acs.chemmater.4c00583

Lori, M. S., Ohadi, M., Estabragh, M. A., Afsharipour, S., Banat, I. M., & Dehghannoudeh, G. (2021). pH-sensitive polymer-based carriers as a useful approach for oral delivery of therapeutic protein: A review. *Protein and Peptide Letters*, *28*(11), 1230–1237. DOI: 10.2174/0929866528666210720142841 PMID: 34303327

Low, P. S., & Kularatne, S. A. (2009). Folate-targeted therapeutic and imaging agents for cancer. *Current Opinion in Chemical Biology*, *13*(3), 256–262. DOI: 10.1016/j.cbpa.2009.03.022 PMID: 19419901

Lu, Q., Cui, Z., Wu, Y., Wang, H., Zhang, S., Guan, J., & Mao, S. (2022). Challenges and Strategies to Enhance the Systemic Absorption of Inhaled Peptides and Proteins. *Pharmaceutical Research*, 40(5), 1037. DOI: 10.1007/s11095-022-03435-3

Lu, B., Luo, D., Zhao, A., Wang, H., Zhao, Y., Maitz, M. F., & Huang, N. (2019). pH responsive chitosan and hyaluronic acid layer by layer film for drug delivery applications. *Progress in Organic Coatings*, *135*, 240–247. DOI: 10.1016/j.porgcoat.2019.06.012

Lu, H., Cai, Z., & Hu, P. (2024). Recent Advances in Polymeric Delivery Vehicles for Controlled and Sustained Drug Release. *Pharmaceutics*, *16*(9), 1184. DOI: 10.3390/pharmaceutics16091184 PMID: 39339220

Luo, Y., Teng, Z., & Wang, Q. (2016). Development of calcium-fortified nanostructured lipid carriers for oral delivery of curcumin. *International Journal of Pharmaceutics*, *507*(1-2), 61–70. DOI: 10.1016/j.ijpharm.2016.04.033 PMID: 27154254

Luo, Y., & Wang, Q. (2014). Alginate-based hydrogel beads for drug delivery: Preparation, properties and applications. *Biomacromolecules*, *15*(9), 3211–3224. DOI: 10.1021/bm5005636

Maddiboyina, B., Desu, P. K., Vasam, M., Challa, V. T., Surendra, A. V., Rao, R. S., Alagarsamy, S., & Jhawat, V. (2022). An insight of nanogels as novel drug delivery system with potential hybrid nanogel applications. *Journal of Biomaterials Science. Polymer Edition*, *33*(2), 262–278. DOI: 10.1080/09205063.2021.1982643 PMID: 34547214

Maderuelo, C., Zarzuelo, A., & Lanao, J. M. (2011). Critical factors in the release of drugs from sustained release hydrophilic matrices. *Journal of Controlled Release*, *154*(1), 2–19. DOI: 10.1016/j.jconrel.2011.04.002 PMID: 21497624

Magli, S., Rossi, L., Consentino, C., Bertini, S., Nicotra, F., & Russo, L. (2021). Combined Analytical Approaches to Standardize and Characterize Biomaterials Formulations: Application to Chitosan-Gelatin Cross-Linked Hydrogels. *Biomolecules*, *11*(5), 683. DOI: 10.3390/biom11050683 PMID: 34062918

Mahar, R., Chakraborty, A., Nainwal, N., Bahuguna, R., Sajwan, M., & Jakhmola, V. (2023). Application of PLGA as a biodegradable and biocompatible polymer for pulmonary delivery of drugs. *AAPS PharmSciTech*, *24*(1), 39. DOI: 10.1208/s12249-023-02502-1 PMID: 36653547

Majrashi, M. a. A., Yahya, E. B., & Mushtaq, R. Y., HPS, A. K., Rizg, W. Y., Alissa, M., Alkharobi, H., Badr, M. Y., & Hosny, K. M. (. (2024). Revolutionizing drug delivery: Exploring the impact of advanced 3D printing technologies on polymer-based systems. *Journal of Drug Delivery Science and Technology*, *98*, 105839. DOI: 10.1016/j.jddst.2024.105839

Majumdar, S. (2025). Cellulose-based hydrogels for drug delivery applications. In *Cellulose-Based Hydrogells* (pp. 259–279). Elsevier. DOI: 10.1016/B978-0-443-22049-4.00012-6

Makadia, H. K., & Siegel, S. J. (2011). Poly(lactic-co-glycolic acid) (PLGA) as biodegradable controlled drug delivery carrier. *Polymers*, *3*(3), 1377–1397. DOI: 10.3390/polym3031377 PMID: 22577513

Mandal, A., Ghosh, M., Basu, S. K., & Mukherjee, B. (2021). Redox-responsive and pH-sensitive smart nanoparticles for targeted delivery of anticancer drugs. *European Journal of Pharmaceutical Sciences*, *163*, 105856. DOI: 10.1016/j.ejps.2021.105856

Maniar, K. H., Jones, I. A., Gopalakrishna, R., & Vangsness, C. T.Jr. (2018). Lowering side effects of NSAID usage in osteoarthritis: Recent attempts at minimizing dosage. *Expert Opinion on Pharmacotherapy*, *19*(2), 93–102. DOI: 10.1080/14656566.2017.1414802 PMID: 29212381

Mano, J. F. (2008). Stimuli-responsive polymeric systems for biomedical applications. *Advanced Engineering Materials*, *10*(6), 515–527. DOI: 10.1002/adem.200700355

Martins, A. F., Facchi, S. P., Follmann, H. D., Pereira, A. G. B., Rubira, A. F., & Muniz, E. C. (2009). Biodegradable and injectable hydrogels based on chitosan and poly(vinyl alcohol) crosslinked with glutaraldehyde. *International Journal of Biological Macromolecules*, *65*, 136–145. DOI: 10.1016/j.ijbiomac.2014.11.005

Mashaqbeh, H., Al-Ghzawi, B., & BaniAmer, F. (2024). Exploring the formulation and approaches of injectable hydrogels utilizing hyaluronic acid in biomedical uses. *Advances in Pharmacological and Pharmaceutical Sciences*, *2024*(1), 3869387. DOI: 10.1155/2024/3869387 PMID: 38831895

Massironi, A. (2024). Hybrid organic polymer/inorganic nano-materials for biomedical applications: Where we are and where to go? *Current Nanoscience*, *20*(2), 188–205. DOI: 10.2174/1573413719666230410113733

Matanović, M. R., Kristl, J., & Grabnar, P. A. (2014). Thermoresponsive polymers: Insights into decisive hydrogel characteristics, mechanisms of gelation, and promising biomedical applications. *International Journal of Pharmaceutics*, *472*(1), 262–275. DOI: 10.1016/j.ijpharm.2014.06.029 PMID: 24950367

Mauri, E., Negri, A., Rebellato, E., Masi, M., Perale, G., & Rossi, F. (2018). Hydrogel-Nanoparticles Composite System for Controlled Drug Delivery. *Gels (Basel, Switzerland)*, *4*(3), 74. DOI: 10.3390/gels4030074 PMID: 30674850

McConnell, H. L., & Mishra, A. (2022). Cells of the blood–brain barrier: An overview of the neurovascular unit in health and disease. *The Blood-Brain Barrier. Methods and Protocols*, *2492*, 3–24. DOI: 10.1007/978-1-0716-2289-6_1 PMID: 35733036

McCrudden M., McAlister E., Courtenay A. J., González-Vázquez P., Singh T. R. R., & Donnelly R. F. (2015). Microneedle applications in improving skin appearance. *Experimental Dermatology*, 24(8), 561. DOI: 10.1111/exd.12723

McIntosh, M., Stone, B. A., & Stanisich, V. A. (2005). Curdlan and other bacterial (1→ 3)-β-D-glucans. *Applied Microbiology and Biotechnology*, *68*(2), 163–173. DOI: 10.1007/s00253-005-1959-5 PMID: 15818477

Metaxa, A. F., Efthimiadou, E. K., & Kordas, G. (2014). Cellulose-based drug carriers for cancer therapy: Cytotoxic evaluation in cancer and healthy cells. *Materials Letters*, *132*, 432–435. DOI: 10.1016/j.matlet.2014.06.134

Mikušová, V., & Mikuš, P. (2021). Advances in chitosan-based nanoparticles for drug delivery. *International Journal of Molecular Sciences*, *22*(17), 9652. DOI: 10.3390/ijms22179652 PMID: 34502560

Millotti, G., Laffleur, F., Perera, G., Vigl, C., Pickl, K., Sinner, F., & Bernkop-Schnürch, A. (2014). In vivo evaluation of thiolated chitosan tablets for oral insulin delivery. *Journal of Pharmaceutical Sciences*, *103*(10), 3165–3170. DOI: 10.1002/jps.24102 PMID: 25139279

Mishra, D. K., Pandey, V., Maheshwari, R., Ghode, P., & Tekade, R. K. (2019). *Cutaneous and Transdermal Drug Delivery. Elsevier eBooks, 595*. Elsevier BV. DOI: 10.1016/B978-0-12-817909-3.00015-7

Modi, K., Modi, M., Mishra, D., Mittal, P., Sorathiya, U., & Shelat, P. (2013). Oral controlled release drug delivery system: An overview. *International Research Journal of Pharmacy*, *4*(3), 70–76. DOI: 10.7897/2230-8407.04312

Mohamed, A. A., Kbah, N. Z., & Wennas, O. N. (2023). Poly (lactic-co-glycolic acid) nanoparticles for drug delivery of rupatadine fumarate: Development and evaluation. *Farmatsiia (Sofia)*, *70*(4), 1257–1264. DOI: 10.3897/pharmacia.70.e110191

Mohammed, P. N., Hussen, N. H., Hasan, A. H., Salh, H. J. H., Jamalis, J., Ahmed, S., Bhat, A. R., & Kamal, M. A. (2025). A review on the role of nanoparticles for targeted brain drug delivery: Synthesis, characterization, and applications. *EXCLI Journal*, *24*, 34. DOI: 10.17179/excli2024-7163 PMID: 39967907

Mohanty, D., Sahoo, S., & Khatua, S. (2016). Alginate microbeads coated with chitosan: Formulation, characterization and their application as a controlled drug delivery system. *Carbohydrate Polymers*, *148*, 282–289. DOI: 10.1016/j.carbpol.2016.04.064

Mottaghitalab, F., Yazdi, M. K., Saeb, M. R., Bączek, T., & Farokhi, M. (2024). Green and sustainable hydrogels based on quaternized chitosan to enhance wound healing. *Chemical Engineering Journal, 152288*, 152288. Advance online publication. DOI: 10.1016/j.cej.2024.152288

Mukherjee, C., Varghese, D., Krishna, J. S., Boominathan, T., Rakeshkumar, R., Dineshkumar, S., Brahmananda Rao, C. V. S., & Sivaramakrishna, A. (2023). Recent advances in biodegradable polymers–properties, applications and future prospects. *European Polymer Journal*, *192*, 112068. DOI: 10.1016/j.eurpolymj.2023.112068

Municoy, S., Alvarez Echazu, M. I., Antezana, P. E., Galdopórpora, J. M., Olivetti, C., Mebert, A., & Desimone, M. F. (2020). Stimuli-responsive materials for tissue engineering and drug delivery. *International Journal of Molecular Sciences*, *21*(13), 4724. DOI: 10.3390/ijms21134724 PMID: 32630690

Mura, P., Maestrelli, F., Cirri, M., & Mennini, N. (2022). Multiple roles of chitosan in mucosal drug delivery: An updated review. *Marine Drugs*, *20*(5), 335. DOI: 10.3390/md20050335 PMID: 35621986

Mura, S., Nicolas, J., & Couvreur, P. (2013). Stimuli-responsive nanocarriers for drug delivery. *Nature Materials*, *12*(11), 991–1003. DOI: 10.1038/nmat3776 PMID: 24150417

Murdande, S. B., Pikal, M. J., Shanker, R. M., & Bogner, R. H. (2011). Aqueous solubility of crystalline and amorphous drugs: Challenges in measurement. *Pharmaceutical Development and Technology*, *16*(3), 187–200. DOI: 10.3109/10837451003774377 PMID: 20429826

Murugesan, S., Byran, G., Lakshmanan, K., Karri, V. V. S. R., & Radhakrishnan, A. (2019). Oral Modified Drug Release Solid Dosage Form with Special Reference to Design; An Overview. *Current Drug Research Reviews*, *12*(1), 16–25. DOI: 10.2174/2589977511666191121094520 PMID: 31755398

Muthu, M., Gopal, J., Chun, S., Devadoss, A. J. P., Hasan, N., & Sivanesan, I. (2021). Crustacean waste-derived chitosan: Antioxidant properties and future perspective. *Antioxidants*, *10*(2), 228. DOI: 10.3390/antiox10020228 PMID: 33546282

Nakagawa Y., Ushidome K., Masuda K., Igarashi K., Matsumoto Y., Yamasoba T., Anraku Y., Takai M., & Cabral H. (2023). Multi-armed star-shaped block copolymers of poly(ethylene glycol)-poly(furfuryl glycidol) as long circulating nanocarriers. *Polymers*, 15(12), 2626. DOI: 10.3390/polym15122626

Nance, E., Pun, S. H., Saigal, R., & Sellers, D. L. (2022). Drug delivery to the central nervous system. *Nature Reviews. Materials*, *7*(4), 314–331. DOI: 10.1038/s41578-021-00394-w PMID: 38464996

Narančić, T., Cerrone, F., Beagan, N., & O'Connor, K. E. (2020). Recent Advances in Bioplastics: Application and Biodegradation. *Polymers*, *12*(4), 920. DOI: 10.3390/polym12040920 PMID: 32326661

Narayanaswamy, R., & Torchilin, V. P. (2019). Hydrogels and Their Applications in Targeted Drug Delivery. *Molecules (Basel, Switzerland)*, *24*(3), 603. DOI: 10.3390/molecules24030603 PMID: 30744011

Nayak, A. K., Maji, R., & Das, B. (2010). Gastroretentive drug delivery systems: A review. *Asian Journal of Pharmaceutical and Clinical Research*, *3*(1). https://ajpcr.com/Vol3Issue1/250.pdf

Nguyen, H. T., Chien, T. B., & Cư ng, Đ. X. (2023). Polymer-Based Hydrogels Applied in Drug Delivery: An Overview. *Gels (Basel, Switzerland)*, *9*(7), 523. DOI: 10.3390/gels9070523 PMID: 37504402

Niculescu, A., Bîrcă, A. C., & Grumezescu, A. M. (2021). New applications of lipid and Polymer-Based nanoparticles for nucleic acids delivery. *Pharmaceutics*, *13*(12), 2053. DOI: 10.3390/pharmaceutics13122053 PMID: 34959335

Nie, T., Wang, W., Liu, X., Wang, Y., Li, K., Song, X., Zhang, J., Yu, L., & He, Z. (2021). Sustained release systems for delivery of therapeutic peptide/protein. *Biomacromolecules*, *22*(6), 2299–2324. DOI: 10.1021/acs.biomac.1c00160 PMID: 33957752

Ni, M., Xiong, M., Zhang, X., Cai, G., Chen, H., Zeng, Q., & Yu, Z. (2015). Poly(lactic-co-glycolic acid) nanoparticles conjugated with CD133 aptamers for targeted salinomycin delivery to CD133+ osteosarcoma cancer stem cells. *International Journal of Nanomedicine*, 2537–2554. PMID: 25848270

Nizam, M., Purohit, R., & Taufik, M. (2024). 3D printing in Healthcare: A review on drug printing, challenges and future perspectives. *Materials Today. Communications*, *40*, 110199. DOI: 10.1016/j.mtcomm.2024.110199

Nordin, A. H., Ahmad, Z., Husna, S., Ilyas, R., Azemi, A., Ismail, N., Nordin, M., Ngadi, N., Siti, N., Nabgan, W., Norfarhana, A., & Azami, M. (2023). The State of the Art of Natural Polymer Functionalized Fe3O4 Magnetic Nanoparticle Composites for Drug Delivery Applications: A Review. *Gels (Basel, Switzerland)*, *9*(2), 121. DOI: 10.3390/gels9020121 PMID: 36826291

Ofridam, F., Tarhini, M., Badri, W., Liao, W., Lebaz, N., Gagnière, É., . . . Elaïssari, A. (2020). Stimuli-Responsive Polymer Coatings. In *Polymer Coatings: Technologies and Applications* (pp. 199-225). CRC Press. DOI: 10.1201/9780429199226-11

Ofridam, F., Tarhini, M., Lebaz, N., Gagnière, É., Mangin, D., & Elaissari, A. (2021). pH-sensitive polymers: Classification and some fine potential applications. *Polymers for Advanced Technologies, 32*(4), 1455–1484. DOI: 10.1002/pat.5230

Ossipov, D. A. (2010). Nanostructured hyaluronic acid-based materials for active delivery to cancer. *Expert Opinion on Drug Delivery, 7*(6), 681–703. DOI: 10.1517/17425241003730399 PMID: 20367530

Pahuja, P., Arora, S., & Pawar, P. (2012). Ocular drug delivery system: A reference to natural polymers. *Expert Opinion on Drug Delivery, 9*(7), 837–861. DOI: 10.1517/17425247.2012.690733 PMID: 22703523

Paliwal, S. R., Paliwal, R., Agrawal, G. P., & Vyas, S. P. (2016). Hyaluronic acid modified pH-sensitive liposomes for targeted intracellular delivery of doxorubicin. *Journal of Liposome Research, 26*(4), 276–287. DOI: 10.3109/08982104.2015.1117489 PMID: 26784587

Pal, K., Banthia, A. K., & Majumdar, D. K. (2009). Polymeric Hydrogels: Characterization and Biomedical Applications. *Designed Monomers and Polymers, 12*(3), 197–220. DOI: 10.1163/156855509X436030

Pal, P., Sambhakar, S., Paliwal, S., Kumar, S., & Kalsi, V. (2024). Biofabrication paradigms in corneal regeneration: Bridging bioprinting techniques, natural bioinks, and stem cell therapeutics. *Journal of Biomaterials Science. Polymer Edition, 35*(5), 717–755. DOI: 10.1080/09205063.2024.2301817 PMID: 38214998

Panchpuri, M., Painuli, R., & Kumar, C. (2025). Artificial Intelligence in Smart Drug Delivery Systems: A Step Toward Personalized Medicine. *RSC Pharmaceutics*. Advance online publication. DOI: 10.1039/D5PM00089K

Pandey, R., Sharma, A., Zahoor, A., Sharma, S., Khuller, G. K., & Prasad, B. (2003). Poly(lactide-co-glycolide) nanoparticle-based inhalable sustained drug delivery system for experimental tuberculosis. *The Journal of Antimicrobial Chemotherapy, 52*(6), 981–986. DOI: 10.1093/jac/dkg477 PMID: 14613962

Pang, X., ... (2020). Engineering Thermo-pH Dual Responsive Hydrogel for Enhanced Tumor Accumulation, Penetration, and Chemo-Protein Combination Therapy. https://doaj.org/article/dbd573f72d7c451ab5abeb5271ab8eba

Pan, H., Zhang, X., & Li, Z. (2021). pH-sensitive drug delivery systems for oral administration: A review. *Journal of Drug Delivery Science and Technology, 63*, 102366. DOI: 10.1016/j.jddst.2021.102366

Parashar, A. K., Saraogi, G. K., Jain, P. K., Kurmi, B., Shrivastava, V., & Arora, V. (2024). Polymer-drug conjugates: Revolutionizing nanotheranostic agents for diagnosis and therapy. *Discover. Oncology, 15*(1), 641. Advance online publication. DOI: 10.1007/s12672-024-01509-9 PMID: 39527173

Park, C., Jeong, S., & Kim, Y. (2021). Advanced delivery systems for transdermal applications. *Drug Development and Industrial Pharmacy, 47*(10), 1620–1632. DOI: 10.1080/03639045.2021.1895573

Park, H., & Kim, S. (2022). Biopolymers for sustained release drug delivery. *Journal of Applied Polymer Science, 139*(17), 50785.

Park, H., & Kim, S. (2022). pH-sensitive drug delivery systems for oral applications. *Journal of Applied Polymer Science, 139*(17), 50785.

Park, J., Park, K., & Lee, S. (2017). Multilayered alginate microcapsules for controlled multi-drug release. *Carbohydrate Polymers, 157*, 1363–1371. DOI: 10.1016/j.carbpol.2016.11.070

Park, S., Lee, J., & Kim, Y. (2020). Enhanced skin penetration with lipid nanoparticles for transdermal delivery. *International Journal of Pharmaceutics, 585*, 119440. DOI: 10.1016/j.ijpharm.2020.119440

Parvaresh, A., Izadi, Z., Nemati, H., Derakhshankhah, H., & Jaymand, M. (2023). Redox-and pH-responsive alginate-based magnetic hydrogel:"Smart" drug delivery and protein corona studies. *Journal of Molecular Liquids, 382*, 121990. DOI: 10.1016/j.molliq.2023.121990

Patel, P., Sahoo, S., & Agarwal, R. (2023). Designing "stealth" nanoparticles with stimuli-responsive polymers. *Drug Discovery Today, 28*(5), 890–904.

Patel, S., Kumar, R., & Mehta, S. (2023). Hyaluronic acid-based drug delivery: pH and enzymatic responsiveness for targeted tumor therapy. *Acta Biomaterialia, 155*, 456–469.

Patkar, S. V., & Sherje, A. (2025). Recent Advances in Smart Hydrogel for Drug Delivery in Cancer Therapeutics. *Indian Journal of Pharmaceutical Education and Research, 59*(3), 854–864. DOI: 10.5530/ijper.20250454

Peer, D., Karp, J. M., Hong, S., Farokhzad, O. C., Margalit, R., & Langer, R. (2017). Nanocarriers as an emerging platform for cancer therapy. *Nature Nanotechnology*, *2*(12), 751–760. DOI: 10.1038/nnano.2007.387 PMID: 18654426

Peng, X., Han, Q., Zhou, X., Chen, Y., Huang, X., Guo, X., Peng, R., Wang, H., Peng, X., & Cheng, L. (2022). Effect of pH-sensitive nanoparticles on inhibiting oral biofilms. *Drug Delivery*, *29*(1), 561–573. DOI: 10.1080/10717544.2022.2037788 PMID: 35156501

Peppas, N. A., Bures, P., Leobandung, W., & Ichikawa, H. (2000). Hydrogels in pharmaceutical formulations. *European Journal of Pharmaceutics and Biopharmaceutics*, *50*(1), 27–46. DOI: 10.1016/S0939-6411(00)00090-4 PMID: 10840191

Pérez-Luna, V. H., & González-Reynoso, O. (2018). Encapsulation of Biological Agents in Hydrogels for Therapeutic Applications. *Gels (Basel, Switzerland)*, *4*(3), 61. DOI: 10.3390/gels4030061 PMID: 30674837

Pham, T.-N., Su, C.-F., Huang, C.-C., & Jan, J. (2020). Biomimetic hydrogels based on L-Dopa conjugated gelatin as pH-responsive drug carriers and antimicrobial agents. *Colloids and Surfaces. B, Biointerfaces*, *196*, 111316. DOI: 10.1016/j.colsurfb.2020.111316 PMID: 32827950

Pignatello, R., Stancampiano, A. H. S., Ventura, C. A., & Puglisi, G. (2007). Dexamethasone sodium phosphate-loaded Chitosan based delivery systems for buccal application. *Journal of Drug Targeting*, *15*(9), 603–610. DOI: 10.1080/10611860701502806 PMID: 17968714

Pinto, J. F. (2010). Site-specific drug delivery systems within the gastrointestinal tract: From the mouth to the colon. *International Journal of Pharmaceutics*, *395*(1-2), 44–52. DOI: 10.1016/j.ijpharm.2010.05.003 PMID: 20546856

Piperno, A., Sciortino, M. T., Giusto, E., Montesi, M., Panseri, S., & Scala, A. (2021). Recent Advances and Challenges in Gene Delivery Mediated by Polyester-Based Nanoparticles. *International Journal of Nanomedicine*, 5981. DOI: 10.2147/IJN.S321329

Prabaharan, M., & Mano, J. F. (2004). Chitosan-based particles as controlled drug delivery systems. *Drug Delivery*, *12*(1), 41–57. DOI: 10.1080/10717540590889781 PMID: 15801720

Pradhan, T., Chelike, D. K., Roy, D., Pramanik, T., & Dolui, S. (2025). Stimuli-Responsive Multiacceptor Conjugated Polymers: Recent trend and future direction. *ACS Polymers Au*, *5*(2), 62–79. Advance online publication. DOI: 10.1021/acspolymersau.4c00082 PMID: 40226348

Prasanth, A., Kumar, A., Shruthi, B., & Sindhu, S. (2019). Kinetic study and in vitro drug release studies of nitrendipine loaded arylamide grafted chitosan blend microspheres. *Materials Research Express*, *6*(12), 125427. DOI: 10.1088/2053-1591/ab5811

Prausnitz, M. R., & Langer, R. (2008). Transdermal drug delivery. *Nature Biotechnology*, *26*(11), 1261–1268. DOI: 10.1038/nbt.1504 PMID: 18997767

Prausnitz, M. R., Mikszta, J. A., & Andrianov, A. K. (2004). Microneedle-based vaccines. *Journal of Controlled Release*, *97*(3), 503–511. PMID: 15212882

Prausnitz, M., Langer, R., & Chen, X. (2019). Transdermal drug delivery: Past, present, and future. *Advanced Drug Delivery Reviews*, *64*(1), 40–57. DOI: 10.1016/j.addr.2012.07.008

Premaletha, K., Licy, C. D., Jose, S., Saraladevi, A., Shirwaikar, A., & Shirwaikar, A. (2010). Formulation, characterization and optimization of Hepatitis B surface antigen (HBsAg)-loaded chitosan microspheres for oral delivery. *Pharmaceutical Development and Technology*, *17*(2), 251–258. DOI: 10.3109/10837450.2010.535824 PMID: 21108582

Priya, S., Daryani, J., Desai, V. M., & Singhvi, G. (2024). Bridging the gap in rheumatoid arthritis treatment with hyaluronic acid-based drug delivery approaches. *International Journal of Biological Macromolecules*, *271*, 132586. DOI: 10.1016/j.ijbiomac.2024.132586 PMID: 38795889

Puertas-Bartolomé, M., Mora-Boza, A., & García-Fernández, L. (2021). Emerging Biofabrication Techniques: A Review on Natural Polymers for Biomedical Applications. *Polymers*, *13*(8), 1209. DOI: 10.3390/polym13081209 PMID: 33918049

Qiu, Y., & Park, K. (2001). Environment-sensitive hydrogels for drug delivery. *Advanced Drug Delivery Reviews*, *53*(3), 321–339. DOI: 10.1016/S0169-409X(01)00203-4 PMID: 11744175

Qu, M., Jiang, X., Zhou, X., Wang, C., Wu, Q., Ren, L., Zhu, J., Zhu, S., Tebon, P., Sun, W., & Khademhosseini, A. (2020). Stimuli-responsive delivery of growth factors for tissue engineering. *Advanced Healthcare Materials*, *9*(7), 1901714. DOI: 10.1002/adhm.201901714 PMID: 32125786

Raeisi, A., & Farjadian, F. (2024). Commercial hydrogel product for drug elivery based on route of administration. *Frontiers in Chemistry*, *12*, 1336717. Advance online publication. DOI: 10.3389/fchem.2024.1336717 PMID: 38476651

Rahman, M. M., Roy, S., Das, S. C., Jha, M. K., Begum, T., Ahsan, Q., & Reza, M. S. (2011). Evaluation of various grades of hydroxypropylmethylcellulose matrix systems as oral sustained release drug delivery systems. *Journal of Pharmaceutical Science and Research*, *3*(1), 930.

Rajitha, P., Gopinath, D., Biswas, R., Sabitha, M., & Jayakumar, R. (2016). Chitosan nanoparticles in drug therapy of infectious and inflammatory diseases. *Expert Opinion on Drug Delivery*, *13*(8), 1177–1194. DOI: 10.1080/17425247.2016.1178232 PMID: 27087148

Ranamukhaarachchi, S., Padeste, C., Häfeli, U. O., Stoeber, B., & Cadarso, V. J. (2017). Design considerations of a hollow microneedle-optofluidic biosensing platform incorporating enzyme-linked assays. *Journal of Micromechanics and Microengineering*, 28(2), 24002. DOI: 10.1088/1361-6439/aa9c9c

Rani, E. E., Sanjana, D. S., Karthikeyan, E., & Nandhini, J. (2025). Smart Nanomaterials: Current State and Future Prospects in Drug Delivery and Tissue Engineering. *Biomedical Materials & Devices*, 1–28. DOI: 10.1007/s44174-025-00350-0

Ravindran, R., Mitra, K., Arumugam, S. K., & Doble, M. (2021). Preparation of Curdlan sulphate-Chitosan nanoparticles as a drug carrier to target Mycobacterium smegmatis infected macrophages. *Carbohydrate Polymers*, *258*, 117686. DOI: 10.1016/j.carbpol.2021.117686 PMID: 33593559

Rawal, S. U., Patel, B. M., & Patel, M. M. (2022). New drug delivery systems developed for brain targeting. *Drugs*, *82*(7), 749–792. DOI: 10.1007/s40265-022-01717-z PMID: 35596879

Ren, Y., Li, P., Xie, Y., Xu, J., Luo, Q., Chen, M., Liu, R., Feng, H., Chen, Y., Liu, Y., Bao, C., Duan, J., Li, J., & Lu, W. (2025). Dual-responsive nanoparticles for enhanced drug delivery in breast Cancer chemotherapy. *Journal of Controlled Release*, *377*, 146–161. DOI: 10.1016/j.jconrel.2024.11.026 PMID: 39549730

Rezaei, N., Akbarzadeh, I., Kazemi, S., Montazeri, L., Zarkesh, I., Hossein-Khannazer, N., Hassan, M., & Vosough, M. (2021). Smart materials in regenerative medicine. *Modern Medical Laboratory Journal*, *4*(1), 39–51. DOI: 10.30699/mmlj17.4.1.39

Rigo, D., da Silva, L. M., Fischer, B., Colet, R., Dallago, R. M., & Zeni, J. (2023). Hyaluronic Acid–from Production to Application: A. *Biointerface Research in Applied Chemistry*, *13*(3), 211. DOI: 10.33263/BRIAC133.211

Rijal, G. (2023). Bioinks of Natural Biomaterials for Printing Tissues. In Bioengineering. 10(6): 705. DOI: 10.3390/bioengineering10060705

Rinaudo, M. (2006). Chitin and chitosan: Properties and applications. *Progress in Polymer Science, 31*(7), 603–632. DOI: 10.1016/j.progpolymsci.2006.06.001

Rivera-Hernández, G., Antunes-Ricardo, M., Martínez-Morales, P., & Sanchez, M. L. (2021). Polyvinyl alcohol-based drug delivery systems for cancer treatment. *International Journal of Pharmaceutics, 600*, 120478. DOI: 10.1016/j.ijpharm.2021.120478 PMID: 33722756

Rizzo, F., & Kehr, S. (2020). Recent Advances in Injectable Hydrogels for Controlled and Local Drug Delivery. *Advanced Healthcare Materials, 10*(1), 2001341. Advance online publication. DOI: 10.1002/adhm.202001341 PMID: 33073515

Rodríguez-Fragoso L., Gutiérrez-Sancha I., Rodríguez-Fragoso P., Rodríguez-López A., & Reyes-Esparza J. (2014). Pharmacokinetic properties and safety of cadmium-containing quantum dots as drug delivery systems. *InTech eBooks*. DOI: 10.5772/58553

Roy, D., Semsarilar, M., Guthrie, J. T., & Perrier, S. (2009). Cellulose modification by polymer grafting: A review. *Chemical Society Reviews, 38*(7), 2046–2064. DOI: 10.1039/b808639g PMID: 19551181

Roy, S., & Choudhury, S. (2021). Polymeric pH-responsive drug delivery systems for targeted cancer therapy. *Materials Science and Engineering C, 121*, 111799. DOI: 10.1016/j.msec.2021.111799

Ruiz, M. E., & Scioli-Montoto, S. (2024). Routes of drug administration. In *ADME processes in pharmaceutical sciences: dosage, design, and pharmacotherapy* (pp. 129–172). Springer Nature Switzerland. DOI: 10.1007/978-3-031-50419-8_7

Rwei, A. Y., Wang, W., & Kohane, D. S. (2015). Photoresponsive nanoparticles for drug delivery. *Nano Today, 10*(4), 45. DOI: 10.1016/j.nantod.2015.06.004 PMID: 26644797

Sabu, C., Mufeedha, P., & Pramod, K. (2018). Yeast-inspired drug delivery: Biotechnology meets bioengineering and synthetic biology. *Expert Opinion on Drug Delivery, 16*(1), 27–41. DOI: 10.1080/17425247.2019.1551874 PMID: 30463446

Saeed, R. M., Dmour, I., & Taha, M. O. (2020). Stable chitosan-based nanoparticles using polyphosphoric acid or hexametaphosphate for tandem ionotropic/covalent crosslinking and subsequent investigation as novel vehicles for drug delivery. *Frontiers in Bioengineering and Biotechnology, 8*, 4. DOI: 10.3389/fbioe.2020.00004 PMID: 32039190

Saha, I., & Datta, S. (2022). Bacterial exopolysaccharides in drug delivery applications. *Journal of Drug Delivery Science and Technology*, *74*, 103557. DOI: 10.1016/j.jddst.2022.103557

Sahoo, S., Gupta, A., & Mehta, P. (2023). Recent advancements in multifunctional stimuli-responsive polymers for targeted drug delivery. *Journal of Controlled Release*, *354*(2), 120–134.

Salunke, S. R., & Patil, S. B. (2016). Ion activated in situ gel of gellan gum containing salbutamol sulphate for nasal administration. *International Journal of Biological Macromolecules*, *87*, 41–47. DOI: 10.1016/j.ijbiomac.2016.02.044 PMID: 26899173

Sapra, B., Thatai, P., Bhandari, S., Sood, J., Jindal, M. K., & Tiwary, A. K. (2013). A Critical Appraisal of Microemulsions for Drug Delivery: Part II. *Therapeutic Delivery*, *5*(1), 83–94. DOI: 10.4155/tde.13.125 PMID: 24341819

Sarma, L., Singh, S., Anjali, N., & Dutta, D. (2025). Challenges and future directions for Next-Generation Drug Delivery Systems. In *Methods in pharmacology and toxicology* (pp. 571–596). https://doi.org/DOI: 10.1007/978-1-0716-4554-3_25

Sarmento, B., Ribeiro, A., Veiga, F., Sampaio, P., Neufeld, R., & Ferreira, D. (2007). Alginate/chitosan nanoparticles are effective for oral insulin delivery. *Pharmaceutical Research*, *24*(12), 2198–2206. DOI: 10.1007/s11095-007-9367-4 PMID: 17577641

Savencu, I., Iurian, S., Porfire, A., Bogdan, C., & Tomu ă, I. (2021). Review of advances in polymeric wound dressing films. *Reactive & Functional Polymers*, *168*, 105059. DOI: 10.1016/j.reactfunctpolym.2021.105059

Scheuplein, R. J. (2015). Permeability of the Skin: A Review of Major Concepts. *Current Problems in Dermatology*, 172. DOI: 10.1159/000401285

Seth, A. K., Parmar, G., Aundhia, C., Shah, N., & Gohil, D. (2024). Smart polymer systems: A futuristic approach to enhance therapeutic efficacy. *Current Organic Chemistry*, *28*(15), 1164–1178. DOI: 10.2174/0113852728305580240429100851

Sezer, A. D., Kazak Sarılmışer, H., Rayaman, E., Çevikbaş, A., Öner, E. T., & Akbuğa, J. (2017). Development and characterization of vancomycin-loaded levan-based microparticular system for drug delivery. *Pharmaceutical Development and Technology*, *22*(5), 627–634. DOI: 10.3109/10837450.2015.1116564 PMID: 26607946

Shah, N., Hussain, M., Rehan, T., Khan, A., & Khan, Z. U. (2022). Overview of polyethylene glycol-based materials with a special focus on core-shell particles for drug delivery application. *Current Pharmaceutical Design*, *28*(5), 352–367. DOI: 10.2174/1381612827666210910104333 PMID: 34514984

Shaibie, N. A., Ramli, N. A., Mohammad Faizal, N. D. F., Srichana, T., & Mohd Amin, M. C. I. (2023). Poly (N-isopropylacrylamide)-Based Polymers: Recent Overview for the Development of Temperature-Responsive Drug Delivery and Biomedical Applications. *Macromolecular Chemistry and Physics*, *224*(20), 2300157. DOI: 10.1002/macp.202300157

Shalash, W. K., Al-Ezzy, R. M., & Ahmaed, A. S. (2021). Antitumor activity of dextran produced from localized leuconostocmesenteroides isolates. *Biochemical and Cellular Archives*, *21*(1), 2371–2374.

Shalini, S., & Chaudhury, A. (2021). Biodegradable polymers in drug delivery: A review of clinical applications. *Drug Development and Industrial Pharmacy*, *47*(9), 1387–1399. DOI: 10.1080/03639045.2021.1908191

Sharma S., Hatware K., Bhadane P., Sindhikar S., & Mishra D. K. (2019). Recent advances in microneedle composites for biomedical applications: Advanced drug delivery technologies. *Materials Science and Engineering C*, 103, 109717. DOI: 10.1016/j.msec.2019.05.002

Sharma, V., Agarwal, R., & Sahoo, S. (2022). Advanced drug delivery systems for autoimmune diseases: A focus on stimuli-responsive polymers. *Journal of Pharmaceutical Sciences*, *111*(12), 3425–3441.

Shehata, N. S., Elwakil, B. H., Elshewemi, S. S., Ghareeb, D. A., & Olama, Z. A. (2023). Selenium nanoparticles coated bacterial polysaccharide with potent antimicrobial and anti-lung cancer activities. *Scientific Reports*, *13*(1), 21871. DOI: 10.1038/s41598-023-48921-9 PMID: 38072846

Shi, D., Beasock, D., Fessler, A., Szebeni, J., Ljubimova, J. Y., Afonin, K. A., & Dobrovolskaia, M. A. (2022). To PEGylate or not to PEGylate: Immunological properties of nanomedicine's most popular component, polyethylene glycol and its alternatives. *Advanced Drug Delivery Reviews*, *180*, 114079. DOI: 10.1016/j.addr.2021.114079 PMID: 34902516

Shi, J., Tan, C., Ge, X., Qin, Z., & Xiong, H. (2024). Recent advances in stimuli-responsive controlled release systems for neuromodulation. *Journal of Materials Chemistry. B*, *12*(24), 5769–5786. Advance online publication. DOI: 10.1039/D4TB00720D PMID: 38804184

Shi, Q., Liu, H., Tang, D., Li, Y., Li, X., & Xu, F. (2019). Bioactuators based on stimulus-responsive hydrogels and their emerging biomedical applications. *NPG Asia Materials*, *11*(1), 64. Advance online publication. DOI: 10.1038/s41427-019-0165-3

Shi, S. S., Lei, S. J., & Fu, C. X. (2020). Studies of the properties of CHG-Loaded alginate fibers for medical application. *Polymer Testing*, *83*, 106141. DOI: 10.1016/j.polymertesting.2019.106141

Sill, T. J., & von Recum, H. A. (2008). Electrospinning: Applications in drug delivery and tissue engineering. *Biomaterials*, *29*(13), 1989–2006. DOI: 10.1016/j.biomaterials.2008.01.011 PMID: 18281090

Silvestro, I., Lopreiato, M., Scotto d'Abusco, A., Di Lisio, V., Martinelli, A., Piozzi, A., & Francolini, I. (2020). Hyaluronic acid reduces bacterial fouling and promotes fibroblasts' adhesion onto chitosan 2D-wound dressings. *International Journal of Molecular Sciences*, *21*(6), 2070. DOI: 10.3390/ijms21062070 PMID: 32197295

Singh, A., Sharma, R., & Kumar, R. (2023). Stimuli-responsive polymers for neurological drug delivery: Innovations and clinical prospects. *Expert Opinion on Drug Delivery*, *20*(7), 679–694.

Singh, J., & Nayak, P. (2023). pH-responsive polymers for drug delivery: Trends and opportunities. *Journal of Polymer Science*, *61*(22), 2828–2850. DOI: 10.1002/pol.20230403

Singh, S. A., & Vellapandian, C. (2023). Structure of the blood brain barrier and its role in the transporters for the movement of substrates across the barriers. *Current Drug Metabolism*, *24*(4), 250–269. DOI: 10.2174/1389200224666230608110349 PMID: 37291784

Singh, S., Kumar, A., & Sharma, R. (2022). Nanoparticles in transdermal drug delivery: Current research and future perspectives. *Pharmaceutical Research*, *39*(10), 1562–1577. DOI: 10.1007/s11095-022-03322-z

Siqueira, N. M., Cirne, M. F. R., Immich, M. F., & Poletto, F. (2018). Stimuli-responsive polymeric hydrogels and nanogels for drug delivery applications. Elsevier eBooks (p. 343). https://doi.org/DOI: 10.1016/B978-0-08-101997-9.00017-5

Smidsrød, O., & Skjåk-Bræk, G. (1990). Alginate as immobilization matrix for cells. *Trends in Biotechnology*, *8*(3), 71–78. DOI: 10.1016/0167-7799(90)90139-O PMID: 1366500

Smith, A., Johnson, D., & Roberts, F. (2023). Biodegradable polymers for tissue engineering: The role of PPF. *Advanced Healthcare Materials*, *12*(7), 2230–2240.

Song, B., Wu, M., Qin, L., Liang, W., & Wang, X. (2025). Smart Design of Targeted Drug Delivery System for Precise Drug Delivery and Visual Treatment of Brain Gliomas. *Advanced Healthcare Materials*, *14*(4), 2402967. DOI: 10.1002/adhm.202402967 PMID: 39707642

Song, M., Liang, Y., Li, K., Zhang, J., Zhang, N., Tian, B., & Han, J. (2019). Hyaluronic acid modified liposomes for targeted delivery of doxorubicin and paclitaxel to CD44 overexpressing tumor cells with improved dual-drugs synergistic effect. *Journal of Drug Delivery Science and Technology*, *53*, 101179. DOI: 10.1016/j.jddst.2019.101179

Song, Y. H., De, R., & Lee, K. T. (2023). Emerging strategies to fabricate polymeric nanocarriers for enhanced drug delivery across blood-brain barrier: An overview. *Advances in Colloid and Interface Science*, *320*, 103008. DOI: 10.1016/j.cis.2023.103008 PMID: 37776736

Sonia, T. A., & Sharma, C. P. (2012). An overview of natural polymers for oral insulin delivery. *Drug Discovery Today*, *17*(13-14), 784–792. DOI: 10.1016/j.drudis.2012.03.019 PMID: 22521664

Sorathia, K., Patel, M., Soni, T., Vaghasiya, V., Patel, J., & Suhagıa, B. N. (2020). Effect of essential oils on transdermal permeation of metoprolol succinate. *Bulletin of Pharmaceutical Sciences*, *0*(0), 0. Advance online publication. DOI: 10.21608/bfsa.2020.41240.1028

Sosnik, A., & Seremeta, K. P. (2017). Polymeric Hydrogels as Technology Platform for Drug Delivery Applications. *Gels (Basel, Switzerland)*, *3*(3), 25. DOI: 10.3390/gels3030025 PMID: 30920522

Srivastava, A., Yadav, T., Sharma, S., Nayak, A., Kumari, A. A., & Mishra, N. (2015). Polymers in drug delivery. *Journal of Biosciences and Medicines*, *4*(1), 69–84. DOI: 10.4236/jbm.2016.41009

Steichen, S. D. (2016). Design, development and optimization of methacrylic acid-based pH-responsive hydrogels for the oral delivery of model protein therapeutics (Doctoral dissertation).

Sudimack, J., & Lee, R. J. (2000). Targeted drug delivery via the folate receptor. *Advanced Drug Delivery Reviews*, *41*(2), 147–162. DOI: 10.1016/S0169-409X(99)00062-9 PMID: 10699311

Suhail, M., Rosenholm, J. M., Minhas, M. U., Badshah, S. F., Naeem, A., Khan, K. U., & Fahad, M. (2019). Nanogels as drug-delivery systems: A comprehensive overview. *Therapeutic Delivery*, *10*(11), 697–717. DOI: 10.4155/tde-2019-0010 PMID: 31789106

Suk, J. S., Xu, Q., Kim, N., Hanes, J., & Ensign, L. M. (2015). PEGylation as a strategy for improving nanoparticle-based drug and gene delivery. *Advanced Drug Delivery Reviews*, *99*, 28–51. DOI: 10.1016/j.addr.2015.09.012 PMID: 26456916

Sullivan, S. P., Koutsonanos, D. G., & Del Pilar Martin, M. (2008). Dissolving polymer microneedle patches for influenza vaccination. *Biomaterials*, *29*(13), 1928–1941.

Sumaila, M., Marimuthu, T., Kumar, P., & Choonara, Y. E. (2021). Lipopolysaccharide nanosystems for the enhancement of oral bioavailability. *AAPS PharmSciTech*, *22*(7), 1–25. DOI: 10.1208/s12249-021-02124-5 PMID: 34595578

Sun Y., & Davis E. (2021). Nanoplatforms for targeted stimuli-responsive drug delivery: A review of platform materials and stimuli-responsive release and targeting mechanisms. *Nanomaterials*, 11(3), 746. DOI: 10.3390/nano11030746

Sung MH, Park C, Choi JC, et al. (2014). Hyaluronidase inhibitor containing poly-gamma-glutamic acid as an effective component. US, US8916141.

Susa, F., Arpicco, S., Pirri, C. F., & Limongi, T. (2024). An Overview on the Physiopathology of the Blood–Brain Barrier and the Lipid-Based Nanocarriers for Central Nervous System Delivery. *Pharmaceutics*, *16*(7), 849. DOI: 10.3390/pharmaceutics16070849 PMID: 39065547

Tabernero, A., & Cardea, S. (2020). Microbial Exopolysaccharides as Drug Carriers. *Polymers*, *12*(9), 2142. DOI: 10.3390/polym12092142 PMID: 32961830

Takahama, R., Kato, H., Tajima, K., Tagawa, S., & Kondo, T. (2021). Biofabrication of a hyaluronan/bacterial cellulose composite nanofibril by secretion from engineered Gluconacetobacter. *Biomacromolecules*, *22*(11), 4709–4719. DOI: 10.1021/acs.biomac.1c00987 PMID: 34705422

Tamada, J. A., & Langer, R. (1993). Erosion kinetics of hydrolytically degradable polymers. *Proceedings of the National Academy of Sciences of the United States of America*, *90*(3), 552–556. DOI: 10.1073/pnas.90.2.552 PMID: 8421690

Tanaka, M., & Vécsei, L. (2024). A decade of dedication: Pioneering perspectives on neurological diseases and mental illnesses. *Biomedicines*, *12*(5), 1083. DOI: 10.3390/biomedicines12051083 PMID: 38791045

Tan, Y., Chen, L., Li, K., Lou, B., Liu, Y., & Liu, Z. (2022). Yeast as carrier for drug delivery and vaccine construction. *Journal of Controlled Release*, *346*, 358–379. DOI: 10.1016/j.jconrel.2022.04.032 PMID: 35483637

Tao, Y., Chan, H. F., Shi, B., Li, M., & Leong, K. W. (2020). Light: A magical tool for controlled drug delivery. *Advanced Functional Materials*, *30*(49), 2005029. DOI: 10.1002/adfm.202005029 PMID: 34483808

Thirumalaisamy, R., Aroulmoji, V., Iqbal, M. N., Deepa, M., Sivasankar, C., Khan, R., & Selvankumar, T. (2021). Molecular insights of hyaluronic acid-hydroxychloroquine conjugate as a promising drug in targeting SARS-CoV-2 viral proteins. *Journal of Molecular Structure*, *1238*, 130457. DOI: 10.1016/j.molstruc.2021.130457 PMID: 33867575

Thirupathi, K., Phan, T. T. V., Santhamoorthy, M., Ramkumar, V., & Kim, S. (2022). pH and Thermoresponsive PNIPAm-co-Polyacrylamide Hydrogel for Dual Stimuli-Responsive Controlled Drug Delivery. *Polymers*, *15*(1), 167. DOI: 10.3390/polym15010167 PMID: 36616517

Thu, H. E., Zulfakar, M. H., & Ng, S.-F. (2012). Alginate based bilayer hydrocolloid films as potential slow-release modern wound dressing. In International Journal of Pharmaceutics. 434(1): 375). https://doi.org/DOI: 10.1016/j.ijpharm.2012.05.044

Tian, M. L., Zhou, J., Qi, X., & Shen, R. (2021). Thermo-sensitive hydrogel and their biomedical applications. *IOP Conference Series. Earth and Environmental Science*, *714*(3), 32062. DOI: 10.1088/1755-1315/714/3/032062

Tibbitt, M. W., & Anseth, K. S. (2009). Hydrogels as extracellular matrix mimics for 3D cell culture. *Biotechnology and Bioengineering*, *103*(4), 655–663. DOI: 10.1002/bit.22361 PMID: 19472329

Tomalia, D. A. (2005). Birth of a new macromolecular architecture: Dendrimers as quantized building blocks for nanoscale synthetic polymer chemistry. *Progress in Polymer Science*, *30*(3–4), 294–324. DOI: 10.1016/j.progpolymsci.2005.01.007

Tønnesen, H. H., & Karlsen, J. (2002). Alginate in drug delivery systems. *Drug Development and Industrial Pharmacy*, *28*(6), 621–630. DOI: 10.1081/DDC-120003853 PMID: 12149954

Tran, T.-N. T. (2013). Cutaneous drug delivery: An update. *Journal of Investigative Dermatology Symposium Proceedings*, 16(1). DOI: 10.1038/jidsymp.2013.28

Tran, T. T., & Tran, P. H. (2019). Controlled release film forming systems in drug delivery: The potential for efficient drug delivery. *Pharmaceutics*, *11*(6), 290. DOI: 10.3390/pharmaceutics11060290 PMID: 31226748

Trucillo, P. (2024). Biomaterials for drug delivery and human applications. *Materials (Basel)*, *17*(2), 456. DOI: 10.3390/ma17020456 PMID: 38255624

Tsung, T., Tsai, Y., Lee, H., Chen, Y., & Lu, D. (2023). Biodegradable Polymer-Based Drug-Delivery systems for ocular diseases. *International Journal of Molecular Sciences*, *24*(16), 12976. DOI: 10.3390/ijms241612976 PMID: 37629157

Tundisi, L. L., Mostaço, G. B., Carricondo, P. C., & Petri, D. F. S. (2021). Hydroxypropyl methylcellulose: Physicochemical properties and ocular drug delivery formulations. *European Journal of Pharmaceutical Sciences*, *159*, 105736. DOI: 10.1016/j.ejps.2021.105736 PMID: 33516807

Turgeon, S. L., Beaulieu, M., Schmitt, C., & Sanchez, C. (2007). Protein-polysaccharide complexes and coacervates. *Current Opinion in Colloid & Interface Science*, *12*(4-5), 166–178. DOI: 10.1016/j.cocis.2007.07.007

Tyler, B., Gullotti, D., Mangraviti, A., Utsuki, T., & Brem, H. (2016). Polylactic acid (PLA) controlled delivery carriers for biomedical applications. *Advanced Drug Delivery Reviews*, *107*, 163–175. DOI: 10.1016/j.addr.2016.06.018 PMID: 27426411

Ueno, M. (2007). Molecular anatomy of the brain endothelial barrier: An overview of the distributional features. *Current Medicinal Chemistry*, *14*(11), 1199–1206. DOI: 10.2174/092986707780597943 PMID: 17504140

Ulbrich, K., Hekmatara, T., Herbert, E., & Kreuter, J. (2011). Transferrin- and transferrin-receptor-antibody-modified nanoparticles enable drug delivery across the blood-brain barrier (BBB). *European Journal of Pharmaceutics and Biopharmaceutics*, *71*(2), 251–256. DOI: 10.1016/j.ejpb.2008.08.021 PMID: 18805484

Ullah, F., Othman, M. B. H., Javed, F., Ahmad, Z., & Akil, H. M. (2015). Classification, processing and application of hydrogels: A review. *Materials Science and Engineering C*, *57*, 414–433. DOI: 10.1016/j.msec.2015.07.053 PMID: 26354282

Van, D. B. S. L., Shalgunov, V., & Herth, M. M. (2022). Transport of nanomedicines across the blood-brain barrier: Challenges and opportunities for imaging and therapy. *Biomaterials Advances*, *141*, 213125. DOI: 10.1016/j.bioadv.2022.213125 PMID: 36182833

Varshosaz, J., Taymouri, S., Jafari, E., Jahanian-Najafabadi, A., & Taheri, A. (2018). Formulation and characterization of cellulose acetate butyrate nanoparticles loaded with nevirapine for HIV treatment. *Journal of Drug Delivery Science and Technology*, *48*, 9–20. DOI: 10.1016/j.jddst.2018.08.020

Vasanth, S., Misiriya, A., Thahsin, F., & Priya, S. (2025). Nonmagnetic Inorganic/Organic Core-Shell Nanoconstructs for Cancer Theranostics. In *Core-Shell Nano Constructs for Cancer Theragnostic: Current Scenario, Challenges and Regulatory Aspects* (pp. 365-388). Singapore: Springer Nature Singapore. DOI: 10.1007/978-981-96-3025-7_13

Vashist, R., Sangave, S. D., & Alexander, A. (2024). Surface active ligands for enhanced brain targeting of nanoparticles. In *Application of Nanocarriers in Brain Delivery of Therapeutics* (pp. 241–262). Springer Nature Singapore. DOI: 10.1007/978-981-97-2859-6_9

Vehring, R. (2008). Pharmaceutical particle engineering via spray drying. *Pharmaceutical Research, 25*(5), 999–1022. DOI: 10.1007/s11095-007-9475-1 PMID: 18040761

Visan, A. I., Popescu-Pelin, G., & Socol, G. (2021). Degradation behavior of polymers used as coating materials for drug delivery—A basic review. *Polymers, 13*(8), 1272. DOI: 10.3390/polym13081272 PMID: 33919820

Vossen, M., De Geest, B. G., & De Smedt, S. C. (2017). Biodegradable microbeads for intraperitoneal drug delivery in ovarian cancer: In vivo evidence for a sustained effect. *Journal of Controlled Release, 254*, 83–91. DOI: 10.1016/j.jconrel.2017.03.382

Walvekar, P., Lulinski, P., Kumar, P., Aminabhavi, T. M., & Choonara, Y. E. (2024). A review of hyaluronic acid-based therapeutics for the treatment and management of arthritis. *International Journal of Biological Macromolecules, 130645*, 130645. Advance online publication. DOI: 10.1016/j.ijbiomac.2024.130645 PMID: 38460633

Wang, J., Liu, K., & Chen, X. (2022). Dextran derivatives in sustained drug release: Mechanisms and applications. *Biomaterials Science, 10*(8), 1254–1268.

Wang, C., Li, J., Sun, W., Dong, C., Zhang, L., & Liu, W. (2022). Theranostic alginate-based microbeads: An emerging platform for combined therapy and imaging. *Materials Science and Engineering C, 130*, 112435. DOI: 10.1016/j.msec.2021.112435

Wang, J., Liu, S., Huang, J., Ren, K., Zhu, Y., & Yang, S. (2023). Alginate: Microbial production, functionalization, and biomedical applications. *International Journal of Biological Macromolecules, 242*, 125048. DOI: 10.1016/j.ijbiomac.2023.125048 PMID: 37236570

Wang, T., Hou, J., Su, C., Zhao, L., & Shi, Y. (2017). Hyaluronic acid-coated chitosan nanoparticles induce ROS-mediated tumor cell apoptosis and enhance antitumor efficiency by targeted drug delivery via CD44. *Journal of Nanobiotechnology, 15*(1), 1–12. DOI: 10.1186/s12951-016-0245-2 PMID: 28068992

Wang, W., Meng, Q., Li, Q., Liu, J., Zhou, M., Jin, Z., & Zhao, K. (2020). Chitosan derivatives and their application in biomedicine. *International Journal of Molecular Sciences, 21*(2), 487. DOI: 10.3390/ijms21020487 PMID: 31940963

Wang, Y., Zhang, Y., & Liu, X. (2022). Microneedle-based drug delivery systems: A review. *International Journal of Pharmaceutics*, *617*, 114694. DOI: 10.1016/j.ijpharm.2022.114694

Wang, Z., Zhai, B., Sun, J., Zhang, X., Zou, J., Shi, Y., & Guo, D. (2024). Recent advances of injectable in situ-forming hydrogels for preventing postoperative tumor recurrence. *Drug Delivery*, *31*(1), 2400476. DOI: 10.1080/10717544.2024.2400476 PMID: 39252545

Wei, B., Yang, G., & Hong, F. (2011). Preparation and evaluation of a kind of bacterial cellulose dry films with antibacterial properties. *Carbohydrate Polymers*, *84*(1), 533–538. DOI: 10.1016/j.carbpol.2010.12.017

Wei, P., Cornel, E. J., & Du, J. (2021). Ultrasound-responsive polymer-based drug delivery systems. *Drug Delivery and Translational Research*, *11*(4), 1323–1339. DOI: 10.1007/s13346-021-00963-0 PMID: 33761101

Wells, C. M., Harris, M., Choi, L., Murali, V. P., Guerra, F. D., & Jennings, J. A. (2019). Stimuli-responsive drug release from smart polymers. *Journal of Functional Biomaterials*, *10*(3), 34. DOI: 10.3390/jfb10030034 PMID: 31370252

Wilczewska A. Z., Niemirowicz K., Markiewicz K. H., & Car H. (2012). Nanoparticles as drug delivery systems. *Pharmacological Reports*, 64(5), 1020. DOI: 10.1016/S1734-1140(12)70901-5

Williams J. (2013). Vector design for improved DNA vaccine efficacy, safety and production. *Vaccines*, 1(3), 225. DOI: 10.3390/vaccines1030225

Witting, M., Boreham, A., Brodwolf, R., Vavrova, K., Alexiev, U., Friess, W., & Hedtrich, S. (2015). Interactions of hyaluronic acid with the skin and implications for the dermal delivery of biomacromolecules. *Molecular Pharmaceutics*, *12*(5), 1391–1401. DOI: 10.1021/mp500676e PMID: 25871518

Wong, T. W. (2009). Chitosan and its use in design of insulin delivery system. *Recent Patents on Drug Delivery & Formulation*, *3*(1), 8–25. DOI: 10.2174/187221109787158346 PMID: 19149726

Woodruff, M. A., & Hutmacher, D. W. (2010). The return of a forgotten polymer—Polycaprolactone in the 21st century. *Progress in Polymer Science*, *35*(10), 1217–1256. DOI: 10.1016/j.progpolymsci.2010.04.002

Wu, D., Chen, Q., Chen, X., Han, F., Chen, Z., & Wang, Y. (2023). The blood–brain barrier: Structure, regulation and drug delivery. *Signal Transduction and Targeted Therapy*, *8*(1), 217. DOI: 10.1038/s41392-023-01481-w PMID: 37231000

Wu, Y., Li, P., Jiang, Z., Sun, X., He, H., Yan, P., Xu, Y., & Liu, Y. (2023). Bioinspired yeast-based β-glucan system for oral drug delivery. *Carbohydrate Polymers, 319*, 121163. DOI: 10.1016/j.carbpol.2023.121163 PMID: 37567689

Xu, A. R., Wang, J. J., & Guo, X. (2018). Fabrication of cellulose aerogels using a green/clean procedure. *Journal of Macromolecular Science, Part B: Physics, 57*(1), 1–7. DOI: 10.1080/00222348.2017.1396825

Xu, L., Cao, Y., Xu, Y., Li, R., & Xu, X. (2024). Redox-Responsive Polymeric Nanoparticle for Nucleic Acid Delivery and Cancer Therapy: Progress, Opportunities, and Challenges. *Macromolecular Bioscience, 24*(3), 2300238. DOI: 10.1002/mabi.202300238 PMID: 37573033

Xu, , Liu, Y., Fu, W., Yao, M., Ding, Z., Xuan, J., Li, D., Wang, S., Xia, Y., & Cao, M. (2020). Poly(N-isopropylacrylamide)-Based Thermoresponsive Composite Hydrogels for Biomedical Applications. *Polymers, 12*(3), 580. DOI: 10.3390/polym12030580 PMID: 32150904

Xu, Y., Zhan, C., Fan, L., Wang, L., & Zheng, H. (2007). Preparation of dual crosslinked alginate–chitosan blend gel beads and in vitro controlled release in oral site-specific drug delivery system. *International Journal of Pharmaceutics, 336*(2), 329–337. DOI: 10.1016/j.ijpharm.2006.12.019 PMID: 17223290

Yaghmur, A., & Mu, H. (2021). Recent advances in drug delivery applications of cubosomes, hexosomes, and solid lipid nanoparticles. *Acta Pharmaceutica Sinica B*, 11(4), 871. DOI: 10.1016/j.apsb.2021.02.013

Yallapu, M. M., Reddy, M. K., & Labhasetwar, V. (2007). *Nanogels: Chemistry to drug delivery*. Biomedical Applications of Nanotechnology.

Yang, H., Shan, C., & Kolen, A. F. (2020). Medical Instrument Detection in Ultrasound-Guided Interventions: A Review. *arXiv (Cornell University)*. Cornell University. /arxiv.2007.04807DOI: 10.48550

Yang, H., Wang, H., Wang, Z., & Liu, L. (2019). pH-sensitive alginate microspheres for oral drug delivery: Preparation and characterization. *International Journal of Biological Macromolecules, 125*, 990–997. DOI: 10.1016/j.ijbiomac.2018.12.027

Yang, L., Wang, L., & Zhang, Q. (2021). Nanocarriers for transdermal drug delivery: Opportunities and challenges. *Pharmaceutical Nanocarriers, 17*(3), 245–256. DOI: 10.1016/j.phnan.2021.03.002

Yang, X., Yang, D., Zhu, X., Nie, J., & Ma, G. (2019). Electrospun and photocrosslinked gelatin/dextran–maleic anhydride composite fibers for tissue engineering. *European Polymer Journal, 113*, 142–147. DOI: 10.1016/j.eurpolymj.2019.01.059

Yan, X., Liu, Z., & Zhang, J. (2021). Mechanisms and strategies for improving transdermal drug delivery. *Journal of Controlled Release, 332*, 201–211. DOI: 10.1016/j.jconrel.2021.03.001

Ye, S., He, S., Su, C., Jiang, L., Wen, Y., Zhu, Z., & Shao, W. (2018). Morphological, release and antibacterial performances of amoxicillin-loaded cellulose aerogels. *Molecules (Basel, Switzerland), 23*(8), 2082. DOI: 10.3390/molecules23082082 PMID: 30127283

Yetisgin, A. A., Cetinel, S., Zuvin, M., Kosar, A., & Kutlu, O. (2020). Therapeutic nanoparticles and their targeted delivery applications. *Molecules (Basel, Switzerland), 25*(9), 2193. DOI: 10.3390/molecules25092193 PMID: 32397080

Yin, L., Ding, J., He, C., Cui, L., Tang, C., & Yin, C. (2020). Smart polymeric nano-carriers for therapeutic gene delivery. *Current Pharmaceutical Design, 26*(12), 1353–1366. DOI: 10.2174/1381612826666200128123546

Yoshida, T., Lai, T. C., Kwon, G. S., & Sako, K. (2013). pH-and ion-sensitive polymers for drug delivery. *Expert Opinion on Drug Delivery, 10*(11), 1497–1513. DOI: 10.1517/17425247.2013.821978 PMID: 23930949

Yotsuzuka, F. (2001). Curdlan. In Cho, S. S., & Dreher, M. L. (Eds.), *Handbook of dietary fiber* (pp. 737–757). DOI: 10.1201/9780203904220-41

Younes, I., & Rinaudo, M. (2015). Chitin and Chitosan Preparation from Marine Sources. Structure, Properties and Applications. *Marine Drugs, 13*(3), 1133–1174. DOI: 10.3390/md13031133 PMID: 25738328

Yu, W., Jiang, G., & Li, P. (2015). Gelatin methacryloyl-based microneedles for controlled release. *Acta Biomaterialia, 12*, 220–230. DOI: 10.1016/j.actbio.2015.08.013

Yu, X., & Pishko, M. V. (2011). Release of paclitaxel from pH sensitive and biodegradable dextran based hydrogels. *Soft Matter, 7*(19), 8898–8904. DOI: 10.1039/c1sm05729d

Zamboulis, A., Nakiou, E. A., Christodoulou, E., Bikiaris, D. N., Kontonasaki, E., Liverani, L., & Boccaccini, A. R. (2019). Polyglycerol hyperbranched polyesters: Synthesis, properties and pharmaceutical and biomedical applications. *International Journal of Molecular Sciences, 20*(24), 6210. DOI: 10.3390/ijms20246210 PMID: 31835372

Zare, M., Bigham, A., Zare, M., Luo, H., Rezvani Ghomi, E., & Ramakrishna, S. (2021). pHEMA: An overview for biomedical applications. *International Journal of Molecular Sciences, 22*(12), 6376. DOI: 10.3390/ijms22126376 PMID: 34203608

Zeb, A., Gul, M., Nguyen, T. T. L., & Maeng, H. J. (2022). Controlled release and targeted drug delivery with poly (lactic-co-glycolic acid) nanoparticles: Reviewing two decades of research. *Journal of Pharmaceutical Investigation*, *52*(6), 683–724. DOI: 10.1007/s40005-022-00584-w

Zhang, W., Zhao, Q., & Yuan, J. (2018). Porous polyelectrolytes: The interplay of charge and pores for new functionalities. *Angewandte Chemie International Edition*, *57*(23), 6754–6773. DOI: 10.1002/anie.201710272 PMID: 29124842

Zhang, X., & Luo, X. (2020). Thermoresponsive polymers for drug delivery. *Materials Science and Engineering C*, *111*, 110758. DOI: 10.1016/j.msec.2020.110758

Zhang, Y., & Lee, J. (2022). Amphiphilic polymers in drug delivery systems. *European Journal of Pharmaceutics and Biopharmaceutics*, *174*, 32–45. DOI: 10.1016/j.ejpb.2022.02.006

Zhang, Y., & Liu, Z. (2023). Protein and gene delivery using pH-sensitive polymers. *Journal of Controlled Release*, *344*, 185–200.

Zhang, Y., Wang, L., & Liu, X. (2024). Advances in pH-responsive polymers for gene delivery: The case of PDMAEMA. *Macromolecular Bioscience*, *24*(3), 1203–1218.

Zhang, Y., Yu, J., Bomba, H. N., Zhu, Y., & Gu, Z. (2016). Mechanical force-triggered drug delivery. *Chemical Reviews*, *116*(19), 12536–12563. DOI: 10.1021/acs.chemrev.6b00369 PMID: 27680291

Zhao, J., Li, Y., & Zhang, Z. (2022). Skin penetration enhancement by electroporation for transdermal drug delivery. *Journal of Controlled Release*, *338*, 183–190. DOI: 10.1016/j.jconrel.2022.07.027

Zhao, L., Feng, Y., & Huang, J. (2023). Controlled drug release using poly(sebacic acid) for implantable devices. *Materials Today*, *51*, 125–138.

Zhao, S., Zhu, H., Chen, Z., Shuai, S., Zhang, N., Liu, Y., Rao, Z., Li, Y., Zhao, C., Zhou, K., Ge, W., & Hao, J. (2019). Preparation and properties of a temperature- and pH- responsive polypeptide hydrogel. *Materials Research Express*, *6*(8), 85711. DOI: 10.1088/2053-1591/ab253e

Zhao, W., Xu, X., & Chen, L. (2021). Controlled release formulations for transdermal applications. *International Journal of Nanomedicine*, *16*, 1373–1389. DOI: 10.2147/IJN.S308135

Zhao, X., Wu, H., Guo, B., Dong, R., Qiu, Y., & Ma, P. X. (2019). Antibacterial anti-oxidant electroactive injectable hydrogel as self-healing wound dressing with hemostasis and adhesiveness for cutaneous wound healing. *Biomaterials*, *122*, 34–47. DOI: 10.1016/j.biomaterials.2017.01.011 PMID: 28107663

Zhao, Z., Han, J., Xu, S., Jin, Z., Yin, T. H., & Zhao, K. (2022). Amoxicillin encapsulated in the N-2-hydroxypropyl trimethyl ammonium chloride chitosan and N, O-carboxymethyl chitosan nanoparticles: Preparation, characterization, and antibacterial activity. *International Journal of Biological Macromolecules*, *221*, 613–622. DOI: 10.1016/j.ijbiomac.2022.09.035 PMID: 36089095

Zheng X., Xie J., Zhang X., Sun W., Zhao H., Li Y., & Wang C. (2020). An overview of polymeric nanomicelles in clinical trials and on the market. *Chinese Chemical Letters*, 32(1), 243. DOI: 10.1016/j.cclet.2020.11.029

Zheng, M., & Yuan, J. (2022). Polymeric nanostructures based on azobenzene and their biomedical applications: Synthesis, self-assembly and stimuli-responsiveness. *Organic & Biomolecular Chemistry*, *20*(4), 749–767. DOI: 10.1039/D1OB01823J PMID: 34908082

Zhu, D., Yao, K., Bo, J., Zhang, H., Liu, L., Dong, X., & Leng, X. (2010). Hydrophilic/lipophilic N-methylene phosphonic chitosan as a promising non-viral vector for gene delivery. *Journal of Materials Science. Materials in Medicine*, *21*(1), 223–229. DOI: 10.1007/s10856-009-3849-3 PMID: 19680604

Zhu, J., Tang, X., Jia, Y., Ho, C. T., & Huang, Q. (2020). Applications and delivery mechanisms of hyaluronic acid used for topical/transdermal delivery–a review. *International Journal of Pharmaceutics*, *578*, 119127. DOI: 10.1016/j.ijpharm.2020.119127 PMID: 32036009

Žigrayová, D., Mikušová, V., & Mikuš, P. (2023). Advances in antiviral delivery systems and Chitosan-Based polymeric and nanoparticulate antivirals and antiviral carriers. *Viruses*, *15*(3), 647. DOI: 10.3390/v15030647 PMID: 36992356

Zou, Z., Zhang, B., Nie, X., Cheng, Y., Zhang, H., Liao, M., & Li, S. (2020). A sodium alginate-based sustained-release IPN hydrogel and its applications. *RSC Advances*, *10*(65), 39722–39730. DOI: 10.1039/D0RA04316H PMID: 35515393

About the Contributors

Md. Abul Kalam Azad (A. K. Azad), an expert in Polymers and Biomaterials science and encapsulation technology with a focus on composite materials and surface technologies for targeted drug delivery, is currently the senior lecturer at Department of Pharmaceutical Technology, Faculty of Pharmacy, University MAIWP International, Malaysia. In 2023, Dr. Azad was listed as Top 2% scientist by Stanford University for his extraordinary achievement on polymers related research. He has published over 100 publications, 20 book chapters and several patents. All of his efforts lead to become organizing and scientific committee member of many international conferences including PRISAL-UCMI INTERNATIONAL CONFERENCE. 19th August 2023, (Organizing Committee Member), 4th Malaysian Congress of Toxicology (MYCOT 2023), IIUM, Kuantan, Malaysia, 5th Malaysian Congress of Toxicology (MYCOT 2024), TBC, Malaysia., Secretariate, 2nd AIMST Conference, 2022, AIMST University, Kedah, Malaysia. Dr. Azad also invited as a keynote speaker in national and international conferences on Polymer Nanocomposites. Dr. A. K. Azad obtained his Masters, PhD from the International Islamic University Malaysia and has postdoctoral research fellow for academic & research experience, having been head of the Pharmaceutical Engineering Department at the MAHSA University, Malaysia for 2 years. This is complemented by his long academic and research experience as leading guest editor in Journal of Herbal Medicine, (Q2, impact factor 2.3) Elsevier, Leading guest editor in Polymers, (Q1, impact factor 5.0) MDPI and Editor-In-Chief (2018-present), Research and Review: Drugs and Drugs Development, Editorial Board member (2018), Trends in Pharmaceuticals and Nanotechnology and Editorial Board member (2018), Journal of Pharmacological Research and Developments. He has reviewed more than 200 articles from Elsevier, Springer Nature, Taylor and Francis, Dove Press, Bentham Science, John Wiley & Sons and Nature as well. He also received numerous National and international awards (The Best Award (Open Category) in Innovation and Invention, MTExpo-2018, PWTC, Kuala Lumpur, Malaysia) and gold medal

(MTExpo-2018, PWTC, Kuala Lumpur, Malaysia) for his outstanding research findings. Recently, He awarded US Green Card in the category of EB-1A for his extraordinary contributions in pharmaceutical formulations development.

Mireia Mallandrich Miret is a researcher in topical delivery and nose-to-brain delivery, advancing skin and mucosal permeation evaluations of innovative semi-solid formulations and nanosystems for targeted delivery of anti-inflammatory and immunosuppressant drugs. She is a Tenure-Track Lecturer at the University of Barcelona and co-editor of scientific journals.

Sukalyan Kumar Kundu completed his PhD and postdoc from Japan and Joined Jahangirnagar University (JU) as Lecturer in 2008. He was a Senate Member, JU, Savar, Dhaka, 2024 and also Former Chairman, Department of Pharmacy, JU, Savar, Dhaka, Bangladesh.

***

Ankur Bhardwaj is working as Assistant professor in the Department of Biotechnology, Shri Vaishnav Institute of Sciences, Shri Vaishnav Vidyapeeth Vishwavidyalaya, Indore. He received B.Sc.[HONS] (2009) in Biotechnology from Allahabad Agricultural Institute-Deemed University, Allahabad, Uttar Pradesh; M.Sc. (2011) in Biotechnology from SRM University, Kattankulathur, Chennai, Tamil Nadu and Ph.D. (2021) in Molecular Biology and Microbiology from Mangalayatan University, Aligarh, Uttar Pradesh and received 3 young scientist awards during his Ph.D. tenure. He has more than 10 years of teaching and research experience. He prominently has worked in the field of Plant Biotechnology, Molecular Biology, Immuno-informatics and Microbiology. He has published 10 research papers in peer reviewed Journals, 14 book chapters. He has successfully guided more than 20 academic students in their final year project. Dr. Bhardwaj is a lifetime member of the Scholars Academic and Scientific Society and has more than 05 conference abstracts. Recently, Dr. Bhardwaj received Prof. M.L. Khorana IJPS Best Paper Award in the area of Pharmaceutical & Medicinal Synthetic Chemistry.

Nur Jalal Mondal received his M.Sc. degree from the University of Science and Technology, Meghalaya, India in 2018. He is a Ph.D. research scholar at the Institute of Advanced Study in Science and Technology (IASST), Guwahati, India. His focus area is nanoscience and nanotechnology. He is currently working on Development of efficient heterogenous nanocomposite and doped nanoparticles-based catalyst.

Index

E

H

I

M

N

P

R

S

T

www.ingramcontent.com/pod-product-compliance
Lightning Source LLC
LaVergne TN
LVHW080857300926
849162LV00003B/447

* 9 7 9 8 3 3 7 3 1 0 4 7 3 *